The Arctic and Antarctic: their division into geobotanical areas

XXIX KOMAROV LECTURE PRESENTED 14 OCTOBER 1974

The Arctic and Antarctic: their division into geobotanical areas

V. D. ALEKSANDROVA

V. L. Komarov Botanical Institute of the Academy of Sciences of the USSR

TRANSLATED BY DORIS LÖVE

CAMBRIDGE UNIVERSITY PRESS

Cambridge

London New York New Rochelle

Melbourne Sydney

CAMBRIDGE UNIVERSITY PRESS
Cambridge, New York, Melbourne, Madrid, Cape Town, Singapore, São Paulo, Delhi

Cambridge University Press
The Edinburgh Building, Cambridge CB2 8RU, UK

Published in the United States of America by Cambridge University Press, New York

www.cambridge.org
Information on this title: www.cambridge.org/9780521114264

First published in Russian as
Геоботаническое районирование Арктики и Антарктики.
by Nauka, Leningrad, 1977
English translation first published 1980

This digitally printed version 2009

A catalogue record for this publication is available from the British Library

ISBN 978-0-521-23119-0 hardback
ISBN 978-0-521-11426-4 paperback

CONTENTS

	Translator's foreword	vii
	Preface	xi
	Acknowledgements	xiii
1	Division of the Arctic into geobotanical areas	1
2	The geobotanical regions of the Arctic: the tundra region	19
3	The geobotanical regions of the Arctic: the region of the arctic polar deserts	144
4	Division of the Antarctic into geobotanical areas	169
5	Conclusions	187
	References	191
	List of Latin plant names	225
	Index	239

TRANSLATOR'S FOREWORD

The landmass of the USSR is enormous and its 'Arctic' alone spans over 150° of longitude from the Atlantic to the Pacific and reaches latitude 82° N at its northernmost point on Franz Joseph's Land. It is, however, fairly well explored, from the point of view of both flora and vegetation.

It may come as a surprise to some readers that much of this exploration has been done by dedicated women scientists, who alone or together with male colleagues have braved much danger, hardship and the rigors of extreme climates. One of them is Vera Danilovna Aleksandrova, who has devoted a lifetime to the study and investigation of the flora and vegetation of arctic lands far above the Polar Circle. The present volume is, thus, to a great extent based on her personal experience, but also on her intimate knowledge of the literature on arctic areas inside and outside the USSR, and on the Antarctic. I hope that my translation will be useful to students of arctic and antarctic vegetation, not able themselves to cope with the Russian language.

The translation has been made as faithful to the original as possible. Nothing has been omitted, and only a few minor revisions have been made by the author, who has approved the translation. The Russian units of classification used in this book have been published previously by Aleksandrova (1973).

The transliteration of personal and geographical names as well as the titles in the bibliography follows the recommendations of Gregory Razran (1959) in *Science*, (**129**, 1111–13). It is a simple and straightforward system, as close to the phonetics of the Russian language as English pronunciation can get. It does not have the cumbersome superscripts and ligatures of the Library of Congress system and is therefore easier both to read and to pronounce. It should be noted that the Russian letters ь and ь are given as ′ and ″, respectively. In the few cases where Russian authors have published in books or journals abroad and their names have been spelled differently from my spellings, I have listed both types of names in the bibliography (e.g. Trautfetter *see* Trautvetter, etc), but in the text only Razran's system is used.

The equivalence of Russian terms has been carefully considered. The multilingual *Geobotanichesky Slovar'* (*Geobotanical Dictionary*) by O. S. Grebenshchikov (Nauka, Moscow, 1965) has been of great help, as has the *Entsiklopedichesky Slovar' Geograficheskikh Terminov* (*Encyclopedical Dictionary of Geographical Terms*), ed. S. V. Kalesnik; Sovietskaya Entsiklopediya, Moscow, 1968). I am also indebted to my husband, Dr Á. Löve, and many Russian and American friends and colleagues for assistance in interpreting certain terms.

There are some terms which have no corresponding English expression. One such word, very much used in Russian ecological literature, is *plakor*. It is derived from a Greek word for 'flat, level', but is used in Russian botany to describe a locality or a vegetation relating to a habitat with fine soil, typical of the zone discussed, on level ground, well drained, neither too wet nor too dry, with a moderate snow cover, melting off neither too early nor too late. I have reduced this formula to the short expression 'zonal, mesic habitat, vegetation', etc., and I hope that it adequately covers the Russian term.

The use of the word 'tundra' in the plural may at first seem strange, because we are used to seeing it in a singular form as a vague and wide term for arctic and alpine vegetation in general. But in a country with as much tundra as the Soviet Union, the distinction of the various types of tundra plant associations is much more precise. I have therefore followed the original when using the expression 'tundras'.

In contrast, it seems we have a much more varied nomenclature in English for the different types of wetlands. Where the Russian language uses one word, *bolota*, we have a whole row of distinct expressions: mires, swamps, bogs, mosses, fens, marshes, etc. In this text I have preferred to use the term 'mire' almost exclusively, because under arctic conditions there is actually nothing comparable to a true 'bog'. Also in northern Scandinavia the vegetation type called 'myr' is almost identical to the subarctic and arctic types described here. The Russian terms *kochka* (hummock), *bugor* (mound, hillock, palsa), *poligonal'noye* (polygonal bog or mire) have been translated according to the Estonian Six-language dictionary *Saksa — Inglise — Rootsi — Soome — Eesti — Vene — Sooteaduslik Oskussonastik* (*German — English — Swedish — Finnish — Estonian — Russian* Dictionary of Scientific Peatbog Terminology) by V. V. Masing and the reference book *Permafrost Terminology* (Tech. Mem. 111, Assoc. Comm. on Geobot. Research, Nat. Res. Council, Canada, 1974), prepared by R. E. J. Brown & W. O. Kupsch.

In the Russian language there are two terms, *lug* and *lugovina*, the first of which means straight 'meadow', but the second is translated as 'short-grass meadow' in the dictionaries. Since this term, at least in the USA, is used for certain types of prairie (steppe) vegetation and since the tundra *lugovinas* described by the author are not only graminoid, but some of them are often herbaceous with predominating forbs, I use the term 'meadow-like communities' here.

Finally, I want to express my gratitude to the editors at the Cambridge University Press for valuable advice and assistance during my work.

January, 1979
Doris Löve

PREFACE

In his monograph *Introduction to an Investigation of the Vegetation of Yakutia*, V. L. Komarov (1926) published material which became of fundamental importance for the study of the geobotanical zonation of the arctic area of the USSR from the Khatanga River to the Chaun Inlet. Of great importance is the map which he drew at a scale of 1:4 200 000 of the northern forest limit and which up to recently was the most sharply defined boundary between the tundras and the forested regions in this wide territory.

At the time when Komarov wrote his work on the vegetation of Yakutia, the division of the polar regions into geobotanical areas was just beginning. The first attempts, carried out during the nineteenth and at the beginning of the twentieth centuries, concerned the East European North of the USSR (Trautvetter, 1851; Schrenk, 1854; Pohle, 1910). At the beginning of the twentieth century, there also appeared outlines for a zonal geobotanical division of the West Siberian North (Zhitkov, 1913; Gorodkov, 1916) as well as the works by Passarge (1921), who, having observed the zonation of the landscapes in the polar lands of both the northern and the southern hemispheres, distinguished 'the cold steppes' (the tundras) and 'the cold deserts', the latter called 'polar deserts' by B. N. Gorodkov.

During the fifty years which have elapsed since the appearance of Komarov's book, the study of the vegetation cover in the Arctic and the Antarctic has made much progress. A very large amount of information was gathered during the thirties in the USSR, when investigations were carried out over a wide area of the tundra territories in connection with the needs of reindeer husbandry and the utilization of water, soil and wildlife resources. The accumulation of information has continued through the end of the forties into recent times, when in the USSR and abroad everything has become more adapted to aviation and mechanical means of communication. During the last few years, photographs taken by the aid of satellites (Vinogradov, 1970; Andersson, 1975) have also been utilized. The development of theories on division into geobotanical areas has, during this time, met with great success (Shennikov, 1940; Lavrenko, 1947, 1968; Sochava, 1948*a*, 1952, 1966, 1967, 1972; Schmithüsen, 1968; Karamysheva & Rachkovskaya, 1973; Il'ina, 1974; Karamysheva *et al*., 1975; Lavrenko & Isachenko, 1976; etc.). Outlines for a zonation of the Arctic have been drawn up within the USSR (Gorodkov, 1935*b*; Leskov, 1947; Sochava, 1948*a*; etc.), in the circumpolar area (Polunin, 1951; Yurtsev, 1966, 1974*a*; Aleksandrova, 1969*b*, 1971*b*), and for the high latitudes (Korotkevich, 1967, 1972) as well as for the Antarctic (Greene, 1964*a*; Holdgate, 1964; Korotkevich, 1966, 1972; Kats, 1971; etc.). Gribova (1975) has drawn a circumpolar map of the vegetation in the Arctic.

There is now an interest in using the accumulated material to formulate a division of the polar lands into geobotanical areas and in examining, from a uniform point of view, the vegetation cover of the circumpolar territories of the Arctic northward from the forest limit to the farthest points of the land and of the Antarctic southward from the line of the antarctic convergence.

ACKNOWLEDGEMENTS

During the preparation of the manuscript for printing, I have enjoyed a number of valuable discussions and received much good advice from V. N. Andreyev, M. S. Boch, S. A. Gribova, A. G. Isachenko, A. E. Katenin, N. V. Matveyeva, T. G. Polozova, O. V. Rebristaya, I. N. Safronova and B. A. Yurtsev, and I want to express my deepest gratitude to them.

1

Division of the Arctic into geobotanical areas

Since so much will be said about division into geobotanical areas, it is first and foremost necessary to define my concept in this respect. By division into geobotanical areas I intend to divide the vegetation cover so that the characteristics of the vegetation, that is, the totality of the plant communities distributed over a given territory, can be considered as diagnostic. Other traits should be considered as characterizing.

The first person to suggest a distinction between diagnostic characteristics, according to which groups of objects may be separated, and characterizing ones, according to which the units are further differentiated, was Tuomikoski (1942). In respect to their nature, the diagnostic characteristics are analogous to the indicator characteristics of a correlation swarm (Terent'yev, 1931, 1959). Takhtadzhyan (1966: 41–2) points to their significance for plant taxonomy and Vasilyevich (1964, 1966) as well as Nitsenko (1966) have shown their importance for the classification of vegetation. The application of diagnostic and characterizing traits is expedient also for the botanical–geographical division into areas (Aleksandrova, 1974). (Note: It is suggested that the division of an area based on any botanical characteristics be called botanical – geographical). The first group of characteristics, the diagnostic ones, defines the limits and establishes the rank and configuration of distinct areas. The second group, the characterizing traits, describes the distinguished areas, emphasizes their particularity and confirms the significance and rank of the boundaries between them.

The choice of various diagnostic traits separates the different systems for dividing vegetation cover into areas. According to this criterion, the distinction is first and foremost floristic, when the character of the flora serves as the diagnostic characteristic, or geobotanical, when the character of the vegetation plays this role. Although the flora and vegetation form a single unit, the flora exists in the form of definite plant associations, and the taxa of the flora in a given region appear as the components of the plant communities. There is, however, a distinct difference between divisions into floristic and into geobotanical areas. In the first case, the taxa of the plant systematists are the focus of attention. If, thereby, the very varied characteristics of the plants are considered, including their role in the composition of the plant communities, all these traits can be used as characters, describing given species. In the second case, the central problem for the division of the vegetation into areas is its classification (Shennikov, 1940: 25–30; Sochava, 1948*a*, *b*, 1952: 530; Braun-Blanquet, 1964: 720–56; Schmithüsen, 1968; Karamysheva & Rachkovskaya, 1973: 171–2; etc.). It is fundamentally possible to set up specific, phytocoenological characteristics here which are outside the concepts of the floristics, such as, e.g., the structure of the plant communities.

In addition, supplementary characteristics are added to define a distinct area which, together with the diagnostic traits, plays a role in its description. If a wide variety of traits, both of the flora and of the vegetation, are used as characterizing, we obtain as a result a synthetic (Lavrenko, 1968; and others) or a complex (Yurtsev, 1966; and others) botanical–geographical division into areas. It should be noted that according to the nature of the diagnostic characteristics, this division becomes, when applied by E. M. Lavrenko, a geobotanical one, and when applied by B. A. Yurtsev, a floristic one.

In the Arctic, the principles of the classification of the vegetation and its division into areas have been differently developed within the USSR and abroad. In part, it is explained by the fact that there are different geobotanical provinces in the different parts of the circumpolar Arctic.

Classification and division into areas of the arctic vegetation outside the USSR

In the western hemisphere, the application of basic, diagnostic characteristics for the division of the Arctic into areas in respect to the degree of closedness of the vegetation is very widespread. This goes back to the principles of classification formulated by Warming (1888) who, when describing the vegetation of Greenland, distinguished two vegetation 'formations': the 'Field-Flur' or fell-field (literally: desert-like ground) and 'Heide' or 'heath' (literally: wasteland, in a sense close to what is understood by that term in Shennikov, 1938: 487). According to Warming (1928: 299) the fell-field is the formation where 'plants grow singly often with great intervals, though here and there denser spots of vegetation occur, consisting of one or more species, and it is the substratum (loose soil, rock) which lends its generally grey or greyish black tinge to the surface'. 'Heath' differs from 'fell-field' in that here the vegetation cover is closed. By heath, Warming implied not only the kind of heath where the ericaceous dwarfshrubs form a closed cover with interlacing branchlets and the space between them is filled by mosses, lichens, and grasses (Warming, 1928: 302), but also lichen-heaths and moss-heaths. By the latter, Warming meant mainly forms of *Rhacomitrieta.* In addition, Warming distinguished copses (thickets of shrubs), as well as 'herb-field and grass-field in well-ventilated soil', 'freshwaters vegetation', 'moors and meadows' and 'shore vegetation'. The terms 'heath' and 'fell-field' introduced by Warming as well as some additional ones, e.g., the term 'barrens' for areas with a very disrupt vegetation, are widely used in foreign literature, both for the purpose of vegetation classification (cf. Beschel, 1963*b*; and others) and also as additional concepts to characterize vegetation (Böcher, 1954, 1963*b*; and others). It should be noted here that although some authors imply by the term 'heath' only associations dominated by ericaceous shrubs, this term is, as a rule, applied solely for the purpose of emphasizing the predominance of a closed cover, some small part of which also may be bare ground. Thus, Porsild (1951: 12) calls the dwarfshrub tundra 'dwarfshrub heath' and the lichens and moss tundras 'lichen and moss heaths'. Churchill (1955: 609) uses the term 'heath' for the tussocky tundras of Alaska. Beschel (1963*b*: 102–3) described from Axel Heiberg Island beside 'mesic heath (*Cassiope tetragona, Trisetum spicatum, Potentilla hyparctica*)' also 'dry mesic heath (*Dryas integrifolia, Poa arctica, Thamnolia vermicularis*)', and so on. Often, expressions such as 'Dryas-heaths' and even 'Festuca-heaths', and so on are met with. The vagueness in the use of the term 'heath' can be explained by the fact that, in the English language, this word has two meanings: first wasteland, i.e., useless, 'wasted' land, and second dwarfshrub thickets of the heather family (Ericaceae).

The criterion of a closed vegetation cover, widely applied outside the USSR, forms the basis for the division of the Arctic into three major regions: the low-arctic, the middle-arctic, and the high-arctic. According to Polunin (1960: 382; cf. also Knapp, 1965: 92), a closed cover predominates in the low-arctic. In the middle-arctic, the vegetation cover is closed only over large parts of the lowlands; in the high-arctic, there is a patchy cover only under the most favorable conditions, usually with wide spaces between. More essential characteristics have been added by Rønning, who states that the low-arctic is where the vegetation is closed over large distances and is especially rich in grasses, sedges and shrubs; the middle-arctic is where the vegetation is closed only in lowlands and differs by the absence of such dwarfshrubs as *Betula nana* and species of *Vaccinium, Phyllodoce coerulea*, and others; the high-arctic, finally, is where the vegetation is met with only in patches or as widely separated, single individuals (Rønning, 1969: 29).

It should be noted that the degree of closedness of the cover is, to a considerable extent, related to the local and not to the circumpolar character of the vegetation. Thus, in the basin of the Khatanga and Anabar rivers in eastern Siberia 'spotty' tundras with barren nonsorted circles are distributed along the actual forest limit, and even within the northern edge of the open woodland there are spots of barren ground. Sochava (1933*c*: 361) wrote in a paper on the tundra of the Anabar basin: 'in the dwarfbirch-willow subzone, just as to the north of it, the spotty tundras are widely distributed. There is nothing in this subzone that can be considered less fundamentally distinctive than spotty tundras in the subzone of the arctic tundras'. At the same time, hummocky tundras with a closed vegetation cover have developed at the northern edge of the Yamal Peninsula, and so on. Also within the areas of the polar deserts, it is possible to find extensive patches with a closed cover of crustose, fruticose and foliose lichens, mosses and liverworts, in rare cases associated with flowering plants. Therefore, the extent of barren ground cannot be considered everywhere in the Arctic as a diagnostic characteristic for the distinction of major, zonal subdivisions.

The classificatory units of the Scandinavian school have also been utilized for the study of the vegetation within the limits of the Arctic.

Böcher (1954, 1963*b*; and others), studying mainly the sociations, used two parallel systems for their further unification. As the primary distinction of the higher units, he used principally a physiognomical classification: shrub communities, dwarfshrub communities, and so on (Böcher, 1963*b*: 268–72: etc.). Secondarily, he used groups of indicator species for the unification of sociations: Ar – arealogeographical, Cl – climatic, Hb – ecological, EG – ecological-geographical ones. By means of the indicator species he recognized 'groups of sociations' and 'vegetation types'. On the basis of these, he identified 'vegetation complexes' and 'vegetation areas', appearing similar to regional subdivisions (Böcher, 1954; 11–12). Proceeding from these units Böcher accomplished a profound, basic areal division of the vegetation of southwestern Greenland (Böcher, 1954).

Areal divisions based on physiognomic and floristic – physiognomic indicators have also been published: Porsild (1951: 11) divided the American Arctic into four provinces, Polunin (1951: 310) divided all of the circumpolar Arctic into ten sectors, and so on.

There exist also works on the classification of the arctic vegetation based on the methods of the Braun-Blanquet school. Thus, Hadač (1946) has described a number of associations from Spitsbergen and decided on their position in a system of units of higher rank; he included, for instance, the associations *Carisetum subspathaceae* Hč. and *Puccinellietum phryganodis* Hč. in the alliance *Puccinellion phryganodis* Hč., the order *Puccinellietalia phryganodis* Hč, and the class *Puccinellio–Salicornieta* Topa 1939, and so on. Rønning, when describing the *Dryas* tundras of Spitsbergen, placed them within the alliance *Dryadion* Du Rietz 1942 and the order *Seslerietalia* Br. -Bl. 1951 (Rønning, 1965: 11). Hofmann (1968) studied the vegetation of Spitsbergen with these methods; they have also been applied in Greenland (Molenaar, 1974) and arctic Canada (Thannheiser, 1975) and so on. However, the results of the classification according to the Braun-Blanquet school have not been used for the purpose of division into geobotanical areas.

Classification and division into areas of the arctic vegetation inside the USSR

Within the Soviet Union, the development of principles for the classification of vegetation aiming at a division into areas, has followed along physiognomic (in combination with this or that life form) and floristic-physiognomic lines from the very beginning. Thus, Trautvetter (1851) divided the East European tundras into two 'districts': 'the district of the alpine willows' (the arctic tundras) and 'the district of the low-grown birches' (the subarctic tundras). The southern boundary of the polar (according to Schrenk, 1854), the northern (according to Zhitkov, 1913), or the arctic (according to Pohle, 1910, and Gorodkov, 1916) tundras has been drawn along the northern limit of distribution of shrub thickets. Pohle (1910) distinguished also within the limits he had outlined for the 'subarctic zones' of the tundras in the European North, regions of the type of provinces: 'the western region' (from Norway to the Timan) and 'the eastern region' (from Timan to the Urals).

During the nineteen thirties, more serious contributions to the problems of classification and differentiation of the vegetation were put forward by Sochava (1933*c*, 1934*b*, for the Anabar Basin), Andreyev (1935 and others, for the East-European tundras), and Gorodkov (1935*c*). The latter accomplished a geobotanical areal division of the entire USSR tundra zone. As the highest classificatory unit, Gorodkov used the vegetation type, established on the basis of physiognomic, ecological, and partly also phylocoenogenetic criteria: the vegetation on snow (algae on the melting snow cover), on skeletal soils of the arctic deserts, on boulder fields, tundra-type vegetation, tundra meadows, subalpine shrub thickets, hydrophytic and mesophytic flood-plain meadows and shrublands, shrub thickets, forests, and bogs. His highest unit for the geobotanical division was the zone, and the second rank was the subzone. In addition, he distinguished twelve provinces, dividing all the subzones in a longitudinal direction. This remarkable work by Gorodkov has received wide acclaim and has been used as a basis for the illustration of the arctic territories on small-scale vegetation maps of the USSR (Lavrenko, 1939; Sochava, 1949, 1964*a*; etc.).

Leskov (1947) has suggested another system for geobotanical division. Basically, it differs from the Gorodkov system by the rejection of the concept of zone as a taxonomic category in the system of differentiating units. Guided by the principles advanced by Lavrenko (1946: 63–4; 1947) and differing from Gorodkov in his approach to the classification of the vegetation, Leskov distinguished four regions within the limits of the Soviet Arctic: (1) the high-arctic nival, (2) the arctic tundra, (3) the Euro-Siberian shrub region (including the European-West Siberian and Central Siberian provinces), and (4) the Beringian shrub region. The forest-tundras were included in the latter two. Subsequently, Norin (1957, 1961, etc.) formulated the concept 'forest-tundra type of vegetation' and raised the forest-tundra to the rank of a zone. Included in it was also the southern part of the subzone of the shrub-tundras, *sensu* Gorodkov.

An outline for a division, according to which the boundary between the arctic and subarctic tundras appears as the basic limit of first rank, was developed by V. B. Sochava. While using traits connecting the vegetation with important factors in the environment as well as with phylocoenogenetic criteria as diagnostic characteristics, he distinguished the 'arctic belt', including in it the arctic tundras and the polar deserts, and the 'humid' (Sochava, 1948*a*) or the 'temperate' belt (Sochava, 1952), uniting the subarctic tundras with the adjoining, more southerly areas. In the Soviet Far North three 'geobotanical fields' (subdivisions at the rank of province) were recognized within the limits of the arctic belt. The tundra areas of the temperate belt were divided into seven 'geobotanical fields'. This outline was later used with some alterations by Sochava for making a physical–geographical division of Asia into areas (Sochava & Timofeyev, 1968) and for drawing the northern boundary of the Subarctic (Sochava *et al.*, 1972: 3).

The contribution made by B. A. Yurtsev to the differentiation of the arctic vegetation cover should also be mentioned. Although the complex botanical–geographical differentiation by Yurtsev can be considered as floristic, since the characters of the flora appear as diagnostic traits, his results are of great importance for understanding the geobotanical differentiation, thanks to the fact that they take into account a wide array of descriptive characteristics: ecological, biological, coenological, and phylocoenogenetic ones (Yurtsev, 1968*b*) as well as the characteristics of the landscape related to the vegetation cover (Yurtsev, 1966: 17–19, etc.). Yurtsev distinguishes the botanical–geographical 'hyparctic belt' and the adjacent 'arctic' and 'boreal' belts. In spite of using different arguments, he draws the boundary between the arctic and the hyparctic belts just like Sochava along the southern limit of the arctic tundras, which confirms the great importance of this line as a boundary of highest rank. Yurtsev divides the hyparctic circumpolar belt into five provinces.

The system of regional units, as applied by Lavrenko (1947, 1968), was useful to me (Aleksandrova, 1964, 1971*b*). However I suggest separating vegetation types not according to Lavrenko's ideas but to a complex of characteristics, including the specificity of the structure, the character of the typical synusia, and so on. A reconsideration of the principles for the delimitation of the vegetation types of the tundras and of the polar deserts has resulted in a division of the Arctic into geobotanical regions: the polar deserts and the tundras. The latter region is divided into two subregions: the arctic and the subarctic tundras. My own points of view are expressed in detail below.

The higher taxonomic units for the classification of the arctic vegetation

The difficulty, when classifying the vegetation of the Arctic, is primarily connected with the clearly expressed co-dominance of the majority of the plant associations developed there: very often mosses, lichens, herbs, dwarfshrubs and semiprostrate shrubs co-dominate in the same phytocoenosis. Because of this, the application of the principles widely used in the USSR for differentiating higher units of the vegetation on the basis of the dominating biomorph, meets with considerable obstacles.

Supporters of the distinction of the vegetation types according to the dominating lifeform consider that in the tundra zone there is no single, zonal type of vegetation. They try to distinguish the tundra associations into a few types: the lichen-, the moss-, the dwarfshrub- and the shrub-types (Leskov, 1947, and others). However, as a rule, as was mentioned above, there appear in the tundra associations as co-dominants some life forms among which it is difficult to distinguish the main dominant for which this characteristic is the single one of the essential zonal characteristics for the zonal, mesic associations of the tundra zone.*

* Translator's note. The expression 'zonal, mesic association, habitat, etc.' is here used for the Russian term 'plakor', derived from a Greek word for 'surface, flatland', and is used to describe the zonal type of growth on mesic habitats, neither too wet, nor too

The tundra type of vegetation was identified by Gorodkov (1935*c*, 1946*b*; etc.), but Sochava had already expressed himself categorically on the identification of the tundra type of vegetation. He wrote that 'the tundra is a type of vegetation, – a plant community in the wide sense of that word, characterized by the following properties: it is from time immemorial without trees; it is dominated by arctic–alpine plants (by these are understood also the hyparctic forms; that term was introduced later by Tolmachev, V.A.), or by mosses and lichens; it has a special type of soil formation . . . and some other features' (Sochava, 1931: 127). Concepts of 'tundra types of vegetation' have been formed by many Soviet tundra geobotanists (Andreyev, 1954: 8, 12; Norin, 1966; etc.).

Katenin (1972*a*, *b*) distinguished the tundra associations of the East European forest–tundra into three vegetation types: the shrub-tundra type, the dwarfshrub tundra type, and the herbaceous tundra type according to the lifeforms of the plants which compose the upper tier of the community. The classification by Katenin appears logically well composed and better than many present ones, because of its factual basis. He has many comprehensive tables with complete species lists and each association is based on descriptions of two to sixty stands. However, the classification was worked out for the local conditions in the area which the author studied, and when proceeding northward and into areas with a distinctly continental climate, where the layers degrade and a horizontal mosaic develops, it becomes less applicable.

For the differentiation of vegetation types, I myself take into consideration a complex of diagnostic characteristics, including the combination of definite ecobiomorphs and geographical groups of species, the composition of the characteristic synusia, not necessarily the dominating one, but one closely connected with the type of community under consideration, and the characteristics of the structure. The types and subtypes described below can then be distinguished as somewhat deviating from those in my previous publications (Aleksandrova, 1971*b*). (An *ecobiomorph* is a life-form adapted to certain environmental (ecological) conditions; e.g. hekistothermal mesophilous

---

dry, neither too sheltered nor too exposed, and covered by neither too little nor too much snow.

dwarfshrubs, hekistothermal mosses, etc. Hekistotherms (DeCandolle, 1874) are plants able to grow and propagate in average July temperatures ranging from 2 to 10 °C. Yurtsev (1976) united these and some other groups of plants under the term 'cryophytes'.*)

1. *The tundra type of vegetation.* This type comprises co-dominant communities of hyparctic, arctic and arctic-alpine shrubs, dwarfshrubs, herbaceous perennials, mosses and lichens in various combinations. These plants belong to the hekistotherms and in part to the microtherm-hekistotherms and in relation to the moisture conditions to mesophytes (in a wide sense). The tundra communities have cryogenic mosaic composition and are developed on auto-morphic soils. As characteristic synusiae appear such with hekistothermic, semiprostrate shrubs and dwarfshrubs. Two subtypes can be distinguished: *a*) the *subarctic,* and *b*) the *arctic* one. The presence in the former of a characteristic synusia of semiprostrate, hyparctic shrubs (*Betula nana, B. exilis*, etc.) lacking in the second subtype appears as the basic diagnostic difference. When passing from the subarctic subtype into the subtype of the arctic tundras, the role of the hyparctic elements is sharply reduced, while the role of arctic and arctic-alpine species increases, and as the characteristic synusia there now appears the synusia of arctic–alpine dwarf shrubs. Below, these subtypes will be characterized along with the regional description of the vegetation.

2. *The polar deserts.* These are represented by plant associations of lichens (*Ochrolechia* spp., *Pertusaria* spp., *Toninia* spp., *Collema* spp., *Cetraria* spp., *Stereocaulon rivulorum* etc.), mosses (*Ditrichum flexicaule, Polytrichum alpinum,* etc.), and liverworts (*Cephaloziella arctica,* etc.) together with some hyperhekistothermal, arctic and mainly high-arctic, as well as arctic–alpine herbaceous plants (*Phippsia algida, Poa abbreviata, Papaver radicatum* s.l.), which have developed on auto-morphic high-arctic soils. (Note that I prefer to use the name *Papaver radicatum* s.l. for what in the literature, in relation to the area where it occurs, is called *P. polare* (Tolm.) Perf., *P. Dahlianum* Nordh., etc. and is often united into the collective species *P. radicatum*

* Translator's note: compare Löve & Löve, 1974, 1975.

Rottb. for the entire circumpolar Arctic.) The characteristic synusia is here composed of crustose lichens.

In non-zonal, non-mesic habitats, the following vegetation types may be met with:

3. *Thickets of hyparctic shrubs.* These differ from the shrub-tundras in lacking a cryogenic mosaic and in a number of other characteristics (Gorodkov, 1935*c:* 79; Andreyev, 1954: 12). The thickets of hyparctic shrubs occupy a wide area, reaching from the mountain taiga and northern taiga areas of the hyparctic belt far to the south of the limit of the tundra region. Two subtypes of the hyparctic shrub thickets can be distinguished. *a*) Thickets of hyparctic shrubs on watershed habitats on slopes where the snow is sufficiently deep but melts off early; characteristic components are *Betula nana, B. exilis, B. glandulosa, Salix lanata* s.l., *S. glauca* s.l., *Alnus fruticosa,* and *A. crispa.* (Following Komarov, 1936:308, and Yurtsev, 1974*a*, who take the genus *Alnus* in a wide sense, I accept *Alnus fruticosa* Rupr. as a species of the section *Alnobetula* Pouzar, and not as *Duschekia fruticosa* (Rupr.) Pouzar or *Alnaster fruticosa* (Rupr.) Ldb.) *b*) *Thickets of shrubs in valley habitats*, consisting mainly of willows, *Salix krylowii, S. phylicifolia,* etc., associated with some grasses and herbs.

4. *Hekistotherm meadow-like communities.* This vegetation type includes three subtypes. *a*) *The nival meadow-like communities* with predominantly mesophilous, hekistothermic, arctic–alpine herbaceous species such as *Poa alpigena, Lagotis minor, Astragalus alpinus, Pedicularis oederi,* etc., developed within the subregion of the subarctic tundras in habitats with a snowcover melting about midsummer. If, however, the snow melts much later, these types of meadows do not develop because of the excessively shortened vegetation period. Thus, according to e.g. Andreyev (1954: 12), the meadow-like nival communities do not develop in the East European subarctic tundras, if the snow-melt is delayed to the beginning of August. Some investigators (Korchagin, 1933; etc.) consider the nival meadow-like communities analogous to the meadows on watershed habitats belonging to the forest zone. However, as has been emphasized by Gorodkov (1938: 327), this analogy is quite far-fetched and it is instead necessary to look for a resemblance between the nival meadow-like communities on the

plains of the Arctic and the alpine meadows or meadow patches. *b*). *The meadow-like communities on well drained habitats with a favorable exposure,* in part getting an additional nitrogen–phosphate fertilization from beasts or man, and consisting predominantly of arctic–alpine herbs, mostly dicotyledonous of an extremely variable composition. These are typical for both the subarctic and the arctic subregions. *c*). *The cryoxeromesophilous meadow-like communities* are associated with groups of hekistotherm, xero-mesophilous monocotyledons, mainly arctic–alpine species, such as *Kobresia bellardii, Carex rupestris,* etc., together with occasional herbs, and formed in areas of both subarctic and arctic tundras as a closed turf on some patches of dry habitat, mainly affected by föhn-winds (e.g., in Greenland, on Wrangel Island, etc.). In the hekistotherm meadow-like communities, mosses and lichens are usually weakly developed or lacking. There is also a closed turf.

5. *Tundra–steppe and steppe types of vegetation.* To these belong the relicts from the cryo-arid epoch of the Pleistocene, described from northeast Asia by Yurtsev (1974*b*) and other investigators, and met with in the tundras, both the subarctic and the arctic ones, on dry, south-facing slopes. In these habitats present-day steppe species participate, such as *Carex duriuscula,* which is distributed over the steppes of southern Siberia, Mongolia, and North America, *Helichtotrichon krylovii,* etc., and usually also steppe-meadow species, such as *Carex pediformis, C. obtusata,* races of *Pulsatilla patens* s.l., *Silene repens* s.l., etc. Tundra species, *Dryas punctata,* etc., co-dominate with them in the tundra–steppes. Tundra–steppe communities are known also for Greenland (Böcher, 1954; and others; see below p. 86).

6. *Lichen and moss–lichen vegetation on boulderfields.* This vegetation type is represented by communities, for which the most typical are mosses such as *Rhacomitrium lanuginosum,* and lichens such as the epilithic, crustose *Rhizocarpon* spp., *Lecidea* spp., *Lecanora* spp. and others, the foliose *Umbilicaria* spp., etc., and the fruticose *Alectoria ochroleuca, Cornicularia divergens, Cetraria* spp., *Cladonia* (*Cladina*) spp., *Sphaerocarpus globosus,* etc. The patches of angiosperms vary in relation to subzone or area. This type of vegetation is widely distributed and met with not only in all the areas of the Arctic, where there

are boulder fields, but also outside its boundaries (e.g., in Iceland,* etc.).

7. *Homogeneous herb-moss tundra mires.* Two subtypes are distinguished *a*). *Glycophytic*† *associations* of hydrophilous, hyparctic and arctic herbaceous species (*Carex stans, Eriophorum angustifolium,* etc.) and mosses (*Calliergon* spp., *Drepanocladus* spp., etc.) with a peat layer from 30 to 60 cm thick in subarctic tundras and from 5 to 35 cm thick in the arctic ones. These mires may be considered as a tundra subtype of the widely distributed, homogeneous herb-moss-mire. *b*). *Halophytic mire* [*marsh*] *associations* on muddy, low marine beaches or spits (*Carex subspathacea, C. ursina,* etc.); to these are also referred communities of *Puccinellia phryganodes* and *Stellaria humifusa.* A number of authors call the latter 'meadows', but it does seem ecologically more correct to include them with the types of mires in spite of their lack of peat, as has been done by, e.g. Korchagin (1937) and Böcher (1954: 129).

8. *Hillocky and complex polygonal tundra mires.* Hillocky mires consist of flat-topped mounds, 5–10 m or more in diameter and rarely above 0.5 m tall, interspersed by shallow depressions with plant communities similar in composition to the cover of herb-moss mires. The mounds are genetically related to *palsas.* However, the flat peat-mounds of the tundra are distributed in the belt of 'continuous permafrost', differ in size and in shape, and have thinner peat layers. Characteristic components of the shallow depressions are *Carex chordorrhiza* and *C. rotundata* in the subarctic and *C. stans* in the arctic tundras. The polygonal (tetragonal) mires have three components: the depressed polygonal center (10 m or more in diameter), the raised borders, and the network of shallow depressions (troughs) along frost cracks with active ice wedges (veins). Typical components on the raised borders and on the slopes of the mounds in the subarctic are *Betula nana* and *B. exilis* and in some areas, shrubby willows and hyparctic dwarfshrubs.

* Translator's note: In my opinion, Iceland ought to be included in the subarctic; cf. Löve & Löve, 1979.

† Translator's note: 'glycophytes' are plants unable to thrive on habitats containing more than 0.5% sodium chloride in solution.

9. *High-arctic 'mineral' mires without peat.* Here belong associations of mosses such as *Orthothecium chryseum, Campylium* spp., *Bryum* spp., etc., and scattered occurrences of hyperhekistothermal and hekistothermal high-arctic herbaceous plants. Depending on the area, one or more of the following species may be met with: *Phippsia algida, Alopecurus alpinus, Deschampsia borealis, Eriophorum Scheuchzeri, Saxifraga cernua, S. rivularis,* etc.

10. *Meadows* are met with mainly in the river valleys of the subregion of the subarctic tundras, usually in the form of associations becoming more or less mossy, shrubby, or swampy, and with herbaceous species belonging to boreal, boreal–hyparctic, rarely arctic, grasses, sedges and herbs. The meadows in the tundra zone are a subtype of the boreal type of the meadow vegetation.

11. *Open woodland* with microthermal species of trees such as *Picea obovata, P. glauca, P. mariana, Larix sibirica, Betula tortuosa, Chosenia arbutifolia, Populus suaveolens, P. balsamifera,* etc., penetrating into the tundra region along river valleys, are situated in favorable habitats or met with in the form of relict forest islands, e.g. at Mor e-Yu in the Bol'shezemelya Tundra (Tolmachev & Tokarevskikh, 1968) and at Ary-Mas in eastern Taimyr (Tyulina, 1937; Norin *et al.*, 1971, Norin, 1978).

12. *Krummholtz communities.* Small patches of prostrate *Larix gmelinii* occur in the Khatanga River basin as far north as 72°55′ (Aleksandrova, 1937: 183, 197–8; the latitude given here is from the most recently issued maps). Dwarfed specimens of the montane species *Pinus pumila* penetrate into the subarctic tundras of northeastern Asia.

The distribution of these types of vegetation serves as a basic characteristic for the division of the Arctic into geobotanical areas of major rank (regions, subregions, and, in part, also of provinces).

Units used for the division into geobotanical areas

I have accepted the system of geobotanical division worked out by Lavrenko (1947, 1968): dominion, subdominion, region, subregion, province, subprovince, and district, but with some modifications based on my own experience with the vegetation cover of the arctic area. The ideas of Sochava (1948*a*, 1968, etc.) have also been utilized, especially the principles for the differentiation and co-ordination of the units for the division of the vegetation into areas.

The Arctic belongs to the Holarctic Dominion and to Arctogaea, its subdominion (Lavrenko, 1968).

I distinguish the *Region*, following Lavrenko (1968: 50–51), on the basis of the distribution on zonal, mesic habitats of the fundamental type of vegetation, lacking on zonal, mesic habitats in adjacent regions (although, in the latter, communities of this type may exist in non-zonal, non-mesic localities), and also on the basis of the presence of a specific set of non-zonal, non-mesic vegetation types. The lack of certain types of vegetation specific for the neighbouring regions (cf. Braun-Blanquet, 1964: 756) is also of diagnostic importance. Floristic differences belong to the characterizing traits, such as the basic composition of the flora typical only for the region under consideration, and the presence there of endemic taxa of major rank.

The latter characteristic, endemism, is less prominent in the Arctic than in regions with a richer and older flora. However, in the opinion of Yurtsev *et al.* (1975) 'the endemism in the Arctic region is actually not of such a low order, if you take into account the poverty of the flora (apparently, around 1000 species, not counting species scarcely penetrating into the southern part of the tundra zone from the boreal region) and its youth (probably one to three million years); besides, an important part of the endemic taxa is represented by an active, thriving and widely distributed element in the flora. Strictly endemic genera are: *Dupontia* (3 polyploid taxa), × *Arctodupontia,* × *Pucciphippsia,* and *Parrya* s. str. (according to the treatment by Bochantsev, considered as a monotypic genus of the tribe Arabideae with one species, *P. arctica,* whereas the remaining species of *Parrya* s.l. belong to the genus *Neuroloma* of the tribe *Matthioleae*), *Phippsia* (2 species) and *Pleuropogon* s. str. (1 species), all common to the Arctic, but known outside it only from disjunct localities mainly in subarctic high mountains, as arctic relicts; *Arctophila* is distributed outside the Arctic in the subarctic, herbaceous mires of northeastern Asia and northwestern America (often in permafrost areas). The latter three genera do possibly represent an alpine element in the mountain flora of the Tertiary

Arctic. There are quite a few endemic taxa above specific rank, e.g., in the genera *Taraxacum, Puccinellia, Draba, Papaver, Potentilla, Oxytropis, Cerastium, Gastrolychnis, Ranunculus,* and others. The number of endemic species and other taxa (not including the apomictic "microspecies") does exceed 100; among them, many are well-established taxa, others are very young races. There are also numerous subendemic species and races, met with outside the Arctic only in the high mountains of the subarctic, not rarely as relicts in specific habitats (e.g., *Ranunculus pygmaeus, Saxifraga tenuis, S. hyperborea,* etc.).' Lately, Yurtsev (1977) has convincingly demonstrated that the region in which the arctic flora is distributed should also include the major mountain ranges, mainly of eastern Siberia and northwestern America. This region merges with the cryophyte floras of the Arctic within the unifying contour of a 'meta-arctic flora' and even permits the addition to those mentioned above of a number of other taxa, which should be considered as endemics to the circumpolar region with a cryophytic flora. Among these, Yurtsev includes the genera *Gorodkovia, Ermania* s.str., *Novosieversia, Claytoniella,* section *Arctobia* s.str. of the genus *Oxytropis*, a number of groups in *Artemisia* and many others. In the polar deserts, distinguished by myself as an area at the rank of a region, endemic taxa are lacking, although clearly distinct, endemic plant communities are met with. Instead, the negative characteristics of the flora are sharply expressed, especially its particular composition. Thus, for example on Franz Joseph's Land, the list of families (listed according to the Engler system) stops abruptly at Rosaceae, while *Saxifraga* (22% of all the species in the flora) and *Draba* (17% of the species) appear as the most important genera.

Subregions are distinguished by the development on zonal, mesic habitats of different subtypes belonging to the vegetation type of the region under discussion (the characteristic synusiae being represented by groups of species, differing in biology, ecology, and morphology, and belonging to different geographical elements of the flora and of different origin), but also by a special set of non-zonal, non-mesic types of vegetation. The lack of this or that non-mesic, non-zonal formation may also serve as a diagnostic characteristic, e.g., the lack of shrub thickets, etc., in the subregion of the arctic tundras.

The *Provinces* are distinguished first and foremost by the presence of classes and groups of associations, endemic to the province, which are represented by a zonal, mesic type of vegetation, and also by the character of the set of non-zonal, non-mesic classes and groups of associations (sometimes also by some non-zonal types of vegetation) in the province under discussion. The provinces are major areas, each of which differs by its special late- and post-glacial paleogeography and the special florogenesis and phylocoenogenesis connected therewith, but also by the present major climatic differences (cf. Sochava, 1948*a*). The nature of the flora appears as a distinctly expressed characterizing property in that each geographical province has a well expressed aspect, typical only to itself, consisting of a geographical, 'longitudinal' (amphi-atlantic, Siberian, Beringian, etc.) element of the flora as well as definite complexes of endemic species and races.

The *Subprovinces* are distinguished by the presence of endemic, phylocoenotic units of lower rank (subclasses and groups of associations) and by the extent of the relative participation of this or that class or group of associations. Of importance are those characteristics, among which Karamysheva & Rachkovskaya (1973: 176) include the following: the different structure of zonality, the width of the zonal belts (some of which are sometimes omitted), and the smooth or 'distorted' course of the zonal transition. I myself also take into consideration the particular macrostructure of the vegetation cover as expressed in the distinct mosaic of geobotanical areas, in particular the presence or absence of extra-zonal areas against the zonal background in relation to local climatic deviations (e.g., on the island of West Spitsbergen, on Wrangel Island, etc.). As one of the characterizing subprovincial properties, the alteration of aggressive species should be mentioned (such as the intense aggressivity of *Deschampsia brevifolia* in the Novaya Zemlya–Vaigach subprovince being sharply curtailed in the adjacent subprovinces, etc.).

I distinguish the *District* as the specific combination of groups of associations, typical for the subprovince in question, and the associations formed mainly because of its special orography, type of soil-forming bedrock, and also the particular local, climatic conditions. However, in the districts, it is also possible to observe variants in the floristic composition of the plant associations related to the occurrence in this or that district of species from a different type of area (e.g., of atlantic species in the western Novaya Zemlya district, and of Siberian species in the eastern Novaya Zemlya district). As stated by

Karamysheva & Rachkovskaya (1973: 180), 'most frequently, these are species indicating a connection between this territory and the surrounding ones and appearing as evidence of paleogeographical events'.

I consider the terms of the latitude zonation (zone, subzone, belt) as complementary ones, having no taxonomical ranks within the geobotanical division.

When drawing up the boundaries, I happen to clash with the geographical continuum as one of the expressions of the generally particular continuities inherent in the vegetation cover (cf. Aleksandrova, 1969*a*: 12–22, etc.). The problem of geobotanical boundaries in connection with the continua at the regional level has been discussed by Sochava (1965, etc.). The degree of smoothness in transition from one area to another and the shape of the borders may be entirely different (Karamysheva *et al.*, 1969). In the Arctic, the change in vegetation occurs more smoothly latitudinally, while the transition from one province to another is more 'abrupt'. At the same time, it is necessary to take into consideration that using one characteristic may make the progress look gradual, while another character may permit us to decide at what boundary there is a definite limit. Thus, in general, the zonal, mesic associations in the tundras change more smoothly, and the non-zonal, non-mesic ones considerably more abruptly. The progress from the northern belt of the subarctic tundras to the southern belt of the arctic ones, as described below, pp. 88–90) may serve as an example.

2

The geobotanical regions of the Arctic: the tundra region

In the territories spreading northward from the forest limit, we can distinguish two regions (Fig. 1)–the tundra region, and the region of the polar deserts. Diagnostic characteristics for differentiating these areas are: (1) the development on zonal, mesic habitats of different types of vegetation of tundra and of polar desert type, none of which is met with on zonal, mesic habitats in adjacent regions (but which may be found on non-zonal, non-mesic habitats: tundra associations may occur in especially favorable local conditions in the southern belt of the polar deserts, and polar desert associations in the definitely unfavorable conditions in the northern part of the tundra region); (2) the lack in the region of the polar deserts of the majority of the non-zonal, non-mesic types of vegetation, developed in the tundra region (described above under nos. 3–5, 7, 8, 10–12). As an especially conspicuous characteristic appears the lack in the polar deserts of mires with a peat layer so that in connection with it the polar deserts are situated north of the area of *Carex stans*.

The characterizing properties of the flora confirm the essential differences of the vegetation covers in these regions. The hyparctic element of the flora is lacking in the polar deserts, although it still plays a decisive role in the arctic tundras and is predominant in the tundras of the subarctic. In the polar deserts, a number of families are lacking such as Polypodiaceae, Liliaceae, Betulaceae, Empetraceae, Vacciniaceae, and many others. High-arctic species predominate, such as *Phippsia algida*, *Poa abbreviata*, etc. The sharp decrease in the number of species of vascular plants, which apparently do not exceed 50–60 species, is also typical. In spite of the fact that the floristic difference is of a negative nature, it does represent a complex, qualitatively

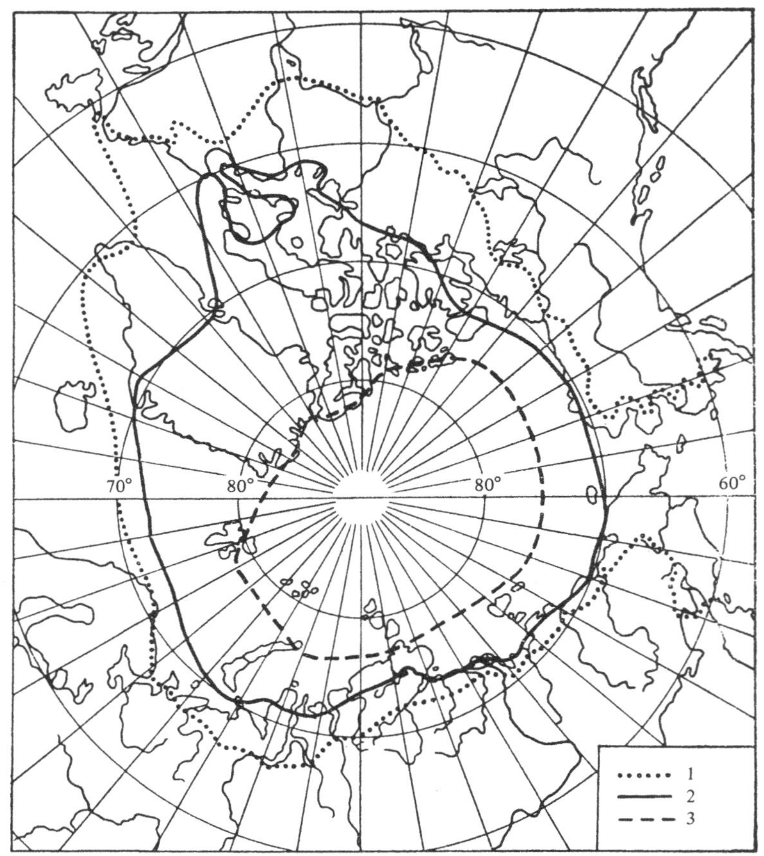

Fig. 1. The geobotanical regions and subregions of the Arctic. The southern boundary of: 1, the tundra region; 2, the subregion of the arctic tundras; 3, the region of the polar deserts.

different from that of the tundra region. The region of the polar deserts agrees most completely with the '1st zone' of the floristic zonation as outlined by Young (1971) and with the 'glacial variant of the subzone of the higharctic tundras' according to the botanical–geographical zonation by Rebristaya & Yurtsev (1973).

This chapter gives a more detailed description of the tundra region and Chapter 3 of the polar desert regions.

The tundra region is distinguished by the development on zonal, mesic habitats of a tundra type of vegetation (cf. p. 10) together with herb-moss, polygonal, and hillocky mires with flat-topped peat mounds, as well as other non-zonal, non-mesic types of vegetation (cf. nos. 3–8, 10–12, pp. 11–14). Open woodlands penetrate as extra-zonal associations from the south, and polar desert-communities from the north.

Two subregions can be distinguished within this region: the subarctic tundras, and the arctic tundras.

It is correct to divide the tundra region into two areas of subregional rank on the basis of a combination of important characteristics, of which the presence on zonal, mesic habitats of particular subtypes of vegetation is decisive. Typical for these subtypes are the different characteristic synusiae, distinguished not only according to species composition, but also in respect to the ecobiomorphs: in the subregion of the subarctic tundras, these synusiae consist of semiprostrate hyparctic shrubs (*Betula nana*, *B. exilis*, *B. glandulosa*, etc.) and dwarfshrubs (*Vaccinium uliginosum* ssp. *microphyllum*, *Empetrum hermaphroditum*, *Ledum decumbens*, etc.), but in the subregion of the arctic tundras, the synusiae are composed of arctic–alpine and arctic dwarfshrubs (*Salix polaris*, *Dryas octopetala*, *D. punctata*, *D. integrifolia*, *Cassiope tetragona*, etc.). The reduction in the number of non-zonal, non-mesic types of vegetation in the subregion of the arctic tundras is another important diagnostic difference: the thickets of shrubs disappear – an especially important character – there are no snow-dependent nival meadow-like communities, and associations of prostrate elfin-wood and open woodland are no longer met with.

A still higher rank is given to the boundary between the arctic and the subarctic tundras by Sochava and Yurtsev, who draw a boundary between different botanical–geographical belts along this line. According to Sochava (1948*a*, *b*, 1952), the arctic tundras belong to the arctic belt, characterized by circumpolar arctic fratria of formations (Buks, 1973; Il'ina, 1975), while the subarctic tundras belong to the temperate, or subarctic belt (Sochava *et al.*, 1972), where different fratriae of formations occur. According to Yurtsev (1966, 1974*a*), the subregion of the arctic tundras belongs to the arctic botanical–geographical belt, while the subarctic (in his terminology: the hyparctic) tundras are part of the hyparctic belt.

To define the transition from one subregion to the other, the principal differences in the history of the formation of the vegetation within the arctic and the subarctic tundras must be considered, as well as the critical decrease in the average amount of heat, which leads to a change in all physico-geographical complexes. As has been pointed out by Tolmachev & Yurtsev (1970), the hyparctic flora, predominating in the subarctic tundras, has developed from components of degenerating forest and non-forest associations from the Pliocene, pre-adapted to the increasingly drier and colder conditions during the late Pliocene and the Pleistocene; these are various communities of shrubs and open woodland, predominantly of deciduous species, such as willows, shrubby birches, and alders, different types of bog-associations, etc. A decisive influence was exerted during the initial formation of the arctic tundras by the alpine tundra (the 'gol'tsy')* of northeastern Asia, particularly in its continental sector. According to Tolmachev & Yurtsev (1970: 90–92) a number of species have penetrated into the Arctic from the 'gol'tsy', such as *Cassiope tetragona*, *Dryas punctata*, *Novosieversia glacialis*, *Erysimum pallasii*, including the definitely arctic species of *Pedicularis*, the precursor of *Braya purpurascens*, and species of *Ranunculus*. Other species originated from close to central Asia. There are also arctic–alpine taxa which now have a circumpolar distribution and some species with an Angaran connection florogenetically, such as *Kobresia bellardii*, *K. simpliciuscula*, *Carex rupestris*, and even circumpolar species of Asiatic origin, such as *Oxyria digyna*, *Alopecurus alpinus*, *Saxifraga cernua*, *Eriophorum scheuchzeri*, *Potentilla hyparctica*, *Eutrema edwardsii* and *Gastrolychnis apetala*; some species represented a special alpine element, derived from the ancient Beringia: *Cardamine bellidifolia*, *Salix arctica*, *Luzula confusa*, and *Poa arctica*. Species can also be traced which have connections with the American mountains (*Poa abbreviata*, *Pleuropogon sabinei*) or xerophytic highlands (*Lesquerella arctica*, etc.) and so on.

The difference between the arctic and the subarctic tundras is enhanced by the fact that the arctic tundras have remained treeless throughout all the periods of the Holocene, a fact confirmed by paleobotanical data (Giterman *et al.*, 1968; etc.). Besides there occurred a repeated expansion of forest into the subarctic tundra territory during periods of warming climate, leading to the enrichment of its

* Translator's note: 'gol'tsy' is a Siberian term for the alpine tundra belt above the timberline.

flora by boreal species as reflected in the composition and structure of its associations.

The boundary between the subarctic and the arctic subregions of the tundra region oscillates around the 6 °C isotherm for July.

The subregion of the subarctic tundras

In zonal, mesic habitats of this subregion, plant communities have developed which belong to the subarctic subtype of the tundra type vegetation. As typical synusiae, hyparctic shrubs and dwarfshrubs dominate or participate. The most typical of these, having a primary diagnostic value, are the polar birches, *Betula nana*, *B. exilis*, *B. middendorffii* and *B. glandulosa*. In non-zonal, non-mesic habitats, different vegetation types are found: herb-moss, lowland, oligotrophic mires with a 30–60 cm-thick peat layer and heterogeneous hillocky and polygonal mires, as the typical components of which appear *Betula nana*, *B. exilis* and other hyparctic shrubs and dwarfshrubs, growing on the sloping parts of frost peat mounds and on raised borders of low-center polygons, thickets of shrubs in the watershed areas and in valleys, meadow associations in river valleys, nival and other hekistothermic tundra-like communities, tundra–steppe and steppe associations (in northeastern Asia and in Greenland), and finally open woodland and prostrate krummholz communities.

The important role played in the vegetation cover by the boreal element of the flora together with the predominating hyparctic element is typical for the subregion of the subarctic tundras. To the boreal element belong first and foremost those arboreal species, the areas of which penetrate far into the subregion of the subarctic tundras (*Larix gmelinii* in prostrate form up to 72°55′ N) and a large number of species of herbs and dwarfshrubs (*Deschampsia flexuosa*, *Calamagrostis langsdorffii*, *Comarum palustre*, *Chamaenerium angustifolium*, *Vaccinium myrtillus*, etc.). At the same time, the participation of arctic species is distinctly expressed, especially of such moderately arctic species as *Carex ensifolia* ssp. *arctisibirica*, *C. lugens*, *Arctagrostis latifolia*, *Arctophila fulva*, etc. The enrichment by boreal species and the generally hyparctic character of the flora is connected with the history of its formation, which has been described above.

Within the limits of the subregion of the subarctic tundras, a number of geobotanical provinces can be distinguished. At the same time, a latitudinal zonality has developed (Fig. 2) in the form of latitudinal belts, which have been described from different parts of the subarctic tundras by a number of authors (Sochava, 1933*c*; Andreyev, 1935, 1954; Gorodkov, 1935*c*; Böcher, 1954; Porsild, 1957*b*; Gribova, 1972; etc.).

The composition of the plant communities within each of the belts is specific for the different provinces. But there are also some general, characteristic changes within the vegetation cover which can be followed around the pole during the progression from south to north. In this respect, the following traits appear as general: the gradually decreasing role in the vegetation cover of hyparctic shrubs, of which the birches play the most important diagnostic role; the change in growth type of the birches in the zonal, mesic phytocoenoses from 'more or less upright' (Polozova, 1966) through 'decumbent' to a 'prostrate' growth form; the decreasing part in the composition of associations of boreal species; the gradual decrease in the vegetation cover of the types of mires with tall flat-topped peat mounds and of river valley meadows; the disappearance from the northern part of the subregion of open woodland and prostrate krummholz formations.

In the subregion of the subarctic tundras, I distinguish three belts (Aleksandrova, 1971*b*), not two, as previously (Aleksandrova, 1970). Mel'tser (1973) and Andreyev (*in* Andreyev & Nakhabtseva, 1974; Andreyev, 1975) have also adopted this idea. The latter author prefers to use the term 'subzone' rather than 'belt', but because both belt and subzone are concepts without rank in the system of units, which I prefer, the question of which of these terms is to be used is of no consequence.

The southern belt of the subarctic tundras

The southern belt is distinguished by the dominance of associations with predominating hyparctic species, of which the most typical are the shrubs *Betula nana*, *B. exilis*, *B. middendorffii*, *B. glandulosa*, *Alnus fruticosa*, *A. crispa*, *Salix glauca*, *S. phylicifolia*, *S. planifolia*, etc. together with a rich variety of boreal species in the flora. Woody species such as *Picea obovata*, *Larix sibirica*, *L. gmelinii*, *L. cajanderi*, *L. laricina*, *Chosenia arbutifolia*, *Populus suaveolens*, *P. balsamifera*, etc., penetrate into this belt along the rivers in the form of patches of

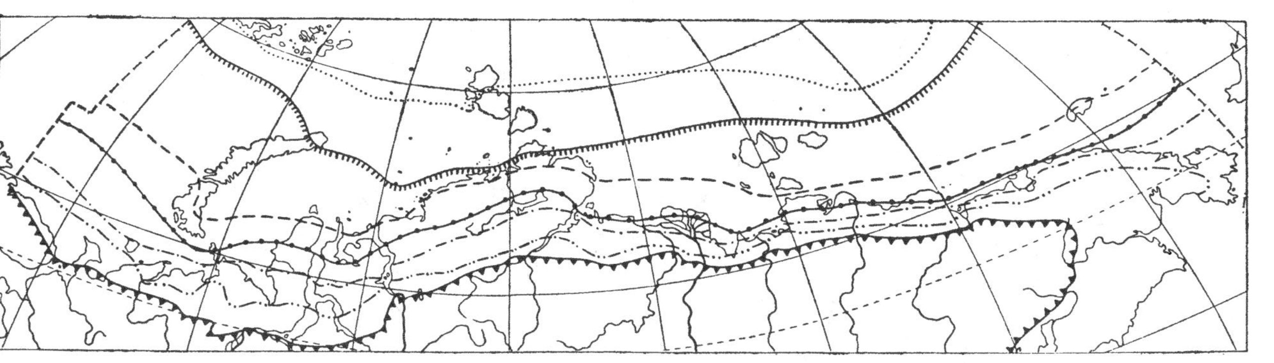

Fig. 2. Latitudinal zonation within the limits of the Soviet Arctic. Southern boundaries of the belts of the Subarctic tundras: (1, of the southern; 2, the middle; and 3, the northern), of the arctic tundras (4, the southern; 5, the northern), and of the polar deserts (6, the southern; 7, the northern).

open woodland, but also as krummholz (prostrate *Larix gmelinii*) or 'crooked' (*Betula tortuosa*) forests.

In the western areas of the USSR tundra zone, the zonal, mesic habitats are in the southern part of the belt occupied by tundras with tall-grown dwarfbirch associations, reaching up to 80 cm, and rarely 1 m, in height, and in the northern part of the belt by tundras with low-grown dwarfbirch associations. On the Kola Peninsula, as well as in Greenland, there are groves of 'crooked' *Betula tortuosa*, which continue to appear all the way to the right-hand banks of the Yenisey River. In the areas of eastern Siberia where there is a continental climate, there appear in the zonal, mesic habitats of the southern subarctic subzone tundras with low-grown associations of *Betula exilis* with a decumbent growth pattern, dependent on the scanty amount of snow cover. The tallest growth occurs among the willows in valleys in species such as *Salix krylovii*, but also in thickets of *Alnus fruticosa*. In Chukotka and Alaska, tussocky tundras of *Eriophorum vaginatum* and *Carex lugens*, and in the Anadyr–Penzhina subprovince also of *C. soczaveana*, with some admixture of birch and willow, predominate in the belt under discussion. Frequently, *Pinus pumila* occurs in the tall-grown shrub tundras together with *Betula middendorffii* in the Anadyr–Penzhina subprovince. In the Canadian Arctic, tundras predominate with a dominance of hyparctic dwarfshrubs together with some *Betula glandulosa*, and in the Greenland province of the subarctic tundras, there are hyparctic dwarfshrub tundras with *Betula nana*. In this belt there are, beside flat-mounded, polygonal and lowlying homogeneous mires, also mires with flat-topped peat mounds, up to 1.5 m tall, on the slopes of which grow shrub thickets.

The middle belt of the subarctic tundras

In the middle belt of the subarctic tundras, the birches, *Betula nana*, *B. exilis*, and *B. glandulosa*, continue, as a rule, to play an important role in the composition of the zonal, mesic phytocoenoses, but they begin to exhibit a creeping growth form, not reaching more than 20–25 cm in height. Among the shrubs, the willows become important compared to species of *Betula* and *Alnus*. There are many hyparctic dwarfshrubs (*Vaccinium uliginosum* ssp. *microphyllum*, *Ledum decumbens*, etc.) in the vegetation cover. The participation of the boreal floristic element in the plant associations is gradually diminishing compared with that in the southern belt of the subarctic tundras while, at the same time, the abundance of the arctic species (*Carex ensifolia* ssp. *arctisibirica*, etc.), in relation to that of the absolute predominance of the hyparctic element, is growing. In this belt, mires are still found with rather tall peat mounds, and rarely, there are open woodland associations and prostrate krummholz; thus, there are relict 'islands' of open woodland at Ary–Mas in the Khatanga basin, which occur within the limits of the middle belt of the subarctic tundras.

The northern belt of the subarctic tundras

In the zonal, mesic habitats of the northern belt, the role of the shrubs is sharply curtailed in the composition of the vegetation cover. Although creeping forms of birches are still found in the zonal, mesic phytocoenoses, they are not universally present. More often, ground-hugging willows such as *Salix pulchra* and *S. reptans* are part of the associations, but even these do not form a pronounced, but rather a fragmentary, tier. *Alnus fruticosa* is not met with any more. Side by side with the hyparctic dwarfshrubs such as *Vaccinium uliginosum* ssp. *microphyllum*, arctic dwarfshrubs begin to appear here, *Dryas* spp., *Cassiope tetragona*, and others. The shrub thickets are represented by associations of willows (*Salix lanata*, *S. lanata* ssp. *richardsonii*, *S. glauca* ssp. *callicarpaea*, etc.).

In the northern belt of the subarctic tundras, the role of the boreal element of the flora is strongly diminished and the importance of the arctic and arctic–alpine element is increased. But, as before, the basic nucleus of the vegetation is formed by the hyparctic species. In this connection, Yurtsev (1966, etc.) considers that this belt, as well as the other two belts of the subarctic tundras, should belong to a 'hyparctic, botanical–geographical belt'. In such a belt the zonal, mesic associations display some characteristics of arctic tundra associations. But the presence of birches in the zonal, mesic habitats, their common occurrence on the slopes of peat mounds and on the raised borders of the low centre polygons in the polygonal mires, and the presence of well developed willow thickets, e.g. in the area of Tareya on the banks of Pyasina River (Taimyr Peninsula) appear as diagnostic characteristics, categorically indicating their relation not to the arctic, but to the subarctic tundras.

The provinces of the subregion of the subarctic tundras

Within the limits of the subarctic tundras, there are clearly expressed provincial differences. These are caused first and foremost by the distinct heterogeneity of the climatic conditions, not only nowadays, but also during the Pleistocene: the presence of an oceanic climate in the atlantic areas, the effect of the East Siberian anticyclone, and so on. The connection between these major climatic subdivisions and the geobotanical provinces has been convincingly demonstrated especially by Sochava (1948*a*). The orographic factor is of importance mainly for the division into districts, but may also affect the appearance of differences at provincial rank in, e.g., the Greenland province.

A major role is played by the essential, paleogeographical differences between the various areas within this territory during the Pleistocene, particularly by the different extent of glaciation and submergence due to marine transgression during various periods of the late Glacial, but also by the presence, on the one hand, of land bridges allowing dispersal of species, and on the other, of barriers preventing it. Side by side with areas fully covered by glaciers, such as eastern Europe, western Europe, western Siberia, and the continental tundras of Canada, there were wide expanses where there was no complete glacier cover, as e.g. in eastern Siberia, Chukotka, and Alaska. Land bridges, over which a floral dispersal was possible, were formed mainly at the time of the maximal emergence of the continental shelf between Chukotka and Alaska (cf. Yurtsev, 1974*a*, 104–110) and, perhaps, between Scandinavia and southern Greenland over Iceland and the Faroe Islands (Heezen & Tharp, 1963 and others). In contrast, the deep penetration into the continent of the aquatic synclynal, the 'Tyrrell Sea' around Hudson's Bay in Canada served as a barrier, preventing floristic exchange.

In the first place, the provinces can be distinguished as geobotanical categories by the presence of endemic classes and groups of plant associations as represented by the zonal, mesic type of vegetation, but also by the characteristic sets of non-zonal, non-mesic communities, not only at the level of classes and groups of communities, but also in a number of cases by the type of vegetation. Thus, the cryo-xerophytic types of vegetation (cf. nos. 4*c* and 5, p. 12) are related to just the kind of definition of the province as follows below. Definitely important characterizing traits are: the expression of floristic differences in the form of definite combinations of geographical elements in the flora; the presence of groups of endemics, specific for each province; as well as the occurrence of differentiating and co-differentiating species (Yurtsev *et al.*, 1975).

Five geobotanical provinces can be distinguished in the subregion of the subarctic tundras: the East European–West Siberian, the East Siberian, the Chukotka–Alaska, the Canadian, and the Greenland provinces.

The East European–West Siberian province of the subarctic tundras

As endemic to the southern part of this province on zonal, mesic habitats there are in its southern part tundras with widely distributed dwarf birch associations, having a closed tier of *Betula nana* and a moss carpet, in which an important part is played by forest mosses, *Hylocomium splendens, Pleurozium schreberi, Polytrichum alpestre*, and others. The distribution over a large part of the province of *Betula nana* associations with an admixture among other species of *Carex ensifolia* ssp. *arctisibirica* is also typical. These types of associations are not met with in the adjacent provinces.

The territories covered by the East European–West Siberian province are mainly represented by a lowland of their own, the surface of which is composed of marine and glacial loamy and sandy deposits. There are some low heights, up to 400 m *s.m.* such as the Timan Ridge, and the Pai-Khoy Highland in the East European North, where the paleozoic bedrock in the majority of cases juts above the marine material, as well as the Byarranga Foothills on the right-hand bank of Pyasina River. (Here, and elsewhere in the text, the elevation is expressed in meters above sea level.) A part of the Ural Mountains protrude from the south into the flat plains, reaching elevations of up to 1100 m within the tundra zone. From the west, the tundra part of the Kola peninsula juts out with its predominant outcrop of bedrock up to 300 m.

The most important condition, both from a paleogeographical and present-day point of view, which causes the originality of the existing flora, is that from the Pleistocene to the present time there has been an intense, cyclonal activity with a constant invasion of atlantic air masses, causing an increased precipitation in comparison with that of the majority of the other provinces and giving rise to a humid climate. One of the consequences is that this province lacks any xerophytic types of vegetation, such as hekistothermal xeromesophytic meadow-like communities and tundra–steppe associations.

During the maximal stage of the Glacial period, the land of this province was completely covered by an ice sheet, and during the maximal marine transgression, a considerable part of it, especially the west Siberian and the Taimyr lowlands, was covered by water. This led to a lack of refugia for a glacial-age flora and to the youthful appearance of the present-day vegetation in its phytocoenotical development in comparison with those parts of the Arctic which have not been subjected to glaciation.

Characteristic of the flora of this province are European species such as *Salix myrsinites, Luzula arcuata, Leymus arenarius, Trollius europaeus*, etc., atlantic and amphi-atlantic ones, such as *Vahlodea atropurpurea, Carex saxatilis* ssp. *saxatilis, Cerastium cerastoides, Chamaepericlymenum suecicum, Diapensia lapponica*, etc., and arctic-alpine and hyparctic species met with also in the mountains of central and southern Europe, such as *Betula nana, Salix hastata, S. herbacea, Saxifraga hieracifolia*, etc. European–west Siberian species not penetrating farther east than the right-hand tributaries of Pyasina River: *Salix phylicifolia, Gastrolychnis angustiflora* ssp. *angustiflora, Salix lapponum, Aconitum septentrionale, Polygonum bistorta*, and others are typical, and species with wider areas, the eastern limit of which (within the limits of the tundra zone) are found also in the eastern part of the Pyasina basin (*Poa alpina, Anthoxanthum alpinum, Diapensia lapponica*, etc.). *Betula tortuosa*, the eastern limit of which reaches to the right-hand bank of Yenisey River, belongs also to this group.

The very high saturation with boreal species such as *Vaccinium myrtillus, Solidago virgaurea, Trientalis europaea, Dianthus superbus, Deschampsia flexuosa, D. caespitosa, Chamaenerium angustifolium, Aconitum septentrionale, Alopecurus pratensis*, etc., is typical for this province in comparison with those adjoining it from the east. The same is valid for the fruticose forms of *Cladonia* (*Cladina*) among the lichens. When approaching the east, these yield space to an increasing amount of *Cetraria cucullata.*

Within the limits of the east European–west Siberian province of the

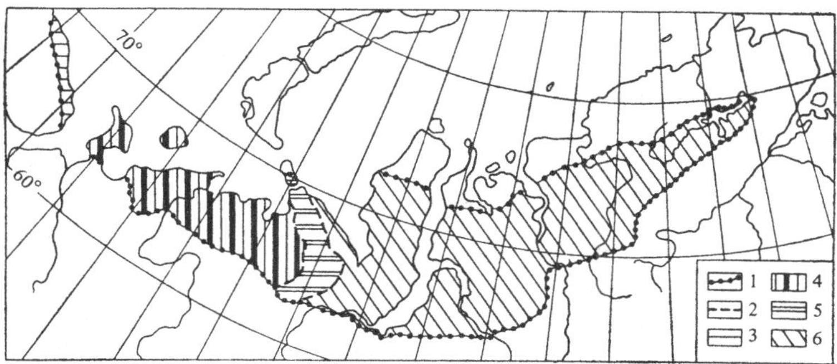

Fig. 3. The east European–west Siberian province of the subarctic tundras. Boundaries: 1, provincial; 2, subprovincial. Subprovinces: 3, the Kola; 4, the east European; 5, the Urals-Pai-Khoy; 6, the Yamal–Gydan–West Taimyr.

subarctic tundras, four geobotanical subprovinces can be distinguished: the Kola, the east European, the Urals–Pai-Khoy, and the Yamal–Gydan–west Taimyr subprovinces (Fig. 3).

The Kola subprovince of the subarctic tundras

The tundras on the Kola peninsula occur in latitudinal (not vertical!) formations only in a narrow coastal belt, the relief of which consists of low heights (up to 300 m.) composed of Precambrian bedrock, and numerous lowlands occupied by lakes and mires. The rock outcrops alternate with areas covered by loose sediments. The Murman Current runs along the coast, contributing to a mild climate of oceanic character; the average yearly temperature is around 0 °C, the temperature of the warmest months is 8–9 °C, and that of the coldest, −9 °C.

The vegetation of this subprovince (cf. Kihlman, 1890; Regel, 1923, 1927, 1928; Tsinzerling, 1932, 1935; Salazkin, 1934; Salazkin *et al.*, 1936; Minyayev, 1963; Breslina, 1969, 1970; Ramenskaya, 1972; Manakov, 1972; and others) is represented by a belt of southern subarctic tundras, where there are patches of 'crooked forest' of *Betula tortuosa* right up to the sea coast. The exceptionally strong concentration of boreal species in the flora served as a basis for Yurtsev (1974*a*:142–3), when he placed the Kola coast tundras in the floristic Boreal region although, according to his system of botanical–geographical zonation, a system in which the degree of activity of the individual species is the important characteristic, this area could also be placed in the Hyparctic belt (Yurtsev, 1966).

The largest area is occupied by petrophytic dwarfshrubs, dwarfshrub–lichen, and shrubby tundra associations.

In the last of these associations, the shrub tier is formed by *Betula nana*, growing up to 70 cm tall, sometimes with admixture of *Salix glauca*. In the group of lichen associations, *Cladonia alpestris, C. mitis* and *C. rangiferina*, and sometimes *Cetraria nivalis*, dominate in the ground cover. In the group of herb-moss associations, there are such mosses as *Pleurozium schreberi, Hylocomium splendens*, and others, and herbaceous species such as *Deschampsia flexuosa, Carex bigelowii*, and a number of other species among which the boreal herbs, *Solidago lapponica, Trientalis europaea*, etc. are noteworthy. On the deeper soils on slopes, *Salix glauca* is mixed into the *Betula nana* thickets.

Willow thickets, growing up to and above 1 m tall on the ledges of slopes and along rivers, consist of *Salix glauca* and *S. lanata* with an admixture of *S. lapponum, S. phylicifolia*, and *S. myrsinites*. Among the willows grow tall herbs, e.g. *Cirsium heterophyllum, Geum rivale, Archangelica norvegica, Filipendula ulmaria, Solidago lapponica*, etc.

The widely distributed dwarfshrub and dwarfshrub–lichen tundras are represented mainly by associations in which *Empetrum hermaphroditum* dominates or is a constant member. On the Kola Peninsula, it forms not only tundras, differing in mosaic composition by the addition of mosses and lichens, but it is also part of the blueberry heaths (Breslina, 1970), which represent a particular hyparctic formation, widely distributed in the meadow-heaths of the Atlantic and Pacific regions of the Holarctic (Lavrenko, 1950; Yurtsev, 1966). In the dwarfshrub tundras of the Kola Peninsula, *Empetrum hermaphroditum* co-dominates often together with *Vaccinium myrtillus* and there are frequently prostrate *Betula nana, Vaccinium uliginosum, V. vitis-idaea, Phyllodoce coerulea*, and other species. Mosses, *Hylocomium splendens, Pleurozium schreberi*, etc., are usually met with in separate spots. In the lichen dwarfshrub associations *Empetrum hermaphroditum* dominates together with an admixture of *Vaccinium myrtillus* but is also associated with some *Arctous alpina*; the *Cladoniae* predominate among the lichens: *C. alpestris, C. rangiferina, C. mitis*, and more rarely there are *Cetraria nivalis* or *Stereocaulon paschale*.

Associations with dominating *Chamaepericlymenum suecicum* are also typical for the Kola Peninsula. This is a hyparctic–boreal species with disjunct amphi-atlantic and amphi-pacific areas.

The forest limit along the Murmansk coast is formed by *Betula tortuosa* and *Pinus lapponica.* Locally, also, *Picea obovata, Betula callosa*, and *B. subarctica* are met with there.

The Kola subprovince of the subarctic tundras also differs from the East European subprovince, besides other characteristics, in the appearance in its flora of amphi-atlantic species such as the sedge *Carex bigelowii* s. str. and by the absence of *Alnus fruticosa.* The tundras of the Kola subprovince are extremely closely related to the mountain tundras of northern Scandinavia and show traces of resemblance also to subarctic Greenland. This may be connected with the supposed existence during the late Pleistocene of a land bridge between Europe and southern Greenland over Iceland and the Faroe Islands (Heezen & Tharp, 1963; Hadač, 1960: 237).

The East European subprovince of the subarctic tundras

The vegetation of the East European subprovince of the subarctic tundras reaching from the White Sea to the Polar Urals and including Kolguyev Island and the southern part of Vaigach Island, has been studied by a number of authors (Trautvetter, 1850; Schrenk, 1854; Tanfil'yev, 1894; Pohle, 1903; Tolmachev, 1923; Perfil'yev, 1928, 1934; Andreyev, 1931, 1932*a*, *b*, 1935, 1954, 1966; Sambuk, 1931, 1933, 1934; Sambuk & Dedov, 1934; Korchagin, 1937; Bogdanovskaya-Gienef, 1938; Smirnova, 1938; Grigor'yev, 1946; Polozova, 1966; Tolmachev & Tokarevskikh, 1968; Gribova & Ignatenko, 1970; Katenin, 1970, 1972*a*, *b*; Norin *et al.*, 1970; Rebristaya, 1970*a*, *b*, 1977; Shamurin, 1970; Rakhmanina, 1971, 1974; Boch & Solonevich, 1972; Gribova, 1972, etc.). Andreyev (1932*b*), Smirnova (1938), Katenin (1972*b*), and Boch & Solonevich (1972) have made complete lists in comprehensive tables on the species composition of the associations there. Gribova (1972), who for some years conducted aerial mapping and ground surveys of key portions of this area (Gribova & Ignatenko, 1970) and used the maps for agricultural purposes, made a new zonation of this area, based on the vegetation maps she had drawn (Gribova, 1972).

Characteristic of the subprovince is the broad belt of 'tall-grown dwarfbirch tundra' (Gribova's term), developed on zonal, mesic habitats in its southern part, and well described by Katenin (1972*b*: 184–231). It consists of a dense and tall-grown tier, from 0.7–0.8 up to 1 m tall under optimal conditions, of *Betula nana* with an admixture of *Salix phylicifolia, S. glauca, S. lanata*, and *S. lapponum*, a thick moss sward of *Pleurozium schreberi, Hylocomium splendens, Polytrichum commune, P. alpestre,* etc., some *Carex globularis*, hyparctic (*Empetrum hermaphroditum, Vaccinium vitis-idaea* ssp. *minus*, etc.) and boreal (*Vaccinium myrtillus*, etc.) dwarfshrubs, and hyparctic (*Calamagrostis lapponica*, etc.) and boreal (*Deschampsia flexuosa, Solidago virgaurea*, etc.) herbs. On light soils, a dwarfshrub–green moss–lichen variant of the tall-grown dwarfbirch tundra has developed with fruticose *Cladoniae* (*Cladinae*) as dominants among the lichens. Where the conditions are more moist, there is a herb–dwarfshrub–green moss–peat moss variant with *Sphagnum girgensohnii, S. nemoreum*, etc. The tall-grown dwarfbirch tundras display a distinctly expressed mosaic due to the cryogenic disruption of the nano-relief (cf. Rakhmanina, 1971, etc.). Leskov and also Nikolayeva (but not Gorodkov and Andreyev!) refer them to the shrub-type vegetation; Gorodkov (1935*c*: 57, 79) and Andreyev (1954:12) put them with the tundras, because of the presence of a cryogenic mosaic and the thick moss cover, which distinguish the tundras from the type of 'tundra shrub' vegetation in their treatment, with which I myself also agree. Farther north, the tall-grown dwarfbirch tundras are exchanged for 'low-grown dwarfbirch associations', in which the height of *Betula nana* does not exceed 0.5 m, and the layer becomes loosely closed (Gribova, 1972: 40). Thus, the southern belt of the subarctic tundras in this province can be distinguished into two belts of secondary rank.

Mires have also developed in the southern belt. These are palsas, mires with low, flat-topped peat mounds and flat-hillocky mires the gentle slopes of which are saturated with water along the troughs (Boch & Solonevich, 1972). There are also shrub-thickets, which are better developed in the river valleys, nival meadow-like communities with some boreal species (*Deschampsia flexuosa, Chamaenerium angustifolium,* etc.), and along the middle course of the river Morye-Yu (at 67° 50′ N), there is a relict forest 'island' of *Picea obovata* together with some *Betula tortuosa* (Tolmachev & Tokarevskikh, 1968).

Typical for the zonal position of the middle belt of the subarctic tundras in this subprovince are the dwarfbirch–willow–moss tundras and the tundras with a sparse growth of willows alone. *Betula nana* is found here in a prostrate form, reaching 10–15 cm tall (Andreyev, 1935: 29). The willows are represented by hyparctic (*Salix glauca*, etc.) as well as arctic (*S. reptans*) species. *Carex ensifolia* ssp. *arctisibirica* should be especially mentioned among the herbaceous species, and the mosses are represented by *Aulacomnium turgidum, A. palustre, Dicranum angustum, D. elongatum*, etc. The flat-hillocky mires with low peat mounds (about 0.5 m high) are the most widely distributed type here.

The northern belt of the subarctic tundras is expressed in this subprovince at Yugorsky Shar in the southern part of Vaigach and the northern part of Kolguyev Island. Typical of this belt are herb–moss tundras with low-grown willows, occasionally also with some *Betula nana*, the development on stony substrates of *Dryas octopetala* association, and an increase in the arctic element of the flora, as the appearance of e.g. *Poa abbreviata.*

The forest limit in the East European subprovince of the subarctic tundras is mainly formed by *Picea obovata.* However, small patches of *Betula tortuosa* 'crooked forests' are also typical. Locally, *Larix sibirica* is also met with there.

In the floristic zonation used by Yurtsev *et al.* (1975), this area is distinguished as the Kanin–Pechora subprovince. Endemic taxa are *Koeleria pohleana* and *Delphinium elatum* ssp. *cryophilum*. Typical differentiating species, penetrating from the south only into that subprovince, are *Fragaria vesca* and some amphi-atlantic species not reaching east of the Urals (cf. p. 30). Together with them, a number of eastern species occur which penetrate from Siberia westward (cf. Rebristaya, 1970*b*, 1977).

Five districts can be distinguished: the Kanin, the Pechora, the Ural foothills, the Kolguyev, and the Yugorsky Shar – south Vaigach districts.

The Urals–Pai-Khoy subprovince of the subarctic tundras

The Polar Urals reach an altitude of 1100 m within the tundra zone. The Pai-Khoy mountains, a remnant of a Hercynian massif with smooth tops, are separated from the Urals by a 45 km broad plain and reach an elevation of not more than 465 m, at Mt Morye-Pai. Together, these mountain areas can be distinguished as a geobotanical subprovince, not only because of the mountain formations with their distinct, vertical belts disrupting the latitudinal zonality, but also because of the special composition of their plant associations. Those most important are: the constant presence in the shrub tier of the dwarfbirch tundras of the Siberian–American species *Salix pulchra*, which in the previously described subprovince is met with only in areas adjacent to the Urals; the occurrence of *Alnus fruticosa* thickets; this species is also found in the preceding subprovince, but is not as active there; and the presence of species endemic to this subprovince as well as some which are not found west of the Urals, or others not found east of these mountains (so-called western and eastern co-differentiating species; cf. Yurtsev *et al.*, 1975).

The vegetation of the Polar Urals and the Pai-Khoi Mountains has been described in papers by a number of authors (Schrenk, 1848, 1854; Gorodkov, 1926, 1935*b*; Sochava, 1930*a*, 1933*b*, 1956*a*, *b*; Andreyev, 1934, 1935; Igoshina, 1933 (and in Andreyev, Igoshina & Leskov, 1935), 1961; Rebristaya, 1970*a*, 1971; Gorchakovsky *et al.*, 1975; etc.). Igoshina (1961) has contributed a detailed botanical–geographical division of this territory.

As stated by Sochava (1956*a*, *b*), the Polar Urals have the lowest situated alpine tundras of all mountain structures in Eurasia, as well as the most widely distributed boulder fields on crests and ridges with a very poor vegetation, indicating the far more severe climate of this area. Due to the low pressure arriving from the Barents Sea, the vegetation period is distinguished by very high humidity and the small amount of atmospheric heat.

The forest limit in the northern Urals is formed by *Larix sibirica* and *Betula tortuosa*; along the upper reaches of the Sob′ River these species reach an altitude of 200–210 m. In the lower belt of the mountain tundra (here and below the term ‘belt’ is applied as a concept without rank), in the parts of the territory of the Polar Urals situated within the limits of the southern belt of the subarctic tundras, dwarfbirch tundras with *Betula nana* have developed, occupying slopes of valleys and replaced on the convex parts half-way up the valleys by sedge–dwarfshrub–moss tundras. The dwarfbirch tundras are close in composition to those of the East European subprovince but, as mentioned above, with one characteristic difference: the presence of an admixture of *Salix pulchra*. In the brook valleys along the water courses, shrub-thickets are met with, consisting mainly of willows: *Salix glauca*, *S. hastata*, *S. lanata*, and *S. pulchra*.

In the middle mountain belt, analogous to the middle belt of tundras on the lowland, sedge–lichen tundras are widely distributed over the more gently sloping surfaces. Typically, lichens predominate and form covers, interrupted locally by stony outcrops and blocks on the surface. Besides lichens such as *Cladonia mitis*, *Alectoria ochroleuca*, etc., and some mosses, like *Rhacomitrium lanuginosum*, there are abundant *Carex ensifolia* ssp. *arctisibirica*, and *C. rupestris*; *Dryas octopetala* and *Vaccinium uliginosum* ssp. *microphyllum* occur; the prostrate shrubs are represented by *Salix arctica* and ground-hugging *Betula nana*. In more moist conditions on relatively gentle slopes there are sedge–dwarfshrub–moss tundras with *Aulacomnium turgidum*, *Tomenthypnum nitens*, *Carex ensifolia* ssp. *arctisibirica*, *C. capillaris*, *Festuca ovina*, *Hierochloë alpina*, *Lloydia serotina*, etc. and prostrate shrubs of *Betula nana*, *Salix pulchra*, *S. arctica* as well as dwarfshrubs such as *Dryas punctata*, *Empetrum hermaphroditum*, *Ledum decumbens*, *Vaccinium uliginosum* ssp. *microphyllum*, etc. On the more convex and stonier slopes, there are lichen tundras (*Alectoria ochroleuca*, *Cornicularia divergens*, etc.) with mosses (*Rhacomitrium lanuginosum*, etc.), arctic–alpine dwarfshrubs (*Dryas punctata*, *Salix polaris*) and some herbs (*Carix ensifolia* ssp. *arctisibirica*, *Luzula confusa*, *Silene acaulis*, *Novosieversia glacialis*, etc.; cf. Sochava, 1956*b*).

In the upper mountain tundra belt, analogous to the arctic tundra lowland, boulder fields with epilithic lichens (particularly the foliose *Umbilicariae* and *Parmeliae*) and mosses predominate (Igoshina, 1961: 186). On skeletal soils, there are *Dicranum*–dwarfshrub tundras with *Salix nummularia* and prostrate *S. pulchra* and *S. arctica*; on skeletal–polygonal substrate, *Dicranum*–dwarfshrub spotty tundras. There are also nival tundras with *Salix polaris*, *Harrimanella hypnoides* and rare herbaceous formations with *Deschampsia brevifolia* and *Ranunculus sulphureus*, rich in liverworts together with *Stereocaulon alpinum* and *Polytrichum alpinum*. Less widely distributed, but typical, are tundras with *Carex ensifolia* ssp. *arctisibirica*, *Salix reticulata*, *S. arctica*, *Aulacomnium turgidum*, and *Tomenthypnum nitens*, sedge–cottongrass mires with *Carex rariflora* and *Eriophorum russeolum*, and *Alectoria* tundras.

At an altitude of 500–600 m in the Polar Urals, a belt begins to form with scattered vegetation (Igoshina, 1961: 188–9), where there is an absolute dominance of boulder fields with epilithic lichens such as *Rhizocarpon*, *Lecanora*, *Caloplaca*, *Lecidea*, *Parmelia*, and other genera. Fragments of scattered vegetation, relating to islands of fine soil, consist of moss and lichen carpets (*Cetraria delisei*, *C. tilesii*, *Solorina crocea*, *Polytrichum hyperboreum*, etc.) and black thin films of liverworts such as *Anthelia juratzkana*, *Gymnomitrium coralloides*, and *Pleuroclada albescens*. Some angiosperms, like *Cardamine bellidifolia*, *Cochlearia arctica*, *Draba* ssp., *Cerastium regelii*, *Gnaphalium supinum*, *Saxifraga oppositifolia*, etc., are also met with.

The Pai-Khoy Mountains, situated within the middle and northern belts of the subarctic tundras and of lower height than the Urals, display two of the mountain belts. The lower one is represented by a particular variant of the zonal subarctic tundras, and the upper one, from about 200 m, is analogous with the arctic tundra lowland.

The floristic originality of this subprovince is evident from the presence of the endemic taxa *Trollius apertus*, *Astragalus gorodkovii*, and *Castilleja vorkutensis* as well as a large number of differentiating and co-differentiating species (Yurtsev *et al.*, 1975).

Two districts can be distinguished within this subprovince: the Polar Urals, and the Pai-Khoy districts.

The Yamal–Gydan–West Taimyr subprovince of the subarctic tundras

The areas of the subarctic tundras in Western Siberia and the western part of the Taimyr lowlands are united into one single subprovince. These territories were subjected not only to complete glaciation, but were also extensively submerged during the marine transgression. This has shaped the distinctive, present relief and the soil-forming bedrocks. The presence in the southern part of this province of dwarfbirch tundras with a closed cover of *Betula nana* with an admixture of *Salix glauca*, *S. phylicifolia*, *S. lapponum*, *S. pulchra* and *S. lanata* is typical; sometimes *Alnus fruticosa* occurs together with these species. The latter does also form independent, small-size stands. At the southern limit of the dwarfshrub tundras, *Carex globularis* participates (Nikolayeva, 1941; Gorchakovsky & Trotsenko, 1974, etc.) and in the major part of that tundra, *Carex ensifolia* ssp. *arctisibirica* is widely distributed. The forest limit is here formed by *Larix sibirica*. *Betula tortuosa* is met with as far as the right-hand bank of Yenisey River.

The appearance on surfaces with a somewhat impeded drainage of tussocky tundras with *Eriophorum vaginatum*, often associated with some *Carex ensifolia* ssp. *arctisibirica* taking part in the tussock formation (Govorukhin, 1933: 77; Vinogradova, 1937: 22–4: Gorodkov, 1944: 18; Polozova, 1970), is a distinguishing characteristic of this subprovince. In the more southerly, tussocky tundras, there is some *Betula nana*, reaching 40 cm in height and mainly situated between the tussocks, as well as *Salix pulchra*, *S. glauca* and *S. reptans* (Gorodkov, 1944) and some dwarfshrubs such as *Ledum decumbens*, *Vaccinium vitis-idaea* ssp. *minus*, *V. uliginosum* ssp. *microphyllum*, *Cassiope tetragona*, etc. In some areas this type of tundra appears to be the essential element of the landscape. Thus, in the eastern part of the Tazov Peninsula (Malyi Yamal), tussocky tundras occupy 'not less than 15–20% of the total surface of the watersheds . . . they are typical also of the lowland tundras, where they occupy peat platforms and large peat-mounds' (Govorukhin, 1933: 77). The tussocky tundras disappear at the boundary with the arctic tundras (Gorodkov, 1944).

Three districts can be distinguished in this subprovince: the Yamal, the Gydan Peninsula (including the Tazov Peninsula) and the West Taimyr districts.

The Yamal district. Here the major part of the surface is a distinct plain with elevations up to 30–50 m, built up of marine, Quaternary deposits, mainly of sand and sandy loam. In the central parts of the peninsula, where the marine sediments are overlaid by moraine formations, the heights reach up to 100 m in elevation. Towards the coast, the land lowers gradually in the form of level terraces.

Information on the vegetation of the subarctic tundras of Yamal, as well as of areas around the lower reaches of the Ob River and on the eastern shore of the Baidaratsky Inlet, has been published in a number of papers (Sukachev, 1911; Zhitkov, 1913; Gorodkov, 1916; Andreyev, 1934, 1938; Igoshina, 1933; Andreyev *et al.*, 1935; Sdobnikov, 1937; Nikolayeva, 1941; Avramchik, 1969; Boch *et al.*, 1971*a*, *b*; Mel'tser, 1973; Gorchakovsky & Trotsenko, 1974; Il'ina, 1975; etc.). The southern tundras have been most intensely studied (Nikolayeva, 1941; a series of publications on the vegetation around the research station 'Kharp' by Gorchakovsky & Trotsenko, 1974; etc.).

The wide occurrence of sandy and sandy-loamy soils in the Yamal district has favored the development of lichen and moss-lichen variants of shrub and herb–dwarfshrub tundras. The predominance of a level relief has strongly promoted the formation of mires and an abundance of lakes. According to data from Boch, Gerasimenko & Tolchel'nikov (1971*a*), there are, in the subarctic parts of Yamal, beside homogenic mires with *Carex stans*, *C. chordorrhiza*, *C. rariflora*, etc., also polygonal mires at various stages of development, as well as swampy and peaty tundras in the troughs of which grow *Carex rariflora*, *Eriophorum angustifolium*, *Drepanocladus revolvens*, and *Sphagnum compactum*, and on the peatmoss hummocks also *Ledum decumbens*, *Vaccinium vitis-idaea* ssp. *minus*, *Dicranum angustum*, *Cladonia rangiferina*, etc. On the raised borders of the low centre polygons in the polygonal mires, *Betula nana*, *Rubus chamaemorus*, hyparctic dwarfshrubs, *Dicranum* and *Sphagnum* mosses predominate. Among the number of characteristics for the vegetation of Yamal, the occurrence of *Dryas octopetala* in the tundras should be noted. Farther east, this species starts to be replaced by *Dryas punctata*.

The Gydan Peninsula district. The relief is more varied; moraine hillocks and ridges are an important part of its formation and loamy soils are more widely distributed here. Information on the vegetation is found in works by Gorodkov (1924, 1927, 1929, 1932, 1944), Govorukhin (1933), Mel'tser (1973), and Il'ina (1975).

In the southern belt, the major part of the surface is occupied by shrub associations with a closed tier of *Betula nana*, *Salix lanata* and *S. pulchra*, growing 50–80 cm tall on the hillsides. Around the lakes and along riverbanks, shrub thickets with dominating *Salix lanata* and an admixture of *Salix pulchra*, *Betula nana*, and sometimes *Alnus fruticosa*, attain a height of 1 m. On zonal, mesic habitats, there are shrub associations with rather dense canopy, formed by *Betula nana* and *Salix pulchra*; among the dwarfshrubs, *Ledum palustre*, *Vaccinium uliginosum* ssp. *microphyllum*, and *Empetrum hermaphroditum* are abundant, *Dryas punctata* occasional. There is much *Carex ensifolia* ssp. *arctisibirica*, and some *C. globularis*. The mosses are represented by *Hylocomium splendens*, *Aulacomnium turgidum*, *Dicranum congestum*, and *Polytrichum alpestre*, the lichens by species such as *Cladonia rangiferina* and *Cetraria cucullata* which are common. Where the drainage is impeded, tussocky tundras of *Eriophorum vaginatum* occur with some *Carex ensifolia* ssp. *arctisibirica*, hyparctic dwarf-bushed, low-grown *Betula nana*, *Salix pulchra* and *Salix glauca*, and mosses, among which there are patches of *Sphagnum* and a small amount of lichens. The mires are represented mainly by low-hillocky mires with flat-topped mounds often overgrown by a dense shrub tier of *Betula nana, Salix lanata, S. glauca, S. pulchra,* etc., reaching a height of 40 cm.

Typical for the middle belt is the decreasing amount of surface occupied by shrub thickets, and the development on zonal, mesic habitats of small hummocky and spotty tundras where the bare spots (non-sorted circles) are surrounded by vegetation cover composed of *Hylocomium alaskanum*, *Aulacomnium turgidum*, *Tomenthypnum nitens*, *Ptilidium ciliare* and other mosses with a small admixture of lichens. The dwarfshrub–herb tier consists of *Carex ensifolia* ssp. *arctisibirica*, *Eriophorum angustifolium*, *Vaccinium vitis-idaea* ssp. *minus* with an admixture of a number of herbaceous species. There are many prostrate shrubs (attaining 20–25 cm in height) of *Betula nana*, *Salix glauca*, *S. lanata*, and *S. pulchra*.

In the northern belt, the quantity of shrubs is sharply reduced, particularly that of *Betula nana*. This species practically peters out at the boundary with the arctic tundras. Among the willows, the role of *Salix reptans* is increasing.

The West Taimyr district. This occupies the rolling plains of the western and northwestern Taimyr lowlands, spreading northwards to the Byrranga Foothills. The similarity between the West Taimyr tundras and those of Gydan is striking, when comparing the descriptions of West Taimyr in the area around Agapa River (Pospelova, 1972, 1974; Pospelova & Zharkova, 1972; etc.) and the lower reaches of Dudypta River (Vinogradova, 1937) with the typical plant associations and in part with the 'climax associations of the watersheds', which were placed by Gorodkov in the middle belt of the Gydan district of the subarctic tundras as briefly described above. Associations (with hummocks about 15–60 cm wide and 15–20 cm tall) have been mentioned from zonal, mesic habitats on a loamy substrate, where the mosses likewise are represented by a mixture of *Aulacomnium turgidum*, *Tomenthypnum nitens*, *Hylocomium alaskanum*, *Ptilidium ciliare*, etc.

Betula nana, not growing taller than 20–25 cm (Vinogradova, 1937: 23; Pospelova & Zharkova, 1972; Pospelova, 1974) together with *Salix pulchra*, *S. glauca* and *S. lanata* forms a micro tier, relating to a network of interhummock low depressions or to the slopes of hummocks, while on the convex nanorelief (the hummocks) dwarfshrub–herb–moss and moss–lichen micro-communities dominate; often some patches of open ground occur. There is an abundance of dwarfshrubs (*Vaccinium uliginosum* ssp. *microphyllum*, *Dryas punctata*, *Cassiope tetragona*, etc.) but also of *Carex ensifolia* ssp. *arctisibirica*. *Eriophorum angustifolium* is also present in the hummocky tundras. On gentle slopes, less well drained, a tussocky cottongrass tundra has developed. The shrub-type vegetation is represented by thickets of *Betula nana*, and more often *Salix lanata*, but also by a few, small stands of *Alnus*. The mires belong characteristically to the type with flat-topped peat mounds. In the troughs between the mounds cottongrasses (*Eriophorum* spp.), *Carex stans*, *Calliergon* and *Drepanocladus* species dominate. On the tops of the flat peat mounds there is much *Betula nana* or willows, *Eriphorum vaginatum*, and hyparctic dwarfshrubs, and among the mosses either peatmosses or species of *Polytrichum* and *Dicranum* predominate.

In the northern belt of the subarctic tundras, the vegetation of which is well known thanks to the work performed at the research station at Tareya (Matveyeva, 1968, 1978; Khodachek, 1969; Polozova, 1970; Aleksandrova, 1971*a*, Matveyeva *et al.*, 1973; Boch & Vasilyevich, 1975, etc.), but also due to other sources (Tolmachev, 1932*a*: 69–89; Serebryakov, 1960, 1963; etc.), there has developed on zonal, mesic habitats a tundra with hummocks about 15–30 cm wide and 10–15 cm high, or a spotty tundra. *Betula nana* in its prostrate form plays a very small role in the composition of the phytocoenoses there (Matveyeva *et al.*, 1973: 20) or is completely lacking. There are only small quantities of willows such as *Salix arctica*, *S. pulchra*, and *S. lanata*. The tier of herbaceous species and dwarfshrubs is dominated by *Carex ensifolia* ssp. *arctisibirica* while *Dryas punctata* appears as a co-dominant. Hyparctic dwarfshrubs such as *Vaccinium vitis-idaea* ssp. *minus*, *Pyrola grandiflora*, etc., participate in small amounts. Among the herbaceous species there are many arctic ones. The mosses *Aulacomnium turgidum*, *Hylocomium alaskanum*, *Tomenthypnum nitens*, and *Ptilidium ciliare* co-dominate; lichens are few; some species of

Cladonia (*C. amaurocrea*, etc.) predominate. Thus, the composition of the associations on zonal, mesic habitats shows characteristics of a transition towards arctic tundras. However, the presence of *Betula nana* on zonal, mesic habitats and its abundant development in tundra–mire complexes and on the raised borders of low centre polygons in polygonal mires (Matveyeva *et al*., 1973: 32), and the distribution of *Salix lanata* shrub thickets, developed on the concave parts of slopes and in habitats sloping down towards the rivers from the watersheds (Aleksandrova, 1971*a*: 186; Matveyeva *et al*., 1973: 32) appear as categorically diagnostic characteristics, placing this area in the subregion of the subarctic tundras.

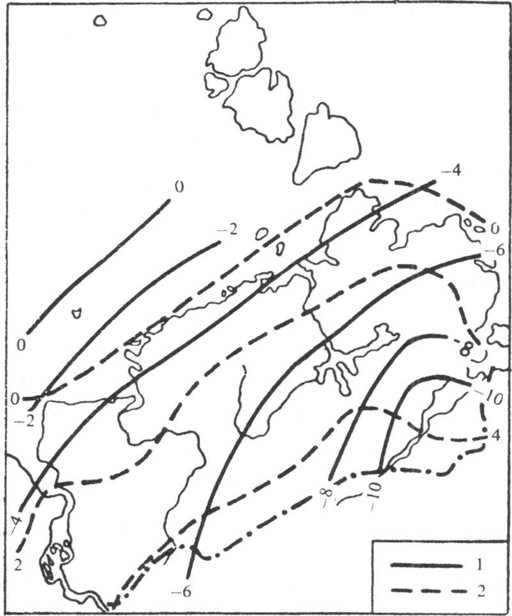

Fig. 4. Iso-anomalies for the January (1) and the February (2) temperatures. After Rubinshtein, 1953.

The boundary with the East Siberian province of the subarctic tundras does not run longitudinally, but from southwest to northeast, which can be explained mainly on climatic grounds (cf. Fig. 4), and cuts the latitudinal belts obliquely. In the south, the area around the middle course of the Dudypta River appears, according to data from Avramchik (1937), to be changing into the East Siberian province, but the vegetation along the southern shores of Lake Taimyr is, according to the description by Tolmachev (1932*a*: 69–89), still similar to the vegetation in the area around Camp Tareya.

The East Siberian province of the subarctic tundras

The transition from the East European–West Siberian province to the East Siberian province is expressed first and foremost in that the basic position of *Betula nana* in the formation of the coenoses is replaced by that of *Betula exilis* and by the disappearance from the tundra of *Betula nana* with its characteristic association of *Salix phylicifolia* and *S. lapponum* (Fig. 5). The boundary between the areas of *Betula nana* (incl. *B. tundrarum*) and *B. exilis* (Fig. 6) runs along the border between the two provinces.

In the southern part of the East Siberian province, the vegetation turns into tundras with low-growing *Betula exilis* associations. The canopy tier of the depressed birches is closed or broken by positive nanorelief, often with patches of bare ground, around the borders of which dwarfshrubs, mosses and lichens predominate. This development is connected with the diminishing thickness of the snow cover, caused by the increasingly continental climate.

Many of the peculiarities typical of the vegetation cover in this province are connected with the high degree of continentality of the climate, sharply expressed in the appearance of frost-wedging and other cryogenic processes. Here we find the spotty tundras developed to various degrees, as well as polygonal mires, 'delli' (peculiar stripe-like phytocoenoses on the slopes; cf. Andreyev, 1971), as well as thermokarst formations with relic, socalled 'cemetery mounds' ('baidzharakhs') surrounded by thermokarst trenches. The important role played by lichen associations with *Cladonia*, typical of many lichen and moss–lichen formations in the East European–West Siberian province, halts here, and *Cetraria cucullata* begins to dominate in the lichen–moss synusia of the dwarfshrub, herb–dwarfshrub, and other types of tundras, and *Alectoria ochroleuca* and *Cornicularia divergens* dominate in the lichen tundras. Around the lower reaches of the Olenek River, cryophytic and steppe species (Yurtsev, 1974*a*, *b*: 18–21; Andreyev & Perfil'yeva, 1975, etc.) already occur in the

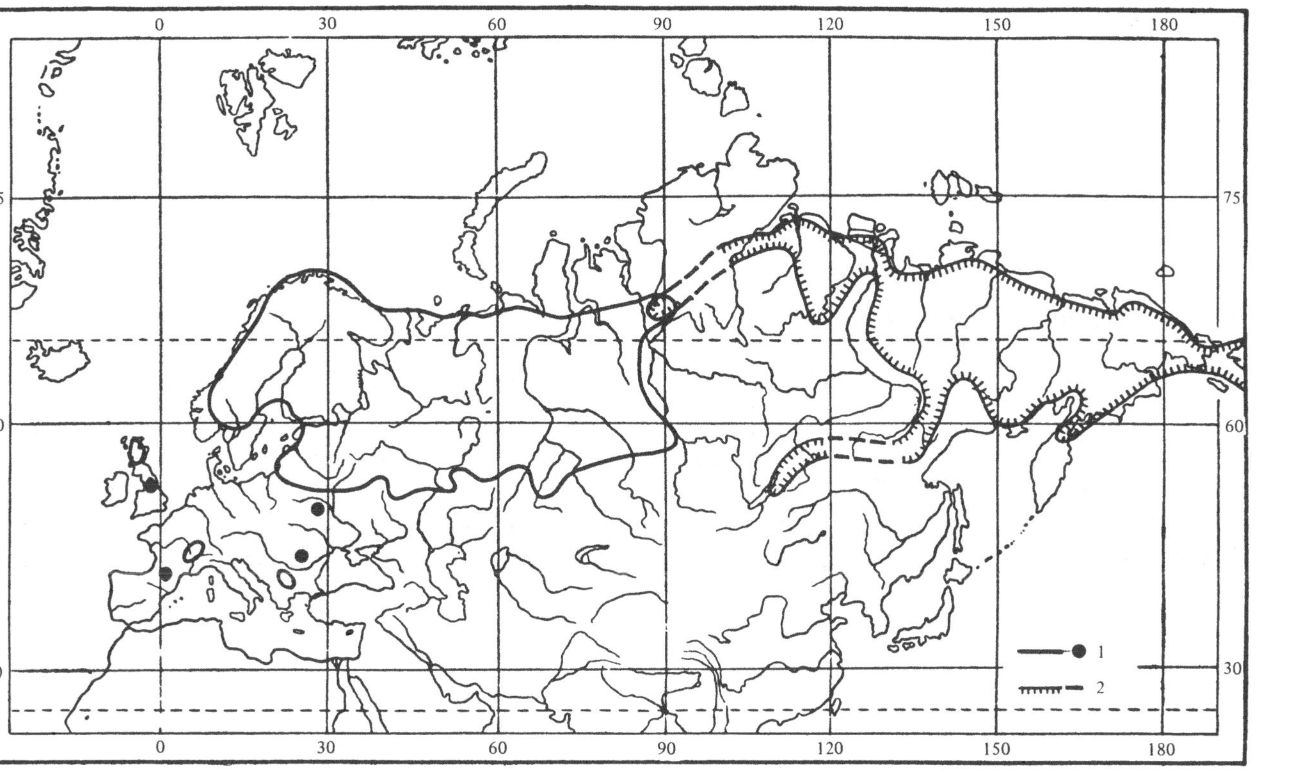

Fig. 5. Areas of *Salix lapponum* L. (1) and *Salix alaxensis* Cov. (2) within Eurasia. According to Skvortsov, 1966: 8.

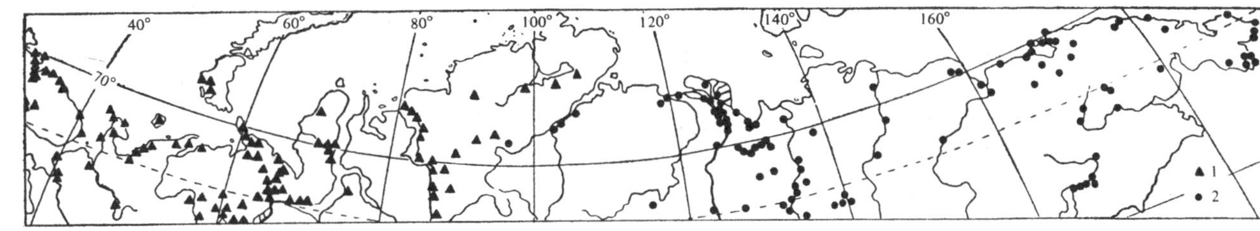

Fig. 6. Distribution of *Betula nana* L. (incl. *B. tundrarum* Perf.) (1) and *B. exilis* Sukacz. (2) within the limits of the Soviet Arctic. According to Cherepanov, 1966.

tundra–steppe communities in corresponding habitats. The strong compression of the latitudinal belts is peculiar for this province. The forest limit, formed in the western parts of *Larix gmelinii* and in the eastern parts by *L. cajanderi*, is strongly displaced towards the north.

The flora has a distinctly expressed east Siberian character. The amphi-atlantic species have disappeared completely and amphi-beringian species begin to mix in (*Salix fuscescens*, *S. alaxensis*, *Diapensia obovata*, and others). The right-hand bank of the Pyasina River appears to be the western limit for the distribution of a number of species growing in the East Siberian province; these are *Betula exilis*, *Salix alaxensis*, *S. boganidensis*, *Polygonum ellipticum*, etc.; and also for *Larix gmelinii*, which is met with far into the tundras in the prostrate krummholz form and as a relict forest 'island' at Ary-Mas. In this province, the number of boreal species decreases sharply, and aggressive arctic species penetrate towards the south (*see below*, p. 51). The boundary between the East European–West Siberian and the East Siberian provinces agrees almost entirely with the western limit of the tundra part of 'Mega-Beringia' according to Yurtsev (1974*a*: 45).

The difference in the vegetation of both these provinces is closely connected with the peculiarities of the climate, i.e., with the cyclonal pressure circulation of the atmosphere over the East European–West Siberian province and with the anticyclonal activity over the East Siberian province.

Due to the continentality of the climate, the deviation of the provincial borders from the longitudinal direction increases from northwest towards southeast (Fig. 4). There is a very important and profound difference in paleogeography between these territories. The importance of the Siberian high pressure area appears to have developed already during ancient eras; it existed as far back as during the Pliocene (Borisov, 1965). The Icelandic low pressure area, imposing a low pressure type of weather on the territories of northern Europe and western Siberia, only developed at the beginning of the Ice Age. Because of the scarcity of precipitation during the winter, no glacial cover was formed over the territories of eastern Siberia as a consequence of the anticyclonal activity except in the far western part, where glaciers formed in the Byrranga Mountains, and only local valley glaciers developed over the rest of the territory. There was no continental icesheet. Instead, the glaciation took the form of a *subsurface glaciation*: the accumulation of an enormous reserve of ice below the ground surface, belonging according to Zubakov's (1965) classification to the family of 'ground-ice formations'. Particularly important appear the enormous ice bodies of the polygonal system of longlasting ice veins, formed during the Ice Age, the genesis of which has been studied by Dostovalov (1960), Romanovsky (1960), Popov (1967, etc.) and others (cf. Brown & Kupsch, 1974). In the area where these occur at present, thermokarst formations appear widely distributed in the form of a network of thermokarst trenches and relic mounds ('baidzharakhs') with peculiar vegetation formations. Thanks to the lack of a glacial cover, the vegetation maintained connection with a succession from the ancient Pliocene sources of the arctic flora and the 'gol′tsy' (alpine tundra) formations of the mountains in northeast Asia, so important for the original formation of the arctic associations.

Three subprovinces can be distinguished: the Khatanga–Olenek, the Kharaulakh, and the Yana–Indigirka subprovinces (Fig. 7).

The Khatanga–Olenek subprovince of the subarctic tundras

This subprovince reaches from the middle course of the Dudypta River and the Khatanga River basin to the lower course of the Lena River, including the Chekanovsky Ridge. It is mainly a rolling plain with elevations up to 40 m, in the southern part up to 100 m, composed of mesozoic sedimentary rocks, overlaid by loam and sandy loam. In the south, the plain changes into the foothills of the Central Siberian Highland, which is already situated within the forest limit. In the northeast of this subprovince there are low mountains: the Pronchishchev (up to 200 m) and the Chekanovsky (up to 500 m) Ridges.

The earliest data on the vegetation (from the lower Olenek River) were collected by Chekanovsky (1896). Later, the tundras in the Anabar River basin were investigated by Sochava (1933*a*, 1934*a*, *b*), at the middle reaches of Dudypta River by Avramchik (1937), along the right-hand bank of Popigay River by Aleksandrova (1937) and in the Nova River basin by Tyulina (1937), Norin *et al.* (1970), Norin (1972, 1974, 1978), Ignatenko *et al.* (1973), Knorre (1974), as well as at the lower reaches of the Olenek River by Yurtsev (1962).

In the southern part of this subprovince, tundras have developed

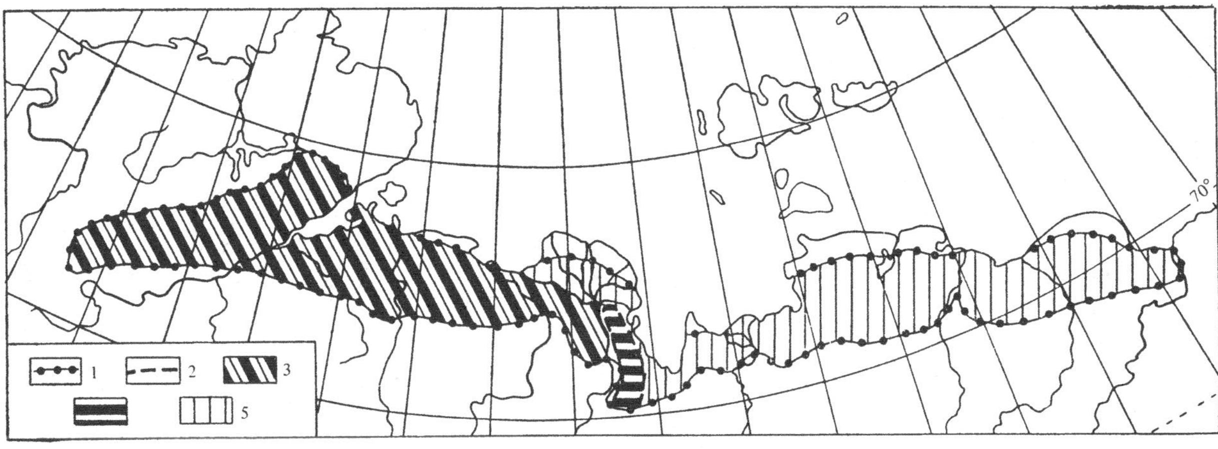

Fig. 7. The East Siberian province of the subarctic tundras. Boundaries: 1, provincial; 2, subprovincial. Subprovinces: 3, the Khatanga–Olenek; 4, the Kharaulakh; and 5, the Yana–Indigirka subprovinces.

with low-grown dwarfbirch associations, often with patches of bare ground, consisting of *Betula exilis*, rarely exceeding 40 cm in height (Sochava, 1934*b*: 292; Avramchik, 1937: 63), with admixture of *Salix pulchra*; mosses are represented by *Aulacomnium turgidum*, etc., and hyparctic dwarfshrubs by *Ledum decumbens*, *Vaccinium uliginosum* ssp. *microphyllum*, etc., associated with *Dryas punctata* and *Carex ensifolia* ssp. *arctisibirica*. Along the very limit of the forest (Sochava, 1933*c*: 360) and in the northernmost part of the open woodland within the polar limit (Tyulina, 1937: 88, 132, 135, etc.), patches of bare ground are typical. Tussocky tundras with *Eriophorum vaginatum* are met with, occupying less well drained habitats with admixture of *Carex ensifolia* ssp. *arctisibirica* and an abundance of low-grown shrubs such as *Betula exilis*, *Salix pulchra*, *S. glauca* and others. On slopes and in river valleys shrub thickets are common, mostly composed of willows, more rarely of dwarfbirch, but sometimes of *Alnus fruticosa*. On sand there are tundras with *Alectoria ochroleuca*. The shrub thickets occupy a considerably smaller area than in the territories to the west of the Pyasina River (Shchelkunova, 1975: 48).

In the middle belt of the subarctic tundras, on zonal, mesic habitats on a loamy substrate, the semi-prostrate, creeping or depressed shrubs (*Betula exilis*, *Salix pulchra* and *S. reptans*), 20–25 cm tall, have a projected coverage of 15% (Aleksandrova, 1937: 190) and are related mainly to shallow depressions in the nanorelief of the spotty tundras. *Dryas punctata* grows mainly on the convex nanorelief around the open patches, other dwarfshrubs are represented by *Cassiope tetragona*, *Vaccinium uliginosum* ssp. *microphyllum*, and others. Among the herbaceous species *Carex ensifolia* ssp. *arctisibirica* predominates. In the moss or moss–lichen carpet are found *Tomenthypnum nitens* and *Hylocomium alaskanum*, etc., and in the Anabar River area *Rhytidium rugosum* is often found (Sochava, 1934*b*: 280). Frequently, there are also tussocky tundras with *Eriophorum vaginatum*. A spotty lichen tundra is widely distributed on sand and sandy loam with predominance of *Alectoria ochroleuca* and *Cornicularia divergens*, but in habitats with a scanty snow cover there are *Dryas* associations with *D. punctata*. At the foot of slopes, where snow accumulates, it is possible to find tundras with dominant *Cassiope tetragona* on screes (Sochava, 1934*b*: 281). In river valleys, shrub associations of *Salix lanata* grow to 70 cm in height, and sometimes groups of *Alnus fruticosa* are observed.

The mires in depressions on the watersheds have mainly flat-topped frost-heaved peat mounds. Typically, *Carex chordorrhiza*, *C. rotundata* and *C. stans* grow in the troughs between the mounds, and on the flat peat mounds there are thickets of *Betula exilis*. On the level and broad expanses of river terraces along major rivers, where the snow blows away during winter and frost-cracking is common, polygonal mires have developed (Aleksandrova, 1937: 201–3).

In the Khatanga River basin the most northerly stand in the world of the woody species *Larix gmelinii* is found. The northernmost phytocoenoses of prostrate krummholz of this larch have been described from the right bank of the Popigay River (Aleksandrova, 1937: 183, 197–9), not, however, from its valley, but rather on a watershed plain about 10 km distant from the river and 6 km north of the middle course of the Sagyr River (Aleksandrova, 1937: 183, Fig. 1) on the top of some low hills. Here, *Larix gmelinii* grows on sandy soil in the form of flat patches with a diameter up to 75 cm. It has short, crooked stems and thick branches, completely flattened to the ground and almost submerged in a moss carpet of *Rhytidium rugosum*, and is surrounded by open formations of chionophobous species. According to recently published maps, the exact latitude of this point is at 72° 55′ N. Similar patches of prostrate krummholz of *Larix* have been described by Yurtsev (1962: 212–15) from the lower reaches of the Olenek River and by Polozova (1961: 292) from the lower reaches of the Lena River. Farther out in the tundras (at 72° 30′ N), there is the most northerly forest 'island' in the world at Ary-Mas on Nova River. At the same time a number of 'settlers from the north' (V. V. Alekhin's term) appear in this subprovince, penetrating towards the south. Thus, e.g., *Dryas punctata* grows not only in the plant associations of the southern belt of the subarctic tundras, but also often enters in considerable abundance into the composition of the ground cover in the open woodland around the Khatanga River (Tyulina, 1937: 92). Also, the northern limits of the areas of many boreal species, which are found in the tundras of the East European–West Siberian province, bend here southwards away from the tundras (cf. e.g., the areas of *Carex aquatilis*, *Alopecurus pratensis*, *Festuca ovina*, *Lycopodium complanatum*, and others in the Arctic Flora of the USSR, I–III, 1960, 1964, 1966). The distinct predominance of eastern components over that of western ones is typical for this province, as is its rapid enrichment by eastern elements of the flora, because of the close vicinity to the arctic Verkhoyansk Mountains (cf. Yurtsev *et al.*, 1975).

Three districts can be distinguished: the Khatanga, the Anabar–Olenek lowland, and the northeastern foothills of the Chekanovsky and Pronchishcheva Ridges.

Yurtsev (1962: 210–12) has described the vertical belts of the Chekanovsky Ridge. In the northern part of this ridge the lower belt (reaching up to 150 m) appears as a mountain variant of the subarctic tundras. It should, however, be mentioned that here a number of continental 'cryophytic steppe species' (in the wide sense of that term) are typical for the *Dryas*–tundras (*Dendranthema mongolicum*, *Thymus extremus*, *Dianthus repens*, *Lychnis sibirica* ssp. *villosula*, etc.). The upper belt is analogous to the plains of the arctic tundras, being devoid of any hyparctic dwarfshrubs. In the southern part of this ridge, the lower belt of the mountain tundras, which is essentially similar to the belt of southern subarctic tundras, occupies elevations from 100–150 to 200–250 m. Above that, the oro-arctic tundras (a term used by Ahti *et al.* (1968), where the prefix 'oro', i.e. 'mountain', indicates a vertical belt) reach to the high summits of the mountain ridge.

The Kharaulakh subprovince of the subarctic tundras

The Kharaulakh Range can be distinguished as a special geobotanical subprovince, not only because it is a mountainous area with elevations up to 1000 m, but also because it is distinguished by the participation in the plant associations of a number of differentiating species from the south (Yurtsev *et al.*, 1975). These species are totally absent in the adjacent western and eastern subprovinces.

Descriptions of the vegetation in the Kharaulakh subprovince have been given by Dushechkin (1937) and Yurtsev (1959). The foothills in the eastern part of the Kharaulakh Range in the vicinity of Tiksi Bay have been treated by Tikhomirov, Petrovsky & Yurtsev (1966) and by Shamurin (1966).

The forest limit in the Kharaulakh mountains is formed by *Larix cajanderi*; a few groups of the dwarftree species *Pinus pumila* are also found there. In the sub-gol'tsy (subalpine) belt between 100 and 150 m altitude, thickets of *Betula exilis* and willows, *Salix pulchra*, *S. glauca* and *S. lanata*, are met with. Tundras with associations of low-growing dwarfbirches, *Betula exilis*, often with an admixture of *Salix pulchra* and rich in hyparctic dwarfshrubs, such as *Vaccinium uliginosum* ssp. *microphyllum*, *Empetrum hermaphroditum*, and *Vaccinium vitis-idaea*, have a thick moss carpet of species like *Hylocomium splendens*, *Aulacomnium palustre*, *Polytrichum alpestre*, and *Dicranum angustum*. The lichens present are *Cetraria cucullata*, *C. crispa*, etc.

Typical for the low-'gol'tsy' (low-alpine) belt in habitats protected from wind are dwarf-shrub mountain tundras, e.g. the *Dryas*–blueberry tundra: *Dryas punctata*, *Vaccinium uliginosum* ssp. *microphyllum* with an admixture of *Betula exilis*, a number of arctic–alpine herbs, but only a small quantity of mosses; *Dryas*–herb tundra: *Dryas punctata*, *Salix nummularia*, various herbaceous species, but lacking in mosses; *Dryas–Cassiope* tundra: *Dryas punctata*, *Cassiope tetragona*, *Salix arctica*, *Arctous alpina*, *Salix polaris*, various arctic–alpine grasses, particularly *Hierochloë alpina*, but with mosses less common, *Hylocomium alaskanum*, *Ptilidium ciliare, Tomenthypnum nitens,* etc. Where more snow accumulates, there are lichen tundras with predominating *Cetraria cucullata, C. crispa*, *C. nivalis*, rarely with *Cladonia sylvatica*, *C. rangiferina*, and *C. alpestris.* In depressions between the mountains, closed towards the north and well heated by the sun, there are also meadow-like communities, very rich in composition, and including 40 or more species (Dushechkin, 1937: 219). Tussocks of *Eriophorum vaginatum* have developed in gently sloping hollows, and tundras with *Alectoria ochroleuca* and black *Cornicularia divergens* are widely distributed on rocky slopes open to the winds.

In the high-'gol'tsy' (high-alpine) belt, there is a very poor vegetation completely lacking in closed formations. It has been described from the Soykudakh Mountains (highest peak 800 m) by Yurtsev (1959). Here, the angiosperms are represented mainly by high-arctic species such as *Alopecurus alpinus*, *Poa abbreviata*, *Draba subcapitata*, etc. A considerable part of the peaks and exposed slopes are totally devoid of any vegetation except for patches of crustose lichens.

In the foothills of the Kharaulakh Range with elevations up to 300–400 m, the relief is divided between stony hills and boggy depressions. There, plant associations appear in the area around Tiksi Bay (cf. Tikhomirov *et al.*, 1966; Shamurin, 1966), such as willow–cottongrass–moss tundras with *Eriophorum vaginatum*, *E.*

angustifolium, *Salix pulchra*, *S. polaris*, *Tomenthypnum nitens*, *Aulacomnium turgidum*, and *Hylocomium alaskanum* which are more closely related to zonal, mesic communities. However, *Dryas*–tundras (with *D. punctata*) and dwarfshrub–herb–moss–lichen tundras (*Vaccinium uliginosum* ssp. *microphyllum*, *V. vitis-idaea* ssp. *minus*, *Ledum decumbens*, *Cassiope tetragona*, *Cetraria cucullata*, *C. crispa*, *Cladonia mitis*, *Hylocomium alaskanum*, *Rhytidium rugosum*, *Aulacomnium turgidum*, etc.), and open formations on bouldery peaks, slopes and plateaus are more widely distributed here, as are also herb–sedge–moss tundras on rolling habitats with *Carex ensifolia* ssp. *arctisibirica* and *Cassiope*–herb–moss tundras. The graminoid tiers of the mires are composed of *Eriophorum angustifolium* or *Carex stans*, *C. rariflora*, *C. chordorrhiza*, and *C. rotundata*, sometimes with an admixture of *Dupontia psilosantha*.

In respect to floristic zonation, the Kharaulakh area has been given the rank of a subprovince (Yurtsev *et al.*, 1975) because of its far-reaching floristic distinctness compared with the areas to the west and east of it. A number of endemics have been reported, such as *Artemisia lagopus* ssp. *abbreviata*, and still not described taxa of *Potentilla*, *Oxytropis*, and *Taraxacum*. But as shown by Yurtsev, 'the not so important racial endemism "sings out" against a background, abundant in differentiating species, mainly arriving from the south, and less often disjunct' (Yurtsev *et al.*, 1975: 19). As reported by Yurtsev, there appear a whole array of representatives for the flora of north-eastern Asia, which is lacking on the lowlands of eastern Yakutia (*Caragana jubata*, etc.). A large number of endemics from the Verkhoyansk Mountains reach up into the Kharaulakhs: *Androsace gorodkovii*, *Gorodkovia jacutica*, *Oxytropis middendorffii* ssp. *orulganica*, and others. Many species penetrating from the south reappear in the Chukotka tundras, e.g. *Helictotrichon dahuricum*, *Carex podocarpa*, *Salix krylovii*, *Claytonia tuberosa*, *Draba kamtschatica*, *Dryas grandis*, *Astragalus pseudadsurgens*, etc. There are also many central Siberian species, which are found in the Kharaulakhs, but are absent east of it.

The Yana–Indigirka subprovince of the subarctic tundras

To the east of the Kharaulakh Range and all the way to the tundras around the Kolyma River, there extends a lowland with widely distributed Quaternary loamy soils, broken here and there by loose eluvium of mesozoic bedrock. The climate is distinguished by its extreme continentality; the winters are not only extremely severe, but also deficient in precipitation. Because of the predominantly calm weather, a thin snow cover is almost evenly spread everywhere. Cryogenic forms of relief are distinctly expressed and so are corresponding vegetation formations, e.g. polygonal mires with raised borders, 'delli' (stripe-like formations on certain slopes), etc. The zonal belts are tightly compressed; in the Indigirka River lowlands the subzone of subarctic tundras is just about 120 km wide, although within this distance all three belts of the subarctic tundras are well developed, according to Andreyev (in Andreyev & Nakhabtseva, 1974).

The vegetation of this subprovince has been described in a number of papers (Komarov, 1926; Skvortsov, 1930; Sheludyakova, 1938; Tyrtikov, 1958; Nosova, 1964; Shchelkunova, 1969; Andreyev, 1971, 1975; Andreyev & Nakhabtseva, 1974; Boch & Tsareva, 1974; Andreyev & Perfil'yeva, 1975, etc.).

It must be mentioned that, in spite of the convincing evidence for the boundaries between the vegetation subzones presented by the above authors (and especially in papers published by Komarov, Tyrtikov, and Nosova, where they are illustrated on maps; cf. also Aleksandrova, 1953), these boundaries are still incorrectly outlined on the geobotanical survey maps (scale 1:4 000 000, etc.); (cf. e.g., *Asia, Vegetation map*, in the book *Bolsh. Sov. Entsikl.*, I, 1970).

Within the subprovince, two districts can be distinguished: the Omoloy–Indigirka, and the Alazeya–Kolyma districts.

The Omoloy–Indigirka district. In this extremely continental district, the vegetation on zonal, mesic habitats in the southern belt of the subarctic tundras is, according to data from several authors (Tyrtikov, 1958; Nosova, 1964; Andreyev & Nakhabtseva, 1974; Boch & Tsareva, 1974; Andreyev, 1975) represented by spotty lichen–moss–dwarfshrub and moss–dwarfshrub tundras with a considerable amount of prostrate *Betula exilis, Salix pulchra* and *Salix glauca*, usually not exceeding 20 cm in height (Tyrtikov, 1958: 73), because of the scanty snow cover. Among the dwarfshrubs *Ledum decumbens* is often predominant, but so are also *Vaccinium uliginosum* ssp. *microphyllum, V. vitis-idaea* ssp. *minus, Dryas punctata, Cassiope tetragona*, etc. Among the graminoids there are *Carex ensifolia* ssp. *arctisibirica*, and frequently also *Eriophorum vaginatum*. The mosses are represented by *Aulacomnium turgidum, Tomenthypnum nitens*, etc., and on small hummocks by *Dicranum* species. *Cetraria cucullata* predominates among the lichens, which often have a coverage of up to 50%. Dwarfshrub, lichen–dwarfshrub and tussocky tundras with *Eriophorum vaginatum* are widely distributed. Low-grown *Betula exilis, Salix pulchra*, and *S. glauca* are common there, and on the low tussocks grow many dwarfshrubs, such as *Ledum decumbens, Vaccinium uliginosum* ssp. *minus*, etc.; sometimes *Carex ensifolia* ssp. *arctisibirica* occurs. The mosses are represented by *Dicranum congestum, Aulacomnium turgidum, Tomenthypnum nitens, Ptilidium ciliare, Sphagnum compactum*, and others. Many lichens are frequently met with, such as *Cetraria cucullata, C. islandica*, and others. Along the rivers, there are thickets of *Salix pulchra* and *S. glauca* reaching 50–60 cm (Nosova, 1964), while the dwarfbirches rarely reach more than 35 cm in height. The thickets of *Alnus fruticosa* grow tallest, because they are able to penetrate above the snow cover without freezing. Mires, mainly polygonal ones, occupy a large area.

In the middle belt of the subarctic tundras (Nosova, 1964; Andreyev & Nakhabtseva, 1974; Andreyev, 1975), tussocky tundras with *Eriophorum vaginatum* and small hummocky ones with hyparctic dwarfshrubs, with sparse and low willows such as *Salix pulchra, S. reptans*, and *S. lanata*, and some prostrate dwarfbirch (*Betula exilis*) associations predominate, but also polygonal mires with *Eriophorum vaginatum* and willows. According to Andreyev, there is in the northern belt of the subarctic tundras a predominance of small hummocky tundras with *Ledum decumbens, Vaccinium uliginosum* ssp. *microphyllum, V. vitis-idaea* ssp. *minus*, and tussocky tundras with *Eriophorum vaginatum*. The forest limit in the Omoloy–Indigirka district is formed by *Larix cajanderi*.

The Alazeya–Kolyma district. A slightly less continental climate is typical of the district. As reported by Andreyev (1975), present permafrost action is less well manifest, but relict permafrost-wedging and ice-wedge formations are widely developed, and hydrolaccoliths or 'pingos' are typical.

Large areas are occupied by polygonal mires with low centre

polygons. The edges of pools or desiccating lakes are overgrown with *Arctophila fulva*. In conjunction with the slight increase in snow cover, the development of the shrubs has improved. In the southern belt of the subarctic tundras *Betula exilis* plays an important role in the plant associations and *Alnus fruticosa* forms typical 'islands'. In the middle subarctic belt *Salix pulchra, S. alaxensis, S. glauca* and *S. lanata* are abundant in river valleys and on zonal, mesic habitats between the small hummocky tundras with hyparctic dwarfshrubs and tussocks of *Eriophorum vaginatum*. Among the cottongrass, *Carex lugens* is beginning to appear, its role increasing towards the east at the same time as *Carex ensifolia* ssp. *arctisibirica* is petering out from the flora. The forest limit is formed by *Larix cajanderi*. The abrupt transition from forest to tundra is typical for the Alazeya–Kolyma district (Andreyev, 1975). Together with *Carex lugens*, other species also typical for the Beringian sector of the Arctic begin to appear here: *Parnassia kotzebuei, Androsace ochotensis*, and others. Tundra-steppe associations with *Carex pediformis* and *C. supina* ssp. *spaniocarpa* have been noted around the lower course of Kolyma (Yurtsev, 1974*b*; Andreyev & Perfil'yeva, 1975).

The flora of this province is also distinguished by its negative properties. More than 60 species common to Beringia and Chukotka Highlands are absent, 36 species do not reach into the Arctic east of the Kharaulakh Mountains, and more than 80 species occur east of the Kolyma river. Differentiating species (*Thellungiella salsuginea, Potentilla viscosa*, etc.) occur in conjunction with the major river valleys (Yurtsev *et al.*, 1975).

The extreme abundance of the peatmosses in the swampy tundras and polygonal mires is one of a number of peculiarities typical of this province.

The Chukotka–Alaska province of the subarctic tundras

In spite of a number of differences between them, I have united Chukotka and Alaska into a single geobotanical province. This has been done mostly on the basis of the character of distribution on zonal, mesic habitats of the groups of associations, which are endemic to these territories, such as the tussocky tundras with *Eriophorum vaginatum* associated with *Carex lugens* as well as the abundance of peatmosses.

These associations dominate on zonal, mesic habitats and disappear a little east of the Mackenzie River.

The general characteristics of the vegetation are connected with the particular paleogeography of the landscapes and their flora and with the many similarities caused by the natural situation. Gorbatsky (1967: 74–93) unites Chukotka and Alaska into a single, physico-geographical region, while emphasizing the major geological unity of these territories with a general similarity in the basic traits of their geography. The Chukotka Highlands and Brooks Range represent individual segments of the same mesozoic formation of folded mountains, etc. To a large extent, their climatic conditions are also similar; both in Chukotka and in Alaska the Pacific Ocean and, particularly, the Bering Sea affect the atmospheric circulation, and the Aleutian low pressure has a very important influence on the weather, especially during the winter.

In the sense used by Yurtsev (1974*a*: 49), both Chukotka and Alaska are situated within the limits of 'Greater Beringia', but this author places them in different floristic provinces. He does, however, mention (Yurtsev, 1974*a*: 104) that although the floristic differences appear well enough defined as far as the mountain elements of the flora are concerned, they are weakly expressed in the composition of the hyparctic tundras on the lowlands. He does therefore consider the possibility of uniting the tundra parts of Chukotka and Alaska into a single geobotanical province (Yurtsev, 1976: 106).

The glaciation was never complete in Chukotka or in Alaska and conditions existed for a dispersal of the flora from the alpine areas of northeastern Asia and northwestern America. The periodic emergence of the Bering shelf favored an exchange of migrants between the two continents. Nevertheless, as Yurtsev has stated, the dispersal of the plants from Asia to America across the Bering landbridge dominated over the dispersal in the opposite direction, so that 'evidently, this accounts for the fact that in northern Asia the process of transformation of the flora in the direction towards an adaptation to a colder and more continental climate surpassed in scale and outstripped in time the analogical transformation of the flora in North America' (Yurtsev, 1974*a*: 109).

One of the differences, for a general comparison, between the vegetation of the tundra areas of Chukotka and Alaska seems to be that the polar forest limit is formed by different tree species: in Chukotka by *Larix cajanderi, Chosenia arbutifolia* and *Populus suaveolens*, in the Anadyr–Penzhina district also by *Betula ermanii* and *Betula cajanderi*, while in Alaska it is formed by *Picea glauca* (Fig. 8), *Larix laricina* (Fig. 9), *Populus balsamifera* (Fig. 10), and locally by

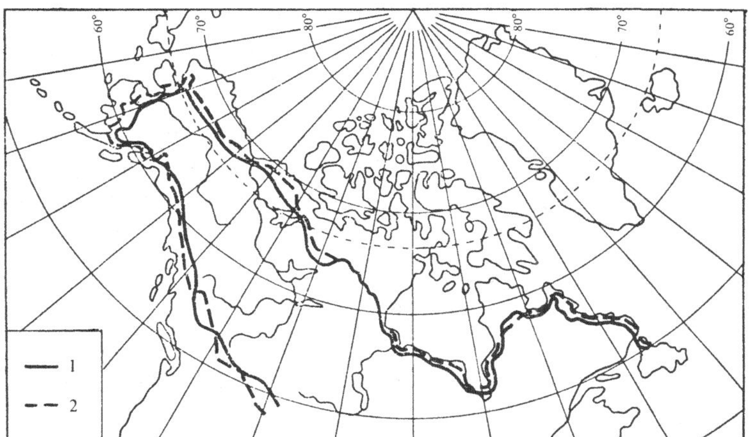

Fig. 8. The northern limit of the distribution of *Picea mariana* (Mill.) Britt., Sterns et Pogg. (1) and of *Picea glauca* (Moench) Voss (2). According to Hustich, 1966.

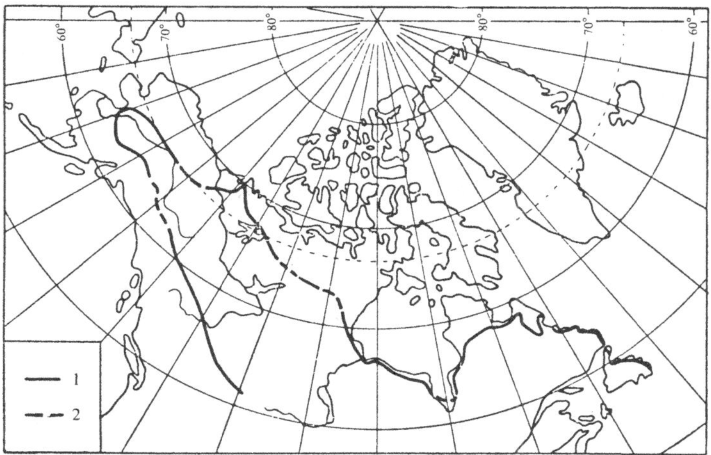

Fig. 9. The distribution of *Larix laricina* (Du Roi) K. Koch. 1, continuous area; 2, disjunct area. According to Hustich, 1966.

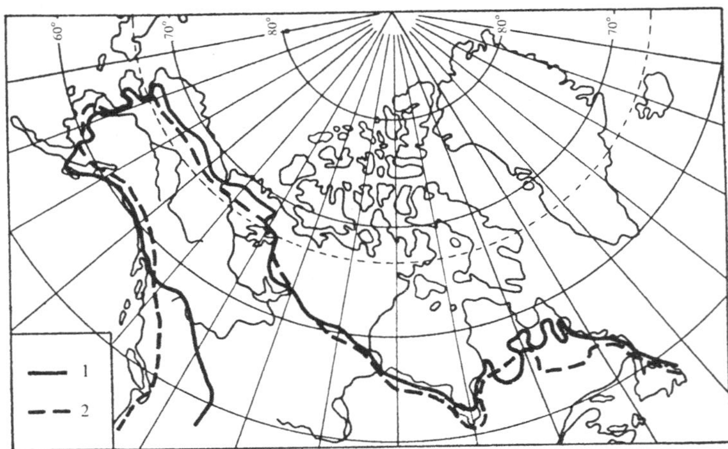

Fig. 10. Northern limit of the distribution of *Populus balsamifera* L. (1) and of *Populus tremuloides* Michx. (2). According to Hustich, 1966.

Populus tremuloides and *Betula papyrifera*. Yurtsev (1974*a*: 106–125) attributes these differences to the peculiarities caused by the Pleistocene history of the Bering landbridge. According to him, the Bering landbridge reached a width of 1500 km when it was broadest during the time of maximum emergence and might therefore have been divided into at least three zones: northern and southern coastal zones, and an interior more continental one. The northern zone could have been characterized by the dry conditions of the arctic tundras. The interior area may have corresponded to the continental variant of the hyparctic tundras. The southern coastal area of the Beringian landbridge, experiencing a softening of the climate due to the air masses from the Pacific Ocean, could as this author suggested have been represented by a zone of non-forested, coastal, hyparctic landscapes and been open to Beringian, north Pacific and other types of species of maritime origin. It is possible that there was a sporadic exchange between the coniferous forest floras of Asia and America (these having already developed independently during the Miocene) during the Pleistocene period, favoring such an exchange. In the opinion of Yurtsev (1974*a*: 121), such an exchange of Asiatic and American species of taiga trees could, however, occur just over the territory limited to the 'neutral' lowland of the Bering landbridge, which separated the mountainous highlands of Asia and America and was periodically destroyed by the transgression of the sea. Another factor was the deterioration of the climate, which resulted in weather conditions unfavorable for the growth of trees. These peculiarities of the paleogeography account, on the one hand, for the similarities in the tundra vegetation of northeastern Asia and northwestern America, and on the other hand, for the complete change in the composition of the woody species which form the polar forest limits in these territories.

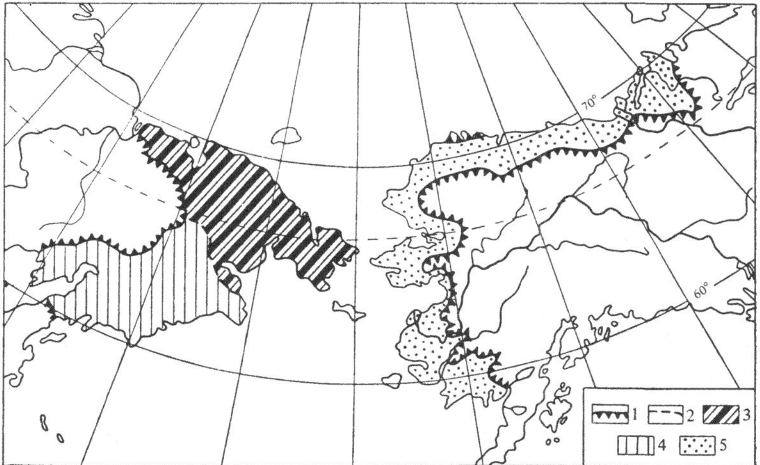

Fig. 11. The Chukotka–Alaska Province of the subarctic tundras. Boundaries: 1, provincial; 2, subprovincial. Subprovinces: 3, the Chukotka; 4, the Anadyr–Penzhina; and 5, the Alaska subprovinces.

Three geobotanical subprovinces can be distinguished within the Chukotka–Alaska province: the Chukotka, the Anadyr–Penzhina, and the Alaska subprovinces (Fig. 11).

The Chukotka subprovince of the subarctic tundras

East of the Kolyma River, the tussocky tundras with *Eriophorum vaginatum* associated with *Carex lugens* assume a zonal, mesic position and become generally the dominating type of tundra on the plains and in the low mountains. As reported by Katenin (1974: 1587), who investigated the vegetation along the middle course of the Amguema River, this is the most widely distributed formation, which is represented by a large number of associations on level or slightly sloping habitats with a loamy substrate. These associations are met with on plateaus, on the ledges of gentle slopes, on the ridges of rows of hills, and are very common on the lowlands as well as on the raised edges of low center polygons in polygonal mires.

The general type of vegetation cover changes abruptly east of the Kolyma River with the transition into different orographic and climatic conditions, and because of the close connection of this flora with the Beringian floristic complexes. In contrast to the plains of the East Siberian province, where the forest limit penetrates far to the north, we enter in Chukotka a mountainous country with a complicated relief. The forest limit east of the Kolyma River turns off towards the south, and the space occupied by tundra vegetation widens noticeably in a latitudinal direction. The mountain structures in Chukotka are represented in its western parts by the Anyusky Highlands, reaching elevations above 1800 m, and in the central and eastern parts by the Chukotka Highlands, also reaching altitudes above 1800 m, stretching out over the Chukotka Peninsula, where the individual peaks reach up to 1500 m elevation. The northern, coastal lowland is mainly represented by a narrow belt, widening around the Chaun and the Kolyuchinskaya Inlets.

According to data available (Gorodkov, 1939, 1946*a*; Derviz-Sokolova, 1964; Yurtsev, 1967, 1970*a*, 1972, 1973*a*: 949–51, 1974*a*: 12–16, 19742b; Reutt, 1970; Katenin, 1974; etc.), tussocky tundras are typical of the southern belt of the subarctic tundras. Together with cottongrass and *Carex lugens, Betula exilis, Salix pulchra, Ledum decumbens, Vaccinium uliginosum* ssp. *microphyllum, V. vitis-idaea* ssp. *minus, Hylocomium alaskanum, Sphagnum* spp., etc. are common, and locally a tier of scattered *Alnus fruticosa* occurs. Some willow brush (*Salix krylovii*, reaching 80–100 cm. and *Salix pulchra*, reaching up to 80 cm in height) and birches (*Betula exilis*, up to 60 cm tall) are associated with tussocky tundra. There are also herb–moss and herb–peatmoss mires. Groves of *Chosenia arbutifolia* and *Populus suaveolens* penetrate deep into the tundras on alluvium along the rivers. There are occasional formations with *Alnus fruticosa* on the slopes.

In the middle belt of the subarctic tundras (when following the zonation by Yurtsev, 1973: 949, this is the 'southern variant of the northern hyparctic tundras'), the tussocky tundras typically display *Eriophorum vaginatum, Carex lugens, Ledum decumbens, Vaccinium vitis-idaea* ssp. *minus* with low-grown *Betula exilis* and constantly present arctic–alpine species, such as *Salix sphenophylla* and others, as well as *Hylocomium alaskanum, Sphagnum* spp., etc. In combination with tundras of low-grown willows and birches (*Betula exilis* up to 30 cm tall), there are also willow thickets in gently sloping depressions in the herb–moss mires. On the slopes of the mountains and on the low plateaus (Katenin, 1974), tundras with different dwarfshrub species (*Dryas*, willows, *Cassiope*, the latter in well-drained, bouldery habitats, where snow accumulates) predominate. Dwarfshrub–sedge tundras with *Carex lugens* are frequent. There are carpet-like formations of dwarfshrubs and epilithic lichens. Many polygonal mires (Reutt, 1970: 282) are found in the low-lying areas, i.e., the northern part of the Kolyma Lowland and the plains of Chaun.

In the northern belt of the subarctic tundras (as defined by Yurtsev, 1973*a*: 949–51, the 'coastal variant of the northern hyparctic tundras'), the species composition of the tussocky tundras becomes impoverished and these tundras occupy less area. The amounts of dwarfshrubs and prostrate bushes diminish, and dwarfbirch associations almost completely disappear, while the position of arctic–alpine species is strengthened (Gorodkov, 1939, 1946*a*; Derviz-Sokolova, 1964; Yurtsev, 1967, 1973*a*).

Tundra–steppe and steppe associations appear in Chukotka. They have been thoroughly described by Yurtsev (1974*a*: 18–21, 1974*b*) and seem to be relicts from the cold and dry Pleistocene epoch. *Carex duriuscula*, otherwise distributed over the steppes of southern Siberia and northern Mongolia as well as on the Canadian prairies, occurs here, as well as *Helictotrichon krylovii* and other steppe, meadow-steppe, and mountain-steppe species.

The forest limit in Chukotka is formed by *Larix cajanderi*, but along the rivers *Chosenia arbutifolia* and *Populus suaveolens* penetrate deep into the tundras.

The vertical belts are complex in Chukotka and their aspect is often confused by temperature inversions. Yurtsev (1974*a*:16) distinguishes in general the following altitudinal belts in the northern part of the Anyusky Range: (1) 'the southern (hyparctic) mountain tundra', where hyparctic shrubs predominate on well-drained habitats and where there are thickets of *Alnus*; (2) 'the typical mountain tundra' with predominantly arctic–alpine and hyparctic species of dwarfshrubs; (3) 'the arctic mountain tundra', where the number of hyparctic species of dwarfshrubs is decreasing; and (4) 'the high-arctic mountain tundra', where the arctic–alpine dwarfshrubs and many herbs are disappearing or rare, where on stony, fine-grained soils the herbaceous species are buried in the carpet of mosses and lichens, and where many arctic species appear, such as *Saxifraga hyperborea, Phippsia algida*, etc., which are lacking below. Above 1300 m, on Mt El'veney, angiosperms are completely lacking. Derviz-Sokolova (1964) described the vertical belts in the easternmost parts of Chukotka. There sedge–cottongrass–willow tundras are characteristic of the lower belt (up to 180–200 m). Mountain tundras with sedge (*Carex lugens*)–dwarfshrub–lichen tundras are typical for the second belt. At an elevation of 250–320 m boulder fields predominate, but in a few places there is a vegetation dominated by crustose lichens and a small quantity of angiosperms, such as *Cardamine bellidifolia, Saxifraga nelsoniana, Artemisia arctica, Luzula nivalis, Oxyria digyna*, etc.

In this subprovince three geobotanical districts can be distinguished: the Lower Kolyma–Chaun, the Amguemsky, and the East Chukotka districts.

In the East Chukotka district, a major role is played by mesophytic meadow-like tundra communities, meadows and graminoid–willow-brush associations. There are no steppe–phytocoenoses, only tundra–steppes occur. The majority of the species composing the associations are co-differentiating and occur also in Alaska, but a number of endemics are also found: *Puccinellia beringensis, Cardamine sphenophylla*, etc. In western Chukotka, the mesophytic meadow-like communities play an insignificant role. Instead there are localities with xerophilous sedge–*Kobresia* associations. Here not only tundra-steppes but also steppe-phytocoenoses are met with. The flora is enriched by Siberian species and, to some extent, by a number of differentiating species, originating from the mountains of eastern Siberia, such as *Carex pediformis, Arenaria tschuktschorum, Phlojodicarpus villosus, Oxytropis ochotensis*, etc. Some of the species occur also in the Kharaulakh Highlands, e.g. *Senecio jacuticus, Astragalus schelichovii*, etc. (Yurtsev *et al*., 1972, Yurtsev, 1973*a*, 1974*a*, *b*).

The Anadyr–Penzhina subprovince of the subarctic tundras

Yurtsev (1974*a*) has placed the Anadyr–Penzhina subprovince in the boreal floristic region, although it belongs to the hyparctic belt. Its vegetation has been described in a number of works (Sochava, 1930*b*, 1932; Gorodkov, 1935*a*, 1938, 1946*a*; Vasil'yev, 1936, 1956; Vikhireva-Vasil'kova *et al.*, 1964; Reutt, 1970; Yurtsev, 1974*a*; etc.). The most extensive information is found in the papers by Gorodkov.

Tussocky tundras appear here as the zonal group of associations, which have developed in the plains between the mountainous areas and are endemic to the subprovince. They consist of *Eriophorum vaginatum*, *Carex soczavaeana* (forming large tussocks) and *Carex lugens*, associated with a considerable amount of low growing *Betula exilis*, *B. middendorffii*, and *Alnus fruticosa.* Odd specimens of *Pinus pumila* occur here. There are many dwarfshrubs: *Vaccinium uliginosum*, *V. vitis-idaea*, and *Ledum decumbens* predominate, less common are *Arctous alpina*, *Empetrum hermaphroditum* and *Diapensia obovata.* Peatmosses are common, *Sphagnum balticum*, *S. lenense*, *S. warnstorfii,* etc., and green mosses, *Dicranum elongatum, Aulacomnium turgidum* and *A. palustre* are also met with. A large area is occupied by tussocky tundras, where *Cladonia rangiferina*, *C. sylvatica*, *Cetraria cucullata* and *C. crispa* are abundant in the ground cover. On gentle mountain slopes and in low passes, not exceeding 150 m in altitude, shrub thickets up to 50 and even 130 cm tall are widely distributed. They have an open canopy of *Betula middendorffii*, *B. exilis*, and *Alnus fruticosa* associated with some *Pinus pumila*, *Salix fuscescens*, and *S. myrtilloides*, and those species of dwarfshrubs, mosses and lichens which also occur in the tussocky tundras. On stony grounds more densely closed dwarfshrub associations are widespread, mainly with species such as *Betula middendorffii* and *Pinus pumila* and an admixture of the shrubs mentioned above, as well as of *Spiraea beauverdiana*, *Rhododendron aureum*, and *Dasiphora fruticosa.* In the gullies thickets of *Pinus pumila* occur up to 150 m altitude.

According to Gorodkov (1935*a*: 70), the sub-gol'tsy (subalpine) shrub belt in the central part of the Penzhina area reaches up to an altitude of 350–400 m above sea level, but lowers towards the north and in the coastal, oceanic area to a maximal altitude of 300–350 m. However, at an altitude of 250 m the shrub thickets are frequently interrupted by boulder fields and mountain tundras with lichens associated with the transition to thickets of *Pinus pumila*. The latter, growing prostrate between mountain tundras with lichens dominated by species such as *Cornicularia divergens* and *Cetraria nivalis*, occurs up to the summits of the mountains as odd specimens at an altitude of 500–700 m.

In this subprovince, mires are found in a few localities but they occupy only limited areas. These mires are situated in depressions where sedge associations of *Carex rotundata* or *C. stans* and mosses and mires with flat-topped peat mounds occur. The latter are overgrown with shrubs, mainly *Betula exilis*, and a large quantity of dwarf-bushes such as *Vaccinium uliginosum*, *Ledum decumbens*, which grow on tussocks formed by *Eriophorum vaginatum*; *Carex rotundata* occurs in herbaceous tier. Mosses such as *Drepanocladus exannulatus* grow in the watery troughs between the mounds. There are also many peat-mosses.

In the western part of the subprovince *Larix cajanderi* forms the forest limit, and in the river valleys stands of *Chosenia arbutifolia* are met with. In addition, there is *Betula cajanderi*, and in the south *Betula ermanii*.

The Alaska subprovince of the subarctic tundras

The associations on zonal, mesic habitats in the subarctic tundras of the Alaska subprovince, including also the part east of Mackenzie River as far as Anderson River, are very similar to those in Chukotka. Here, as there, tussocky tundras with *Eriophorum vaginatum* associated with *Carex lugens* are widely distributed. Hultén (1968) and a number of American ecologists (Churchill, 1955; Britton, 1966, etc.) mention also *E. vaginatum* ssp. *spissum* for Alaska. But I follow Porsild's (1957*a*) and other investigators' (Wiggins & Thomas, 1962; Tolmachev, 1966, etc.) opinion that this eastern American cottongrass does not reach to Alaska.

Carex consimilis is also part of the herbaceous associations, and this variant of the tussocky tundras is endemic to this subprovince. It should be mentioned here that most of the authors dealing with the vegetation of Alaska call this sedge *C. bigelowii* Torr. s.l., while following Hultén (1968) who considered *C. bigelowii* a circumpolar species and *C. ensifolia* Turcz. and *C. consimilis* Holm as its subspecies. However, Porsild (1955, 1957*a*) as well as Wiggins & Thomas (1962:

51) and Hultén in his earlier work (1964: 51) distinguish *C. consimilis* as a species of its own, quite well separated from the amphi-atlantic *C. bigelowii* s.str.

The specificity of this subprovince is also accentuated by the occurrence of the American species *Betula glandulosa* in the shrub thickets and by the participation in a number of associations of American grasses, *Calamagrostis canadensis*, etc., and herbs, *Lupinus arcticus*, etc. There is also, as mentioned above (on p. 59), a complete shift in the woody species forming the forest limit.

The vegetation of the subprovince of Alaska has been described by a number of authors (Palmer & Rouse, 1945; Hanson, 1951, 1953; Churchill, 1955; Bliss, 1956; Bliss & Cantlon, 1957; Britton, 1957, 1966; Viereck & Little, 1972; Corns, 1974, etc.). The most extensive information on the composition of the plant associations is found in the form of comprehensive tables published by Hanson, Churchill and Corns. The tussocky tundras with *Eriophorum vaginatum* and *Carex lugens*, sometimes also with *C. consimilis*, seem to be the most characteristic and the most widely distributed. Occasionally there is an admixture of low-growing shrubs, such as *Betula exilis*, *Salix pulchra*, etc., and, with the exception of the northern variant of the tussocky tundras, also of *Alnus fruticosa*.

(Note: After having studied herbarium material from Alaska, S. K. Cherepanov has suggested that what is found here is not the eastern American *A. crispa* s.str., described from Labrador, but actually *A. fruticosa* (cf. also Yurtsev *et al.*, 1972: 769–70).)

In these tundras there are also hyparctic dwarfshrubs such as *Ledum decumbens*, *Vaccinium vitis-idaea* ssp. *minus*, *V. uliginosum* ssp. *microphyllum*, *Empetrum hermaphroditum*, and others. Between the cottongrass tussocks grow green mosses and peatmosses, and on the edges of the tussocks, lichens.

In the southern belt of the subarctic tundras, beside the omnipresent tussocky tundras on level ground with loamy soils, shrub tundras are met with in areas of low mountains and on stony slopes (Hanson, 1953: 122–3), where low-growing *Betula exilis* with an admixture of *Salix pulchra*, *S. alaxensis*, and *S. lanata* ssp. *richardsonii* predominate and *Spiraea beauverdiana* is occasionally met with. In the ground tier there are many dwarfshrubs, such as *Vaccinium uliginosum* ssp. *microphyllum*, *V. vitis-idaea* ssp. *minus*, *Empetrum hermaphroditum*, *Ledum decumbens*, *Loiseleuria procumbens*, *Salix phlebophylla*, *S. reticulata*, etc. Typical among the graminoid species are *Carex podocarpa*, *Calamagrostis canadensis*, etc., among the mosses *Hylocomium splendens*, *Drepanocladus uncinatus*, *Ptilidium ciliare*, and among the lichens *Cladonia rangiferina*, *C. sylvatica*, *C. pleurota*, and *Cetraria cucullata*.

On slopes varying from gentle to steep, dwarfshrub tundras are observed up to 200 m altitude with *Vaccinium uliginosum* ssp. *microphyllum*, *Empetrum hermaphroditum*, *Ledum decumbens*, *Vaccinium vitis-idaea* ssp. *minus*, *Loiseleuria procumbens*, *Arctous alpina*, and an admixture of low-growing *Betula exilis*, and *Salix pulchra*, mosses and lichens. On slopes of increasing steepness at low elevation, dwarfbirches, *Betula exilis* and *B. glandulosa*, as well as their hybrids, are widely distributed, forming associations of various density, from open to closed thickets, and height from 0.7 to 1.3 m, associated with willows such as *Salix lanata* ssp. *richardsonii*, *S. pulchra*, and others, which are usually 30–50 cm taller than the birches. The moss carpet is formed mainly by *Hylocomium splendens*. There are many dwarfshrubs: *Vaccinium uliginosum* ssp. *microphyllum*, etc. In openings between them lichens have often developed, e.g. *Cladonia rangiferina*, *C. sylvatica*, *C. cornuta*, *Cetraria cucullata* and *C. islandica* (Hanson, 1953: 119–23).

In brook and river valleys, thickets of *Salix alaxensis*, *S. glauca*, and *S. fuscescens* grow together with some grasses (*Calamagrostis canadensis*, *Festuca altaica*, etc.), herbs, and dwarfshrubs. On well drained slopes, associations of *Alnus fruticosa* with a considerable abundance of *Calamagrostis canadensis* are found. In the more open associations of alder there is an admixture of birches, willows, *Spiraea beauverdiana*, a number of herbaceous species, dwarfshrubs, mosses and lichens. In the southern part of the subprovince, localities are found with tall thickets of *Betula glandulosa*, associated with small amounts of *B. exilis*, alder, willows, and *Spiraea*. *Dryas* tundras are typical of the upper parts of the slopes and on mountain tops at elevations up to 500 m. Some arctic and hyparctic species are also present here.

The mires along the edges of lakes, rivers and brooks and in moist draws are represented by herb–moss associations with *Carex aquatilis*, *C. stans* and *C. rotundata*. Flat-mounded and polygonal mires with large hummocks, overgrown with *Betula exilis*, *Rubus chamaemorus*, willows, and hyparctic dwarfbushes, peatmosses and other mosses, are widely distributed in depressions.

The major part of the area investigated by Corns (1974) does also belong to the southern belt of subarctic tundras. His area is situated east of the Mackenzie River delta, but does not include the lowland on its northeastern side or the Tuktoyaktuk Peninsula. Beside the tussocky tundras, which here are at the eastern limit of their distribution area and already begin to constitute a lesser part of the vegetation, a major role in the vegetation cover is played by tundras with low-growing birches on the summits and gentle slopes of hills. Here, low-growing *Betula exilis* associated with *Salix glauca* s.l., *S. pulchra*, and occasionally with *Alnus fruticosa* dominate, and there are many hyparctic dwarfshrubs as well as mosses and lichens. Along the rivers there are tall (1.5–2 m) shrubs, such as *Salix lanata* ssp. *richardsonii*, *S. alaxensis*, *S. pulchra*, and *Alnus fruticosa*. The latter reaches up to 3–4 m in height in favorable habitats.

The Umiat region described by Churchill (1955) belongs to the middle belt of the subarctic tundras. It covers a rolling plain, stretching northwards from Brooks Range. Here tussocky tundras predominate, with *Eriophorum vaginatum*, usually associated with *Carex lugens*, but sometimes with *C. consimilis*. Churchill (1955) uses the epithet *E. spissum* Fern. (=*E. vaginatum* ssp. *spissum* (Fern.) Hult.) for the cottongrass; the latter is, however, as mentioned above, not found in Alaska. There are also many dwarfshrubs, as e.g. *Vaccinium vitis-idaea* ssp. *minus* and *Ledum decumbens*, often mixed with *Cassiope tetragona*, and plenty of low-grown willows, *Salix glauca* s.l., *S. pulchra*, etc. Prostrate *Betula exilis* is a constant, and sometimes there are scattered specimens of *Alnus fruticosa*. Besides *Cassiope tetragona* and *Salix phlebophylla*, there are admixtures of such arctic species as *Luzula confusa*, *L. nivalis*, *Hierochloë alpina*, and *Poa arctica*. Along brooks and rivers and in gullies on watersheds, thickets of willows, and rarely of alders, are met with, but together with the latter species much *Arctagrostis latifolia* occurs. On hills and the upper part of the slopes on ridges in localities strongly exposed to winds during the winter spotty tundras with *Dryas integrifolia* have developed, in which there is some admixture of low-grown willows such as *Salix pulchra* and *S. glauca* s.l. Polygonal mires are widely distributed.

The absolute elevation of the landscape lowers towards Point

Barrow, where the relief becomes almost flat and the number of lakes and swampy areas increases. With the exception of the area immediately adjacent to Point Barrow, which is referred to the subregion of the arctic tundras where there are no more tussocky cottongrass tundras and *Betula exilis* is lacking, the coastal lowlands belong to the northern belt of the subarctic tundras. There is no more *Alnus fruticosa*, the role of the low-grown shrub thickets in the tussocky tundras is diminished, and the variability and number of the plant associations are decreased (Britton, 1966: 31), while the role of the arctic species is increased.

The fact that it is correct to distinguish Alaska as an area at the subprovincial level is emphasized by its floristic uniqueness. There are a number of endemics and subendemics (*Artemisia comata*, *Erigeron muirii*, etc.), eastern co-differentiating species (*Lupinus arcticus*, etc.) and a large number of differentiating cordilleran species (cf. Yurtsev, 1974*a*).

The forest limit is formed by *Picea glauca* and *P. mariana*, and there are also found *Larix laricina (L. alaskensis)*, *Betula papyrifera*, *Populus balsamifera* and *P. tremuloides*.

The Canadian province of the subarctic tundras

The Canadian province of the subarctic tundras begins east of Anderson River, at the easternmost limit for the distribution of the tussocky cottongrass tundras with *Carex lugens*. The tundras with low-growing *Betula glandulosa* and the hyparctic dwarfshrub tundras with an admixture of this species may be considered endemic for this province. Although *B. glandulosa* is found also in Greenland, it occurs only in the small, southwestern part of the subarctic tundras there, adjoining the southern tip of Greenland, which is situated outside the tundra region. In the dwarfshrub tundras it is, as a rule, substituted by *B. nana*.

The western boundary of this province coincides almost completely with the eastern limit of 'Greater Beringia' in Yurtsev's sense (1974*a*: 49). It is also an important boundary from the point of view of orography and climate as well as paleogeography in connection with the major differences in the Quaternary history of the landscape. The complicated relief of Alaska – the result of mesozoic orogenesis and subsequent processes of uplift, erosion, and partial denudation, glacial excavation, and accumulation – changes east of the Mackenzie River into the region of the Canadian Shield. Along its western edge the mesozoic folds of the Alaskan area turn southeastwards. On the surface of the Shield appear Precambrian, crystalline bedrocks. In its form the Shield reminds one of a gigantic bowl, the rims of which are raised, particularly at the eastern edge, forming the northern part of Labrador, Baffin Island, and the southern part of Greenland, while the central part is depressed, forming the Hudson's Bay synclinal. The western part of the Shield consists of the Laurentian Upland with elevations from 150 to 500 m and a strongly levelled surface as a result of protracted denudation and scouring by glaciers, which has left here not so much accumulated debris as eroded forms of relief. Along the northern coast and around Hudson's Bay there are marine, postglacial sediments. Except for the northern tip of Labrador the climate is strongly continental, although somewhat milder than in eastern Siberia.

While neither Chukotka nor Alaska were completely covered by glaciers during the maximal phase of the Glacial Age, the territories of the Canadian Shield were fully enveloped by an icesheet. According to Ignat'yev (1963: 241), who has reviewed data from foreign literature, the Cordilleras in the west and Baffin Island and Labrador in the east served as centers for the formation of the glacial icesheets. The Cordilleran Icesheet, about 3 km thick, did not spread very far east of the mountain ranges and was built up mainly by the transfer of moisture from the Pacific Ocean. In contrast, the eastern Laurentian Icesheet received much moisture from the atmospheric airmasses originating over the Atlantic Ocean. It therefore spread out extensively. At the period of maximum extension, the two icesheets coalesced, forming a glacial sheet reaching from the Cordilleras to the eastern coast of the continent. The total glaciated areas exceeded by 2.5 times the areas under glacial cover in Europe. The last of the icesheets, the Wisconsin, reached its maximum 20 000–40 000 years B.P.; it disappeared from the continent about 6500 years B.P. The Greenland Icesheet and other major glaciers of the Arctic islands remain as relicts of it. During its retreat the icesheet split into two parts, the Labrador and the Keewatin Icesheets, separated by Hudson's Bay; along the edges enormous ice-dammed lakes were formed, as relics of which the large Canadian lakes still exist. At the same time the streams of meltwater raised the surface of the oceans, which in many places submerged the continent.

After the liberation from the icesheet the territories of the Canadian Shield began to be reoccupied by vegetation. In the parts situated below the tundra zone the action of the icesheet had destroyed not only all previously existing flora, but had also removed the layer of fine soil. The return of the vegetation started from the west, the south, and the east and is apparently not yet completed. The sources of dispersal were Beringian and Siberian–Beringian species from the west, and from the east amphi-atlantic species. The waterbody of Hudson's Bay, which during the period of maximal transgression by the sea extended far to the south (Fig. 12), presented an obstacle for the exchange of the flora, and below we shall see the rather clearly outlined border, a little to the west of Hudson's Bay, between the western and the eastern parts of this province of the subarctic tundras in Canada. The western part is rich in Beringian and Siberian–Beringian species, the eastern part in amphi-atlantic ones. The border is also reflected in the botanical–geographical division of the circumpolar Arctic made by Polunin (1951) and the floristic division seen by Yurtsev (1974*a*: 142). It agrees approximately with the extension of the Yurtsev 'Mega-Beringia' (1974*a*: 145). To the group of western species, not extending to Hudson's Bay, belong such important association-formers from northeast Asia and northwest America as *Eriophorum vaginatum*, *Carex lugens*, *Salix pulchra*, and a whole row of other species such as *Kobresia sibirica*, *Salix polaris*, *Caltha arctica*, *Parrya nudicaulis*, *Lupinus arcticus*, *Nardosmia frigida*, *Senecio frigidus*, *Artemisia tilesii*, *Saussurea angustifolia*, etc. Others, such as *Carex rotundata*, *Salix alaxensis*, *S. lanata* ssp. *richardsonii*, *S. arbusculoides*, *Ranunculus gmelinii*, *Astragalus arcticus*, etc., reach as far as the western shore of Hudson's Bay, but do not extend to its eastern side. To the group of species met with only east of Hudson's Bay and not reaching to its western side belong *Deschampsia alpina*, *Carex saxatilis* ssp. *saxatilis*, *Salix uva-ursi*, *Ranunculus allenii*, *Draba allenii*, *D. norvegica*, *Potentilla crantzii*, and others.

Three geobotanical subprovinces can be distinguished: the West Canadian, the Hudson's Bay, and the Labrador subprovinces (Fig. 13).*

* Translator's note: Although the original text uses the epithets 'Hudsonian' and 'Labradorian', these expressions have been avoided here in order to exclude any confusion with those terms, which are used in a different sense in American biogeography.

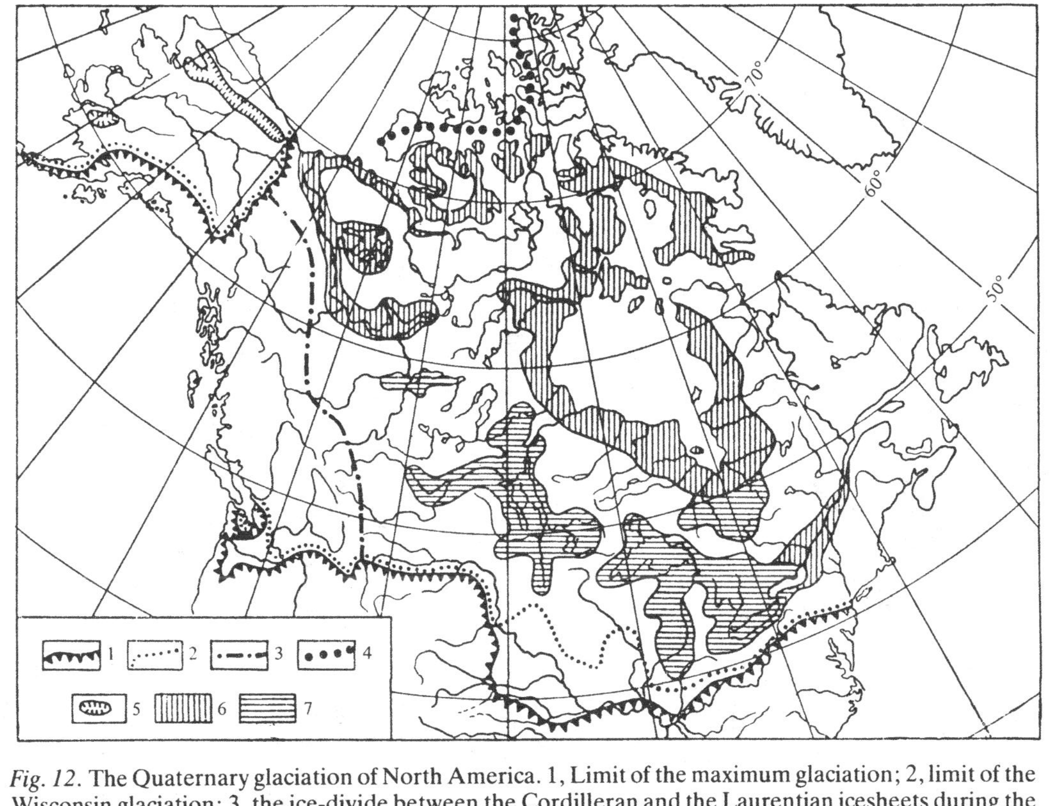

Fig. 12. The Quaternary glaciation of North America. 1, Limit of the maximum glaciation; 2, limit of the Wisconsin glaciation; 3, the ice-divide between the Cordilleran and the Laurentian icesheets during the Wisconsin glaciation; 4, the northern limit of the extension of the Pleistocene glaciation; 5, mountain glaciation; 6, postglacial marine transgression; 7, postglacial lakes. According to Ignat'yev, 1963.

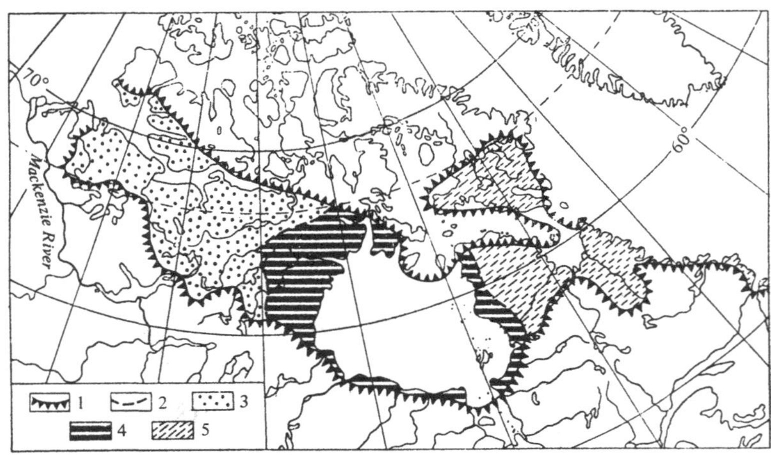

Fig. 13. The Canadian Province of the subarctic tundras. Boundaries: 1, provincial; 2, subprovincial. Subprovinces: 3, the west Canadian; 4, the Hudson's Bay; and 5, the Labrador subprovinces.

The West Canadian subprovince of the subarctic tundras

The continental tundras of Canada from Anderson River in the west to the Hudson's Bay lowland in the east and the southernmost parts of Banks and Victoria Islands belong to the West Canadian subprovince. The vegetation cover, mainly that of the tundras, has been described by Porsild, but also by a number of other investigators (Porsild, 1951, 1955, 1957*a*,*b*; Lindsey, 1952; Banfield, 1954; Ritchie, 1960; Larsen, 1965; Maini, 1966; Porsild & Cody, 1968; Scotter & Telfer, 1975; Thannheiser, 1975, etc.). The flora is considerably better known than the vegetation.

The continental part of the West Canadian subprovince extends over the Laurentian Upland with its uniform relief. It is dominated by flat-topped, rolling hills and ridges, deposited by the glaciers, and plateau-like formations of unbroken outcrops of Precambrian granites and gneisses and, in the western part of the subprovince in rare cases, of Paleozoic, sedimentary bedrock. In between the outcrops are shallow depressions with lakes, mires and tundra vegetation. The best presentation of the character of this landscape is found in the photographs published by Porsild (1951: 18).

The predominance of a stony substrate has led to the extensive distribution of a petrophytic vegetation. North of the Great Bear Lake, according to Porsild (1951: 16), the most common of the tundra associations have developed on the level habitats and consist of low-growing dwarfbirch tundras. These are represented by *Betula glandulosa* together with *Ledum decumbens*, *Rhododendron lapponicum*, *Arctous alpina*, *Cassiope tetragona*, *Vaccinium uliginosum* ssp. *microphyllum* and *Vaccinium vitis-idaea* with a ground cover of mesophytic mosses and lichens. Along brooks, on slopes, and on steep banks around lakes, willow thickets have developed with *Salix arbusculoides*, *S. lanata* ssp. *richardsonii*, *S. glauca* s.l., *S. planifolia* and *S. alaxensis*, and close to the forest limit also with *Alnus crispa*. The rocky surfaces of the wind-exposed summits of gentle hills, composed of bedrocks, are covered by epilithic, crustose lichens, species of *Rhizocarpon*, *Lecidea*, *Lecanora*, *Buellia*, etc., and big patches of black foliose lichens, *Umbilicaria* ssp. On slopes with habitats of fine-grained erosion material associations of lichens, *Cladonia*, *Cetraria*, etc., and dwarf-bushes have developed. They are dominated by *Vaccinium uliginosum* ssp. *microphyllum*, *V. vitis-idaea* ssp. *minus*, *Arctous alpina*, etc., and in addition, willows and birches, represented here by trailing forms. In the thoroughly wet depressions, mires and swampy tundras with *Eriophorum spissum* (in the western part of the subprovince, however, with *E. vaginatum*) have formed. In wet troughs with herbaceous vegetation one can find among other species, *Arctagrostis latifolia*, which is typical of the Canadian tundra territory (in Eurasian tundras *Arctagrostis latifolia* grows only in mesophytic habitats).

The forest limit is formed by *Picea mariana*, *P. glauca*, *Larix laricina*, and locally, *Populus balsamifera*.

Typical for this subprovince is the participation of species from the Beringian area (*Salix alaxensis*, *Carex lugens*, etc.), rapidly diminishing in quantity towards the east. *Parrya arctica* appears to be subendemic for this subprovince.

The Hudson's Bay subprovince of the subarctic tundras

During the last stages of the Ice Age, when the Laurentian Icesheet was melting, the area around Hudson's Bay was flooded by a marine transgression, forming the 'Tyrrell Sea' (Lee, 1960) reaching far towards the south. Later, as a result of isostatic uplift, the sea receded, baring the lowlands around the present Hudson's Bay. Particularly along its southwestern coast, where the elevation as a rule does not exceed 25–50 m, the thickness of the clayey sediments covering the crystalline rock base reaches 15–25 m (Hustich, 1957). In the northwestern and eastern parts of the coast, the alluvial sediments are interrupted more often than in the southwestern parts by outcrops in the form of low, flat, or slightly domed stony mounds, rarely by stony ridges.

The vegetation of this subprovince has been described in a number of works (Marr, 1948; Polunin, 1948, 1960; Hustich, 1957; Porsild, 1957*a*; Schofield, 1959; Porsild & Cody, 1968, etc.).

Dwarf-shrub tundras with *Vaccinium uliginosum* ssp. *microphyllum*, *V. vitis-idaea* ssp. *minus*, *Empetrum hermaphroditum*, *Loiseleuria procumbens*, and *Arctous alpina* are typical for the stony habitats which retain a sufficient quantity of fine-grained material in the southern belt of the subarctic tundras (Hustich, 1957; Ritchie, 1960). In their ground cover lichens predominate (*Cladonia rangiferina*, *C. rangiformis*, *C. mitis*, *Cetraria nivalis* and *C. islandica*) associated with prostrate *Betula glandulosa*. Locally, special associations are formed with *Diapensia lapponica*, *Phyllodoce coerulea* and *Rhododendron lapponicum*. On the lower parts of slopes and along rivers thickets of *Salix planifolia* and other willows, and *Betula glandulosa*, as well as *Alnus crispa*, have developed. In the vegetation cover, occupying large areas of mires, *Carex rotundata*, *C. magellanica*, *C. stans*, *Eriophorum spissum* and *Trichophorum caespitosum* play a large role. *Sphagnum* species, *Andromeda polifolia*, *Rubus chamaemorus*, *Vaccinium uliginosum* ssp. *microphyllum* and *Ledum decumbens* grow on the peat mounds.

Apparently, the vegetation described by Polunin (1948: 262–79) from the western shores of Hudson's Bay around Chesterfield Inlet belongs to the middle belt of the subarctic tundras. Here, on level parts of the surface formed by marine sediments, but also on surfaces of eroded bedrock, large areas are occupied by dwarfshrub tundras in which *Vaccinium uliginosum* ssp. *microphyllum*, *Empetrum hermaphroditum*, and *Ledum decumbens* dominate. There is much *Arctous alpina*, *Vaccinium vitis-idaea* ssp. *minus*, *Salix herbacea*, *Carex bigelowii*, and *Luzula confusa*. Willows, *Salix arctica*, etc., may be found, and sometimes prostrate *Betula glandulosa*. Mosses, such as *Rhacomitrium lanuginosum*, and lichens, *Cetraria ssp.*, *Cladonia* spp., *Dactylina arctica*, *Lobaria linita*, *Peltigera leucophlebia*,

Sphaerophorus globosus, and *Stereocaulon alpinum* form a well developed ground cover, among which there are sometimes small groups of bushy *Betula glandulosa* and *Salix planifolia*. On gneissic outcrops, the surface is covered by black foliose lichens such as *Umbilicaria cylindrica*, *U. hyperborea* and *U. proboscoidea*, but also by species of *Lecanora*, *Lecidea*, *Parmelia* and *Rhizocarpon*. In crevices and hollows *Rhacomitrium lanuginosum* and some other mosses may be met with in association with a number of angiosperms, especially *Hierochloë alpina*, but also with *Luzula confusa*, *Festuca brachyphylla*, *Silene acaulis* and others. Mires occupy large areas. In the mires described by Polunin the graminoid species consist of *Carex stans*, *Dupontia fisheri*, *Carex chordorrhiza*, *C. saxatilis* ssp. *saxatilis*, *C. saxatilis ssp. laxa*, *Hierochloë pauciflora*, *Arctagrostis latifolia*, and others. *Salix arctophila* occurs there, too, and the mosses *Calliergon sarmentosum*, *Drepanocladus revolvens*, *Meesia triquetra* and *Sphagnum squarrosum* are mentioned. On the peat mounds, where peat-mosses are abundant, grow *Andromeda polifolia*, *Carex rariflora*, *Eriophorum spissum*, *Pedicularis sudetica*, *Ranunculus lapponicus*, *Rubus chamaemorus*, and other species.

Southampton Island, situated in the northwestern part of Hudson's Bay, belongs to the northern belt of the subarctic tundras. There, according to Polunin (1948: 248–62), the dwarfshrub tundras are best distributed on plains with relatively well drained habitats. They display much *Dryas integrifolia* associated with *Arctous erythrocarpa*, *Cassiope tetragona*, *Empetrum hermaphroditum*, *Ledum decumbens*, *Rhododendron lapponicum*, *Salix* spp., *Vaccinium uliginosum* ssp. *microphyllum*, *V. vitis-idaea* ssp. *minus* (this species may predominate locally), and an admixture of such arctic species as *Poa arctica*, *Polygonum viviparum*, *Salix arctica*, *S. reticulata*, *Luzula confusa*, *Saxifraga oppositifolia*, and others. The ground cover is formed by tundra species of mosses and lichens. There are also herb–moss mires with *Carex stans*, *C. membranacea*, *Arctagrostis latifolia*, *Eriophorum angustifolium*, etc. On the raised borders of the low center polygons in the polygonal mires there are thickets of *Salix arctophila*, etc.

The floristic specificity of the Hudson's Bay subprovince is inherent in the fact that it is a transitional phytochore with strongly expressed west–east and east–west gradients, the former more clearly manifest. Areas of western and eastern species overlap here, e.g. those of *Carex rotundata* and *C. miliaris* (Fig. 14), of *Eriophorum vaginatum* and *E. spissum* (Fig. 15), and so on. Two subendemics are found here, *Oxytropis bellii* and *O. hudsonica*.

The forest limit is formed by *Picea mariana* and *P. glauca*.

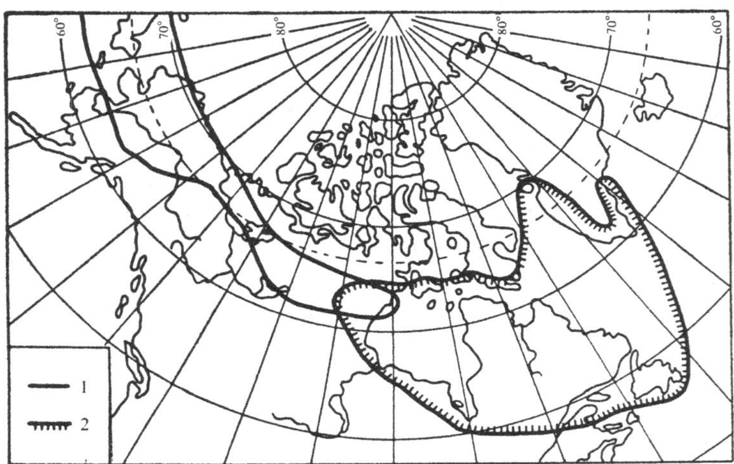

Fig. 14. Distribution of *Carex rotundata* Wahlenb. (1), and of *Carex miliaris* Michx. (2) in North America. According to Hultén, 1964.

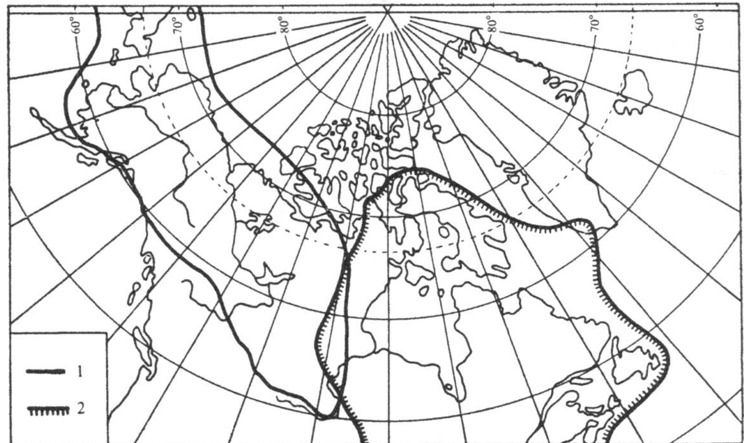

Fig. 15. Distribution of *Eriophorum vaginatum* L. ssp. *vaginatum* (1) and of *Eriphorum vaginatum* L. ssp. *spissum* (Fern.) Hult., (2) in North America. According to Porsild, 1957*a*.

The Labrador subprovince of the subarctic tundras

The northern part of Labrador (actually the Ungava and Labrador Peninsulas of the northern Quebec Province, cf. Fig. 13) and the southwestern parts of Baffin Island belong to the subarctic tundras. These areas cover the steeply rising eastern edge of the Canadian Shield. On the Ungava Peninsula in the eastern part of the Labrador subprovince, where the relief is shaped by glaciers, the elevation reaches 830 m, but east of there, on the Labrador Peninsula, it rises rapidly above 1000 m and up to 1670 m. Here the Atlantic Ocean has cut a multitude of long and narrow, deep fjords with promontories up to 1000 m tall. On the northern Labrador Peninsula, coastal cliffs with 300–700 m tall precipices are typical. The climate of this subprovince has lost the traits of continentality, which are characteristic of the Laurentian plateau, and approaches instead an oceanic climate. At high elevations there are remnants of relic glaciers, now fast disappearing.

The vegetation of the Labrador subprovince has been described by a number of authors (Soper, 1930*a*, *b*, 1936; Abbe, 1938; Polunin, 1948; Hustich, 1949, 1957, 1966; Hare, 1955; Harper, 1964; Rousseau, 1968, etc.). The most complete data are furnished by Polunin (1948).

The zonal vegetation on level parts and gentle slopes is represented by dwarfshrub tundras with predominating *Empetrum hermaphroditum*, *Vaccinium uliginosum* ssp. *microphyllum*, *V. vitis-idaea* ssp. *minus*, mixed with some *Diapensia lapponica*, *Pyrola grandiflora*, *Salix herbacea*, *Ledum decumbens*, *Arctous alpina*, *Cassiope tetragona*, *Rhododendron lapponicum*, *Loiseleuria procumbens*, *Phyllodoce coerulea*, *Harrimanella hypnoides*, as well as *Carex bigelowii*, *C. scirpoidea*, and others. In the northern parts *Dryas integrifolia* is admixed. The latter species dominates (Polunin, 1948: 139) in associations developed on the Paleozoic, crystalline limestone found on Baffin Island. The dominating rock type in this subprovince is otherwise diabases. Locally, in the dwarfshrub tundras, there are low-grown willows, *Salix uva-ursi*, *S. arctica*, prostrate *S. glauca* ssp. *callicarpaea*, and so on, as well as creeping *Betula glandulosa* (Polunin, 1948: 140, 192, 213, etc.). Mosses and lichens form a very well developed ground cover. In favourable conditions, shrub thickets have developed; *Betula glandulosa* and *Salix glauca* ssp. *callicarpaea* up to 50 cm, rarely above 80 cm tall, predominate.

Mires are widely distributed in depressions. So are lakes. In grassy mires, *Eriophorum angustifolium*, *E. scheuchzeri*, *Arctagrostis latifolia*, *Carex membranacea* and *Dupontia fisheri* dominate, associated with *Carex stans*, *C. miliaris*, *C. gynocrates*, *Trichophorum caespitosum*, *Cardamine pratensis*, and others. Mosses are represented by *Calliergon sarmentosum*, *Drepanocladus revolvens*, *Aulacomnium palustre*, *Sphagnum teres*, etc. It is peculiar that *Eriophorum scheuchzeri* participates (as it does also in Greenland) in these subarctic mires. In other areas of the Arctic this species is abundant only at high latitudes. It is also interesting to watch the widening amplitude of *Arctagrostis latifolia* here, where it is reported not only as an associate, but even as one of the dominant species, filling wet hollows and low-lying mires (Polunin, 1948: 146). In the majority of the arctic areas this species is a component of mesophytic tundra associations.

The rather low (up to 150 m high) slightly rounded or flat-topped hills, mostly of diabase, carry a fairly well developed vegetation on their summits. The surface of the rocks is completely covered by crustose lichens and black, foliose *Umbilicaria* spp. Sometimes it is also over-grown by the moss *Rhacomitrium lanuginosum*. In hollows, crevices, and where there are small boulders and gravel, fragments of dwarfshrub tundras have developed with *Salix uva-ursi*, *Vaccinium uliginosum* ssp. *microphyllum*, *Diapensia lapponica*, *V. vitis-idaea* ssp. *minus*, *Cassiope tetragona*, *Carex bigelowii*, *Hierochloë alpina*, *Luzula confusa*, and other species (Polunin, 1948: 136). In gullies and along the edges of brooks Soper (1936: 435) found thickets of willows up to an elevation of 250 m in the interior of southern Baffin Island.

Above 250–300 m a belt of oro-arctic tundras is situated with large areas of denuded surface. There, together with crustose and epilithic, foliose lichens and mosses, mainly *Rhacomitrium lanuginosum* (the surface of which is often colonized by *Cetrariae* and *Cladoniae*, and other lichens) and scattered specimens of *Carex rupestris, Cerastium alpinum, Eutrema edwardsii, Luzula confusa, Potentilla hyparctica*, etc., fragments of associations have developed with *Diapensia lapponica, Dryas integrifolia, Arctous alpina, Hierochloë alpina, Luzula nivalis, Papaver radicatum* s.l., *Saxifraga oppositifolia, Silene acaulis*, and other species (Polunin 1948: 211–12, etc.).

The flora of this subprovince differs in the presence of eastern co-differentiating species such as *Deschampsia alpina, Cerastium cerastoides*, etc., and in the absence of a number of species typical of other subprovinces in the Canadian province of the subarctic tundras, not penetrating east of Hudson's Bay (see above, p. 72).

The forest limit is formed by *Picea glauca, P. mariana*, and *Larix laricina*.

The Greenland province of the subarctic tundras

The specific character of the vegetation of Greenland is connected with the presence of the enormous icesheet, covering 84% of the surface of the country, and with its distinctly mountainous relief in the non-glaciated coastal belt. The glacial shield, in places reaching a height of 3150 m and a depth of 3400 m, exerts an enormous effect on the development of the climate. Because of it there is a constant flow of cold air down over the coastal districts. Above the icesheet an anticyclonal régime develops, which is more stable over the northern part of the country than over its southern parts. From the icesheet valley glaciers lead off, in many places reaching the sea and calving icebergs. Ice-free land is represented either by narrow chains of coastal cliffs and nunataks, or by areas widening out to more or less broad, coastal mountainous districts. These may be 20–30 km up to 200 km in width. The relief consists of steep ridges and mountain massifs, cut by numerous fjords, the depth of which often exceeds 1000 m. On the east coast the mountains are very high (Mt Gunnbjörn, 3700 m), a little less so on the west coast. An important role is played locally by the adiabatic föhn winds which often occur in the valleys of the major fjords. Due to their effect a cold and dry (cryo-arid) climate has developed in the interior areas of the major fjords situated close to the icesheet. Nordenskiöld was startled, when in 1874 in a valley between Egedesminde and Godthaab he discovered a wide area with saline lakes without outlet, salt flats and an appearance of extreme aridity in every character of the landscape (*in* Warming, 1928: 296). The cryo-arid vegetation from the interior areas of Greenland has been described by later authors (Böcher, 1954, etc.).

At the time of maximal glaciation some parts of Greenland remained unglaciated, but the vegetation existing in these refugia may have been very poor. For the postglacial development of a vegetation cover on the tundras of subarctic Greenland the connection both with America (especially Baffin Island – Labrador) and with northern Europe was of great importance. The latter connection may have been particularly active during the maximum emergence of the shelf, where, as suggested by Heezen & Tharp (1963), a landbridge existed via Iceland and the Faroe Islands (see also Hadač, 1960: 237).

A little less than half of the territory of Greenland belongs to the subregion of the subarctic tundras, i.e. the area south of latitude 69–71° N. However, the very southernmost tip, where the climate shows clear signs of oceanity and there are many boreal species and no *Betula nana*, does not belong to the tundras, but rather to the 'north atlantic grassland' region Lavrenko, 1950: 536) or 'ericaceous heath grassland' region (Yurtsev, 1966). For the Greenland province of the subarctic tundras the development of a dwarfshrub tundras is typical of the kind of habitats, which in mountains may be called 'analogous to zonal, mesic habitats' (Sochava, 1956*a*: 527). Most frequently *Empetrum hermaphroditum* and *Vaccinium uliginosum* ssp. *microphyllum* predominate with a typical admixture of prostrate *Betula nana*, the appearance of which separates fundamentally the Greenland tundras from the Canadian ones. Characteristic for this province is also the increased activity and correspondingly wider ecological amplitude of *Salix herbacea*, as well as the fact that *Cassiope tetragona* does not form associations in areas with a long-lasting snowcover. *Salix herbacea* has taken over that role. But, when forming a part of the dwarfshrub tundras, *Cassiope tetragona* passes from a subdominant and even to a dominant in the more northerly variant of this type of tundra and at high altitudes in the mountains. The presence of a number of endemic species in the meadow-like communities is also conspicuous, as is the specific vegetation of the mires and other peculiarities in the vegetation cover.

There are two subprovinces in the Greenland province: the West Greenland and the East Greenland ones (Fig. 16). The difference between them is determined first and foremost by the climate. Along the east coast runs the cold East Greenland Current carrying ice from the Polar Basin. The west coast is subjected to the moderating influence of the warm West Greenland Current, branching off the Gulf Stream which, skirting southern Greenland, enters Baffin Strait along the west coast of Greenland. The floristic factor is very important: in

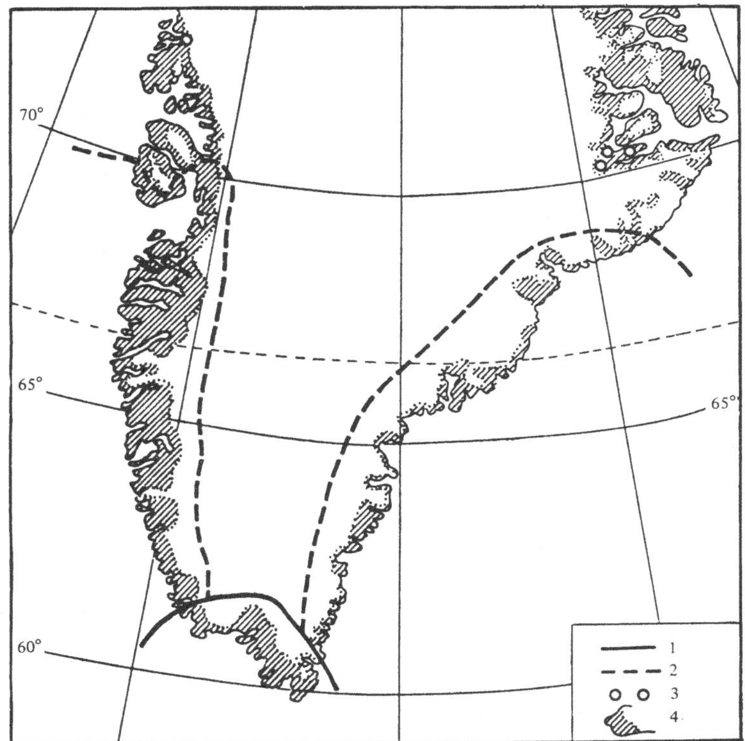

Fig. 16. The Greenland province of the subarctic tundras. 1, the southern boundary of the tundra region and the northern limit of the Atlantic region of the boreal ericaceous heaths and meadows; 2, the boundaries of the west Greenland and east Greenland subprovinces; 3, extrazonal areas of subarctic tundras within the limits of the subregion of the arctic tundras; 4, essentially non-glaciated areas.

the flora of the West Greenland subprovince there are a number of species with mainly American areas, part of which do not occur on the east coast (*Betula glandulosa, Alnus crispa,* etc.). The east coast has also its own co-differentiating species.

The West Greenland subprovince of the subarctic tundras

Information on the vegetation of the area under discussion is found in a number of publications (Porsild & Porsild, 1920; Holttum, 1922; Warming, 1928; Trapnell, 1933; Böcher, 1948, 1950, 1952, 1954,

1959, 1963*a*, *b;* Gelting, 1955; Fredskild, 1961, etc.). Among these, the most valuable are the works by Böcher, in which are found detailed lists of the complete, specific composition of the plant sociations, including also the cryptogams, as well as analyses and data derived from them by two parallel methods of classification. One is based on the dominant life-form, the other on a complex method, according to which the geographical, climatic and ecological ties of the components in the sociations are taken into consideration (see above p. 5). Using the latter method Böcher has carried out a complex, botanical–geographical division of the territories in southwestern Greenland situated between latitudes 60° and 70°N (Böcher, 1954).

The considerable widening of the ice-free coastal area appears to be a consequence of the effect of the West Greenland current. Between Godthaab and Disko Bay (latitude 64°–69°N), this area reaches 100–200 km in width. At the same time, the ice-free areas on the east coast are extremely fragmented and usually do not exceed 30–40 km in width. The boundary of the vegetation subzone on the west coast is pushed northwards. Shrub thickets of *Salix glauca* ssp. *callicarpaea,* described from the northern extreme of the subarctic tundras, are found on the west coast at latitude 71°N with some extrazonal occurrences even farther north (Böcher, 1933*a:* 24, 1938: 305). On the east coast, however, they only reach to latitude 68°N. In the southern part of the West Greenland subprovince there are fragments of 'crooked forests' with *Betula tortuosa. Sorbus groenlandica* reaches from the south as far north as latitude 63°N. Both these species are lacking on the east coast. On the west coast *Juniperus sibirica* extends in prostrate form to Disko Island (latitude 70°N), but on the east coast only to latitude 68°N. A whole row of boreal species, a number of which are found in thickets of *Salix glauca* ssp. *callicarpaea,* such as *Gymnocarpium dryopteris, Lycopodium complanatum, Equisetum sylvaticum, Listera cordata, Linnaea borealis* ssp. *americana,* etc., are either lacking on the east coast or extend only to a lower latitude than on the west coast.

The occurrence of latitudinal zonality in the West Greenland subprovince is strongly disturbed, not only as a consequence of the mountainous relief and the vertical zonation connected with it, but also because of the climatic differences between the coastal and the inland areas. The warm current affects the coastal belt, where, therefore, at latitude 66°n, there is a 'low-arctic, oceanic régime' (Böcher, 1954: 18–19) with an increased amount of precipitation. The interior, cryoarid areas are under the influence of the föhn winds, especially in the valleys of the major fjords ('a continental régime', according to Böcher).

The widespread dwarf shrub tundras are mainly represented by communities with dominating *Empetrum hermaphroditum* and *Vaccinium uliginosum* ssp. *microphyllum* and with admixture in various relationships of species such as *Loiseleuria procumbens, Phyllodoce coerulea, Diapensia lapponica* and *Vaccinium vitis-idaea* ssp. *minus* (these species sometimes assume the position of dominants), *Cassiope tetragona* (in more northerly areas and at higher elevations in the mountains, often co-dominant or even dominant), *Rhododendron lapponicum, Pyrola grandiflora, Ledum decumbens* (the latter often in unimportant quantities), and *Salix herbacea.* Communities with dominating *Arctostaphylos uva-ursi* are occasionally met with (Böcher, 1954: 220). The small admixture of *Betula nana* and *Salix glauca* ssp. *callicarpaea* and of herbs such as *Carex bigelowii* and *Polygonum viviparum* is typical. In mesic environments mosses are well developed, e.g. *Dicranum fuscescens, Drepanocladus uncinatus, Aulacomnium turgidum, Ptilidium ciliare.* In drier conditions there are many lichens, e.g. *Cetraria crispa, C. cucullata, Alectoria ochroleuca, Stereocaulon alpinum, Peltigera malacea,* etc.

Locally, *Betula nana* forms tundras of low-growing dwarfbirch associations, with well developed moss cover in which *Aulacomnium turgidum* is dominating, and there is a characteristic admixture of *Peltigera aphthosa* as well as of herbs and dwarfshrubs, such as *Calamagrostis groenlandica, Pyrola grandiflora, Empetrum hermaphroditum, Vaccinium uliginosum* ssp. *microphyllum,* and sometimes of *Ledum decumbens.* In the southern part of the subprovince there are also dwarfbirch tundras with *Betula glandulosa,* sometimes mixed with *Salix glauca* ssp. *callicarpaea* in a moss carpet of *Pleurozium schreberi, Dicranum* spp. and in the cover some *Ledum decumbens, Empetrum hermaphroditum, Phyllodoce coerulea, Vaccinium uliginosum* ssp. *microphyllum,* and *Carex bigelowii.*

The shrub thickets are mainly represented by widely distributed willow associations with *Salix glauca* ssp. *callicarpaea* reaching 80 cm in height. In the more southerly willow associations there is a typical addition of boreal, herbaceous species: *Calamagrostis langsdorffii, Archangelica norvegica, Chamaenerium angustifolium, Deschampsia flexuosa, Linnaea borealis* ssp. *americana,* etc. (Böcher, 1963*b:* 163). *Alnus crispa* is found up to 250 m altitude and as far as latitude 69°N. It forms dense thickets, alone or associated with *S. glauca* ssp. *callicarpaea.* In the alder thickets boreal ferns are often found, such as *Gymnocarpium dryopteris.* Patches of *Juniperus communis* are found as far north as latitude 67°N, with individual specimens up to latitude 70°N, in the form of small bushes or krummholz, sometimes associated with *Salix glauca* ssp. *callicarpaea.*

Meadow-like communities rich in forbs species (*Alchemilla alpina, A. glomerulans, Taraxacum croceum, Sibbaldia procumbens, Coptis trifolia,* and others) are typical for localities with an oceanic climate on slopes with a favorable exposure. Sometimes *Salix herbacea* is a participant in these meadow-like associations. In the cryo-arid areas of the interior parts of the country xerophytic and xero-mesophytic *Kobresia–Carex* associations are met with ('xerophilous grassland'). These associations are composed of species such as *Carex rupestris, Kobresia bellardii, Festuca brachyphylla,* and others. Especially interesting are the tundra–steppe associations, where a cryophilous–steppe species such as *Carex supina* spp. *spaniocarpa* dominates.

In areas with surplus moisture, mires have developed, occupying small areas. Tussocky mires with *Trichophorum caespitosum* are typical especially in the southern areas, although they are met with as far north as Disko Island (Böcher, 1963*b*: 246). In herb–moss mires, the mosses are represented by *Drepanocladus revolvens, Calliergon* spp., etc., the herbs mainly by sedges, *Carex bicolor, C. microglochin, C. gynocrates, C. rariflora,* in the northern part also by *C. stans,* but also by cottongrasses, *Eriophorum scheuchzeri* and *E. angustifolium.* Rarely there are patches of *Salix arctophila.* In the peatmoss mires with *Sphagnum girgensohnii* and other *Sphagna* there is much *Carex rariflora, Vaccinium uliginosum* ssp. *microphyllum, Empetrum hermaphroditum, Oxycoccus palustris* ssp. *microphyllus* and *Betula nana.*

Vegetation under long-lasting snow cover consists typically of communities with dominant *Salix herbacea* and an admixture of *Harrimanella hypnoides.*

Dryas tundras are met with at altitudes of 400–700 m in the mountains as well as impoverished tundras with *Vaccinium uliginosum* ssp.

microphyllum and *Cassiope tetragona.* In all the territory of this subprovince open formations occupy the majority of the area between boulders.

On the basis of its floristics the West Greenland subprovince can be distinguished, not only by its co-differentiating species, but also by endemics and subendemics, such as *Calamagrostis lapponica* ssp. *groenlandica, Puccinellia groenlandica, P. rosenkrantzii, Sisyrinchium groenlandicum* and *Antennaria glabrata. Dryas punctata* is lacking here, but appears in East Greenland.

The East Greenland subprovince of the subarctic tundras

Due to the cooling effect of the East Greenland current, this subprovince is displaced towards the south compared to the West Greenland subprovince. The vegetation is less diversified here. There are no thickets of *Alnus crispa* or any dwarfbirch tundras with *Betula glandulosa.* In its southern parts there are no *Betula tortuosa* and no dwarfshrub tundras where *Arctostaphylos uva-ursi* and *Vaccinium vitis-idaea* ssp. *minus* dominate. The latter species is completely lacking on the east coast. No *Sorbus groenlandica* occurs here, and a number of the boreal species mentioned above and met with in West Greenland are not found here. However, some species, such as *Dryas punctata,* are not found at the same latitude on the western coast.

The vegetation of this subprovince has been described in a number of works (Hartz, 1895; Warming, 1928; Böcher, 1933*a*, *b*, 1963*a*; Schwarzenbach, 1960; Molenaar, 1974; etc.). Dwarfshrub tundras dominated by *Empetrum hermaphroditum* and *Vaccinium uliginosum* ssp. *microphyllum* are the most characteristic here, mainly with the same components as found on the west coast, although *Salix herbacea* is more often found. The shrub thickets are represented essentially by willow associations with *Salix glauca* ssp. *callicarpaea.* The northernmost locality of this willow species is situated in the interior part of Scoresby Sound, cutting deeply inland, where it has been observed at a distance of 175 km from the ocean shore (Hartz, 1895, Böcher, 1933*b*: 24, 1938: 30). Thickets of this species are, however, not met with north of Mikisfjord (latitude 68°N) and only at Angmagssalik, and farther south they become common. In certain willow associations some boreal species (*Chamaenerium angustifolium,* etc.) are met with, but fewer than on the west coast. In the less dense willowbrush there is much *Empetrum hermaphroditum* and other hyparctic dwarf-shrubs. On slopes in areas with a 'suboceanic' régime, meadow-like communities are often observed, which have an extremely variable composition (*Alchemilla glomerulans, A. filicaulis, A. alpina, Taraxacum croceum, Polygonum viviparum,* and many others). *Salix herbacea* is common there. Nival meadow-like communities are represented by associations with *Sibbaldia procumbens, Cerastium cerastoides, Gnaphalium supinum,* and others associated with *Salix herbacea,* which dominates in habitats with late-melting snow. *Cassiope tetragona* does not occur in such localities, but *Harrimanella hypnoides* is often found there.

As a consequence of the strongly dissected relief, mires are not often found and occupy only small areas. In the herb–moss associations *Eriophorum scheuchzeri* is usually the dominant herb, sometimes associated with *E. angustifolium. Calliergon stramineum* is the most common moss. The mires consist mostly of flat-topped peat mounds, where the moss–turf of *Sphagnum subnitens, Aulacomnium* ssp., etc. is covered by thickets of *Salix glauca* ssp. *callicarpaea,* or has a cover of hyparctic dwarfshrubs, such as *Vaccinium uliginosum* ssp. *microphyllum, Empetrum hermaphroditum, Phyllodoce coerulea,* etc., with an admixture of *Salix glauca* ssp. *callicarpaea, Carex bigelowii* and other herbaceous species as well as lichens, *Cetraria nivalis, Cladonia mitis,* etc. The cover is not infrequently formed of *Salix herbacea* with which usually *Carex bigelowii* and *Polygonum viviparum* are associated.

Characteristically, *Salix herbacea,* which as already mentioned is more aggressive here, is present in almost all the mire associations. I also refer to the wetlands the halophytic associations at the seashore composed of *Puccinellia phryganodes, Carex glareosa, Stellaria humifusa* or *Carex subspathacea* (Böcher, 1933*b*; Molenaar, 1974, etc.).

Cryophytic associations in the inland areas exposed to föhn winds are formed by *Carex rupestris, Kobresia bellardii,* etc., as described by Schwarzenbach (1960).

The subregion of the arctic tundras

The subregion of the arctic tundras has been distinguished as a special botanical–geographical region by all the investigators of the Arctic. In his zonation of European Russia Trautvetter (1851) named this subregion 'the area of the alpine willows' (he called the subarctic tundras 'the area of the lowgrown birches'). Pohle (1910) used the epithet 'the arctic zone of the tundra region'. Sochava (1933*c*) called it 'the subzone of the arctic non-fruiting dwarfshrub tundras' and Leskov (1947) used the name 'the belt of arctic tundras'. Sochava included this territory in a region of higher rank, 'the arctic belt'. In a physicogeographical zonation he later used the term 'the Arctic' (Sochava & Timofeyev, 1968) for the land area situated north of 'the Subarctic' (Sochava *et al.*, 1972). 'The Arctic' is, according to Sochava, characterized mainly by special circumpolar fratriae of formations (Buks, 1973, 1974; Il'ina, 1975), while the territories of the subarctic tundras are included in different areas characterized by different fratriae. Yurtsev (1966, etc.) has also separated the arctic and the subarctic tundras by a boundary of higher rank, dividing the 'arctic belt' and the 'hyparctic' one. As a basis for this, he used the different origins of the floras in the arctic belt and the hyparctic one (cf. above, p. 22). The boundary between the arctic and the subarctic tundras has been designated by Böcher (Böcher *et al.*, 1968: 11) as the southern limit of the 'higharctic belt'. He stressed the lack of shrub thickets in the latter belt. Rønning (1969: 29) called it the border between the 'low arctic' and the 'middle arctic', and Young (1971) the border between floristic zones '3' and '4' (with some exceptions).

The border between the arctic and the subarctic tundras oscillates around the 6° C July isotherm. Grigor'yev (1946:6, 70–3) considered that along this border there occurs an essential change in the character of the atmospheric circulation and the radiation balance, which affects all components in the physico-geographical complex. It is also important that when the polar forest border oscillated in postglacial time, it never crossed this border. Paleobotanical data confirm this (cf. Giterman *et al.*, 1968: 235, etc.) and accordingly, the territories of the arctic tundras have remained treeless throughout the Holocene.

The vegetation cover at the transition from the subregion of the subarctic tundras to the subregion of the arctic tundras is subject to essential changes, namely, the type of arctic associations on zonal, mesic habitats differs sharply from the subarctic one, thickets of hyparctic shrubs disappear in the arctic subregion as do nival, meadow-like communities, floodplain meadows, birchbrush on the peat mounds and raised borders of the low centre polygons in polygonal mires. No more open woodland or prostrate krummholz associations are met with. The most important character is the disappearance of the shrub thickets. This trait is well correlated with all the rest and may be considered as the fundamental diagnostic character for drawing the line between the arctic and the subarctic subregions within the tundra region.

Although a number of scientists have drawn boundaries of higher rank between the Arctic and the Subarctic – Pohle and Young between zones, Sochava and Yurtsev between botanical-geographical belts – I shall distinguish the arctic tundras at the level of a subregion of the geobotanical tundra region, where on zonal, mesic habitats the tundra type of vegetation is represented by its arctic subtype as described above (on p. 21). The complete disappearance from the composition of the zonal arctic associations of any admixture of birches is considered as the basic, diagnostic difference from the Subarctic. *Betula nana* (*B. tundrarum* according to the treatment by Perfil'yev and Cherepanov) does occur in some non-zonal, non-mesic habitats within the area of the arctic tundras, but only as a rare, relict plant of no importance for the composition of the vegetation cover. Thus, during my three years of work on the South Island of Novaya Zemlya, I came upon *Betula nana* (*B. tundrarum*) only once on a gentle, southeast exposed slope on the mainland shore of the Rogachev River delta. It grew in the form of a compact, flat cushion on the ground, measuring 20–25 cm in diameter and 5–6 cm in height, and was densely leafy and sterile.

In spite of the fact that the bareness of the ground is always more intense in the subregion of the arctic tundras than in that of the subarctic tundras (cf. the widely developed 'spotty tundras' and – in habitats with a scanty snow cover – the strongly developed polygonal tundras), there are also in this subregion in certain areas tundra associations with a continuous vegetation cover, as for instance in northern Yamal (Mikhailichenko, 1936) and in northwestern Taimyr (Tikhomirov, 1948*b*), etc. In the zonal, mesic phytocoenoses of the arctic, spotty tundras (as also in those tundras in the Subarctic), the root systems of the plants under the bare spots are close under the ground surface (Aleksandrova, 1962, cf. Fig. 19). This appears to be one of the important, diagnostic characters, distinguishing these phytocoenoses from those in the polar deserts. The first one to point out this character was Perfil'yev (1928).

In the subregion of the arctic tundras the occurrence of the boreal element of the flora is strongly reduced, and so is that of the hyparctic element, at least to a certain degree, although hyparctic species do occur (*Eriophorum angustifolium, Valeriana capitata, Nardosmia frigida,* and others), which penetrate into this zone and are participants in the composition of the plant associations. A number of high-arctic species, which are lacking farther south, appear here, such as *Poa abbreviata, Puccinellia angustata, Ranunculus sabinei, Draba subcapitata, D. oblongata, Saxifraga platysepala,* etc.

Within the limits of the subregion of the arctic tundras, it is possible to distinguish two latitudinal belts, the southern, and the northern belts (cf. Fig. 2).

The southern belt of the arctic tundras

Except where there is a spatial separation in the form of an island (e.g. Novaya Zemlya, etc.), the southern belt of the arctic tundras makes direct contact with the northern subarctic tundras. In the transition zone their associations on zonal, mesic habitats are often similar in composition and structure, although birch species are essentially lacking in the arctic zonal, mesic communities. To a certain extent low growing willow brush is still present, mainly consisting of arctic species such as *Salix reptans, S. arctica,* and Siberian arctic–alpine ones such as *S. sphenophylla,* but also a few hyparctic ones, usually in the form of prostrate, single individuals of, e.g. *Salix pulchra.* Although shrub thickets are completely lacking, it is possible to come across some single low shrubs of species such as *Salix lanata, S. lanata* ssp. *richardsonii,* and *S. arctophila* in swampy localities. Some of the hyparctic dwarfshrubs continue to play an episodic role in the vegetation cover. Thus, e.g., *Vaccinium vitis-idaea* ssp. *minus* is found on Novaya Zemlya as single, sterile individuals in some lichen tundras, and in the Alazeya–Kolyma district it is part of the associations in an arctic variant of the tussocky tundras. *Vaccinium uliginosum* ssp. *microphyllum* penetrates far into the arctic tundras of the eastern Canadian Arctic Archipelago and is also found in arctic Greenland. The peatmosses are well developed in the mires of the southern belt of the arctic tundras; *Sphagnum fimbriatum* is capable of developing big, flat hummocks, *S. squarrosum* of forming carpets, 3–4 m in diameter. In the Kolyma River district of this belt, the peatmosses dominate on the raised borders of the low center polygons in tetragonal bogs (Perfil'yeva & Rykova, 1975). In the Atlantic areas, e.g. on the southern part of Novaya Zemlya, the arctic variant of mires with flat-topped peat mounds is well developed, where on top of the mounds cloudberry brush (*Rubus chamaemorus*) may be found, but most years in a sterile condition (Aleksandrova, 1956).

The northern belt of the arctic tundras

The transition to the northern belt is recognized by the following, basic characteristics: (1) in the zonal, mesic habitats with moderate snow cover associations with semi-prostrate willows (*Salix reptans,* etc.) have disappeared and only dwarfwillows remain, such as *Salix polaris* the stems of which are buried in the moss carpet, as well as a few prostrate types like *Salix nummularia, S. reticulata,* etc. In the northern Canadian Arctic Archipelago and Greenland there is also some creeping *S. arctica;* (2) In non-mesic habitats with scanty snow cover polygonal tundras are widely distributed, where the bare surface (spots, polygons) occupies a considerably larger area than that covered by vegetation, which is confined to net-like strips along the cracks between the polygons. (3) The mires with flat-topped peat mounds have disappeared on Novaya Zemlya, and throughout this circumpolar belt homogeneous herb–moss mires and/or tetragonal mires are absolutely predominant. Besides, a special type of soil ('arctic tundra, slightly gleyified soils'; cf. Karavayeva, 1962: 112, 126–7) has developed under the associations on zonal, mesic habitats. It is different from the more southern types of lightly developed gleyification. Here it is slightly acid (close to neutral) in reaction and has a number of other characteristic traits (Aleksandrova, 1963: 22).

The provinces of the subregion of the arctic tundras

When delimiting the geobotanical provinces in the subregion of the subarctic tundras, a leading role is played by the distribution of associations dominated or characterized by hyparctic or hyparctic–alpine species such as *Betula nana, B. exilis, B. middendorfiii, B. glandulosa, Salix phylicifolia, S. planifolia, S. pulchra, S. alaxensis, S. glauca,*

S. glauca ssp. *callicarpaea, Alnus fruticosa, A. crispa, Eriophorum vaginatum, E. spissum, Carex soczavaeana,* and others. In the subregion of the arctic tundras, however, the decisive role is played by the distribution of associations with dominating or characterizing arctic or arctic–alpine species, such as *Salix polaris, S. arctica, Dryas octopetala, D. punctata, D. integrifolia, Alopecurus alpinus, Deschampsia brevifolia, Cassiope tetragona,* and so on. Although a part of these species have a circumpolar distribution area, their role in the composition of the vegetation cover differs considerably in certain parts of this subregion.

The differences between the provinces in the subregion of the arctic tundras are less well expressed than in the subregion of the subarctic tundras. While in the latter it is possible to distinguish five geobotanical provinces and 15 subprovinces, in the arctic tundras we can only distinguish four provinces and six subprovinces. This major similarity in the vegetation is also reflected in the geobotanical division made by Sochava, who characterized the arctic belt as a single arctic 'fratria of formations' (Il'ina, 1975, etc.), while the subarctic tundras were separated into different regions characterized by different 'fratriae'. The decrease in provincial distinctiveness is, to a considerable extent, related to the fact that the higher the latitude the more uniform is the circumpolar flora (Young, 1971).

In this subregion four geobotanical provinces can be distinguished: the Novaya Zemlya–West Siberian–Central Siberian, East Siberian, the Wrangel Island–West American, and the East American–Greenland provinces (Fig. 17).

Because the vegetation cover at the northern rim of the continents has been studied only to a limited degree and is still insufficiently known, my division into provinces should be considered as preliminary in character.

The Novaya Zemlya–West Siberian–Central Siberian province of the arctic tundras

This province reaches from Novaya Zemlya to the western side of the Olenek Bay. In spite of the differences in the vegetation cover, which increase from west to east, it is united by the wide distribution, especially on zonal, mesic habitats in the southern belt of the subregion, of the arctic variant of the tundras, endemic to this province with its

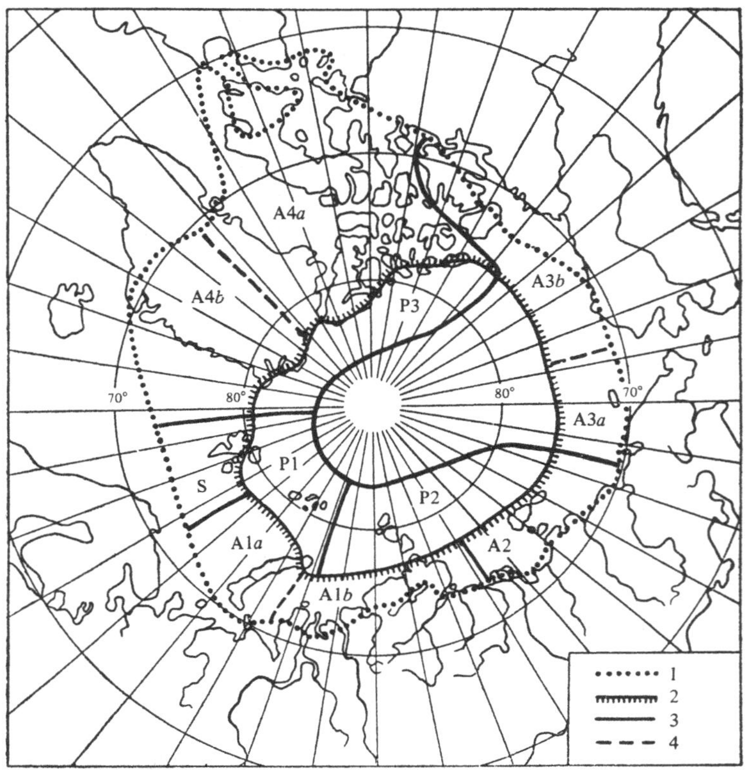

Fig. 17. The geobotanical provinces and subprovinces of the subregion of the arctic tundras and the region of the polar deserts. Southern boundaries of: 1, the subregion of the arctic tundras, and 2, the region of the polar deserts. 3,–, provincial boundaries, 4,– – –, subprovincial boundaries. Provinces of the Arctic tundras: A1, the Novaya Zemlya–West Siberian–Central Siberian province with: A1*a*, the Novaya Zemlya–Vaigach, A1*b*, the Yamal–Gydan–Taimyr–Anabar subprovinces. A2, the East Siberian province; A3, the Wrangel Island–West American province with: A3*a*, the Wrangel Island, A3*b*, the Cape Barrow and the southwestern Canadian Arctic Archipelago subprovinces; A4, the northeastern Canadian–North Greenland province with: A4*a*, the northeastern Canadian-northwestern Greenland, A4*b*, the northeast Greenland subprovinces. S, the Spitsbergen autonomous district. Provinces of the Polar Deserts: P1, the Barents; P2, the Siberian; P3, the Canadian.

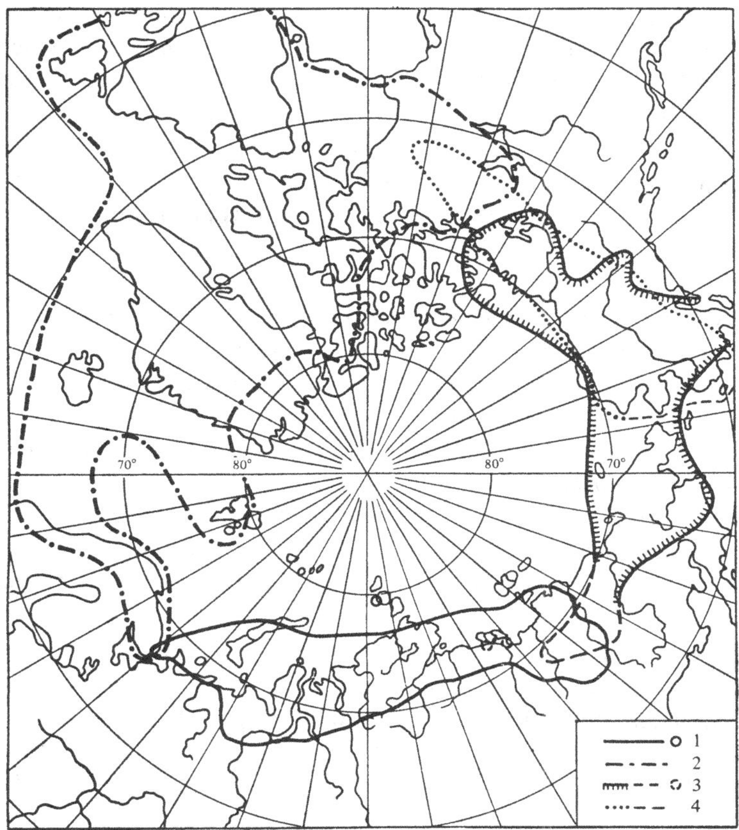

Fig. 18. The distribution areas of 1, *Carex ensifolia* ssp. *arctisibirica* Jurtz., 2, *C. bigelowii* Torr. s.str., 3, *C. lugens* Holm, and 4, *C. consimilis* Holm. According to Porsild, 1957*a*, Hultén, 1964:51, Yegorova, 1966.

abundance of *Carex ensifolia* ssp. *arctisibirica*. It differs from the subarctic form of the tundras in the absence of *B. nana*, or *B. exilis*. Towards the west, starting in Spitsbergen, associations with *Carex ensifolia* ssp. *arctisibirica* disappear as this species drops out of the flora (Fig. 18). In the East Siberian province of the arctic tundras *Carex ensifolia* ssp. *arctisibirica* continues to occur as far east as the Indigirka River, but its activity at that latitude decreases sharply, although farther south it continues to be abundant as far as the Lena River and, locally, in the mountain areas of northern Verkhoyansk Mountains.

The most characteristic element of the flora is the Siberian species, which increase in number from Novaya Zemlya, where they represent about 12% of the flora, towards the east as a consequence of the increasingly continental climate and the approach to the Asiatic centers of origin of the initial Arctic flora.

Within this province two subprovinces can be distinguished: the Novaya Zemlya – Vaigach, and the Yamal–Gydan–Taimyr–Anabar subprovinces.

The Novaya Zemlya – Vaigach subprovince of the arctic tundras

This subprovince is distinguished by some endemic traits, typical of its vegetation cover: the presence of groups of associations on zonal, mesic habitats, dominated by *Deschampsia brevifolia* associated with *Salix polaris, Carex ensifolia* ssp. *arctisibirica,* and mosses, such as *Hylocomium alaskanum, Tomenthypnum nitens* and others, as well as tussock associations with *Deschampsia brevifolia* (Aleksandrova, 1956: 262–4), and the existence of arctic mires with flat-topped peat mounds. There are no polygonal mires. *Cassiope tetragona* is not found in this flora.

The special character of the vegetation is to a major extent related to the natural conditions on Novaya Zemlya. These twin islands are 925 km in length and up to 145 km wide. Except at the southernmost part they are covered by a narrow mountain range. This range is a part of the Novaya Zemlya – Ural Hercynian mountain complex, has elevations up to 1590 m, and is situated between the Barents and the Kara Seas. On the Barents Sea side, along the western coast, the warm Kolguyev–Novaya Zemlya and the West Novaya Zemlya currents flow, while the Kara Sea is cold and ice-bound. Therefore, a front forms above the Novaya Zemlya Range between the Barents Sea and the Kara Sea airmasses, and causes marked differences in the climate on the western and the eastern coasts of Novaya Zemlya as well as the famous Novaya Zemlya's gales, 'Novaya Zemlya Bora' (Vize, 1924, etc.). When the barometric gradient falls towards the Icelandic low-pressure area, air masses swoop down over the western coast, bringing local winds of hurricane force, up to and above 60 m/s, and, as a consequence of the adiabatic expansion, falling temperatures.

The climate is oceanic, cryo-humid, especially on the west coast. Because of this, Novaya Zemlya is at present strongly glaciated; small glaciers and firn formations appear already at latitude 71°30′ N. At 72°50′ valley glaciers appear and from 74°45′ there is an icesheet, where the only unglaciated parts are isolated nunataks and narrow coastal areas, separated by glaciers pushing out into the sea. The winter temperatures are not very low (the average temperature of the coldest month at Malye Karmakuly is −16°), the maximum temperature remaining positive in all months. Nevertheless, the weather is extremely severe because of the very strong winds. The vegetative season on Novaya Zemlya is prolonged, because of the lingering autumn, when snowfall and frosts alternate with relatively warm periods of sleet and rain. Thus, according to my own data, the vegetative period on the South Island lasts about 145 days (cf. Aleksandrova, 1960*a*: 216) and according to Tolmachev (1929) about 140 days at Matochkin Shar, while in Siberia on the Big Lyakovsky Island (Aleksandrova, 1960*a*: 216) it lasts only 90–95 days and around Tiksi Bay (Shamurin, 1960) about 100 days.

Typical for the flora of Novaya Zemlya is the inclusion of European (2%) and amphi-atlantic (8%) species such as *Deschampsia alpina, Carex parallela, Luzula arcuata, Cerastium alpinum, C. arcticum, Arenaria pseudofrigida, Silene acaulis, Saxifraga aizoides, Campanula uniflora,* etc., as well as Siberian and Siberian–American species (12%) such as *Hierochloë pauciflora, Carex saxatilis* ssp. *laxa, Salix pulchra, Minuartia macrocarpa, Erysimum pallasii, Novosieversia glacialis, Artemisia tilesii, Senecio atropurpureus, S. resedifolius,* etc. It should be noticed that part of the European species reach only the western coast of Novaya Zemlya (*Carex parallela, Puccinellia coarctata,* etc.) while some of the Siberian ones only occur on the eastern coast (*Erysimum pallasii, Novosieversia glacialis,* etc.).

In spite of the rather large number of amphi-atlantic species they do not play an important role in the composition of the plant associations. Only *Silene acaulis* is of some importance there. The dominants forming the vegetation cover together with mosses and lichens are, besides the circumpolar *Deschampsia brevifolia* s.l., *Eriophorum angustifolium, Dupontia fisheri,* and such widespread species as *Dryas octopetala, Eriophorum medium,* and *Arctophila fulva*, the mainly Siberian species *Salix polaris* ssp. *polaris* dominating in associations on zonal, mesic habitats, and *Carex ensifolia* ssp. *arctisibirica,* its co-dominant. In the area around Savina River on the east coast of Novaya

Zemlya such Siberian species as *Senecio atropurpureus* and *Novosieversia glacialis* are important components in the tundra associations (Aleksandrova, 1956) and in the area around Matochkin Shar, *Erysimum pallasii* as well (Tolmachev, 1929).

Information on the vegetation of Novaya Zemlya is found in a number of papers (Baer, 1838; Ekstam, 1897; Sokolovsky, 1905; Perfil'yev, 1928; Steffen, 1928; Tolmachev, 1929; Zubkov, 1932, 1934, 1935*a*, *b*; Regel, 1932, 1935; Aleksandrova & Zubkov, 1937; Aleksandrova, 1945, 1946, 1956; etc.).

The vegetation of Vaigach has been much less explored. Among the papers of special geobotanical importance, which do not only list data on the vegetation, are short essays by Tolmachev (1923), Steffen (1928), and Yesipov (1933) as well as incidental remarks in connection with the description of the soils by Ignatenko (1966).

In the southern belt of the arctic tundras, occupying the southern and southwestern parts of Novaya Zemlya and the northern part of Vaigach, the zonal, mesic phytocoenoses on loamy soils, described in detail from Gusinaya Land, are represented by spotty tundras with co-dominant *Salix polaris, Deschampsia brevifolia, Carex ensifolia* ssp. *arctisibirica, Hylocomium alaskanum,* and *Tomenthypnum nitens.* The large number of herbaceous species is typical: in the detailed lists from 18 localities described (Zubkov, 1932: 80–2; Aleksandrova, 1956: 234–9) there are two species of prostrate shrubs (*Salix reptans* and *S. arctica*), two dwarfshrubs (*Salix polaris* and *Dryas octopetala*), eight grasses, nine sedges and rushes (*Carex* and *Juncus* spp.), 43 kinds of herbs, 28 mosses, and 18 species of lichens (fruticose, tubular, and foliose ones). However, the estimated coverage of the herbs amounted on an average to only 2%, but for *Salix polaris* to 15%, while the total coverage of the vegetation was estimated at around 50%, or from 40 to 75%. An impoverished variant of this association, typical of the northern belt of the arctic tundras and described by me from the Kara Sea side of the South Island, differs in the absence of *Salix reptans* and *S. arctica* and in its generally limited species composition.

Because of the humid, oceanic climate water collects in every hollow, also in zonal, mesic habitats where mire species such as *Calliergon sarmentosum* may be found. Due to the saturation with moisture and the fact that the permafrost thaws down to a considerable depth (80–100 cm or more on the South Island), cryogenic processes are always active, such as soil sorting, leading to an enormous variety of patterned formations, stone rings, stone nets, and stone stripes, on which often very colorful vegetation complexes have developed.

Owing to the absolute predominance of stony habitats on Novaya Zemlya and Vaigach, formations of petrophilous vegetation dominate: *Dryas* tundras with predominance of *D. octopetala* (note that in contrast to the Siberian tundras, the *Dryas* tundras are not found on loamy soils here); moss–lichen associations with a predominance of *Rhacomitrium lanuginosum* and lichens, either *Cladonia mitis* or *Cetraria nivalis;* moss–lichen associations with *Cladonia mitis, C. rangiferina, Dicranum elongatum* in habitats with a deeper snow cover; moss–lichen associations with an abundance of *Cetraria nivalis* and *Sphaerophorus globosus* in strongly wind-exposed habitats and *Cladonia mitis* or *Cetraria delisei* where snow accumulates but meltwater does not stagnate, but rather runs off between the scattered rocks and small boulders. The composition of the angiosperms in the moss–lichen and lichen associations varies in relation to the snow régime and the moisture conditions (cf. Aleksandrova, 1956: 242–58). The most common component is *Luzula confusa.*

Thanks to the oceanic climate (the average temperature for January and February equals −12° C, for March −16° C) as well as the deep snow cover, there is no frostcracking in the swampy depressions where the strong winds accumulate the snow and, thus, no polygonal mires. In the southern belt of the arctic tundras, arctic mires with flat peat mounds are widely distributed (Zubkov, 1932: 86–91; Aleksandrova, 1956: 266–79). There, in depressions with graminoid–moss associations (*Carex stans, Eriophorum medium, Dupontia fisheri, Calliergon sarmentosum, Drepanocladus exannulatus,* etc.) with 20–35 cm thick turf, the flat peat mounds occur, up to 10–20 m in diameter and about 50 cm high. They owe their formation to a number of thin ice-veins and small lenses of ice, enclosed in the loamy soil under the peat (cf. Aleksandrova, 1946: 71–6, 1956: 284, etc.); (a similar type of formation of flat peat mounds in a high-arctic variation has been described by Eurola (1971*b*: 90, 99–100) from Spitsbergen). On the mounds peatmosses such as *Sphagnum fimbriatum* may dominate, but they are most often met with around the edges of the mounds. Other dominants on the flat mounds are *Polytrichum alpestre* together with liverworts, or *Cladonia* *mitis* associated with *Dicranum elongatum*. Some mounds are completely overgrown by *Rubus chamaemorus* (sterile). Finally, there are mounds with a dying-off vegetation and white spots of *Ochrolechia* spp. on the peat surface. Homogeneous herb–moss mires are also widely distributed. The vegetation is similar to that of shallow depressions of the mires with flat-topped peat mounds.

On the South Island of Novaya Zemlya there are three mountain belts of vegetation (Aleksandrova, 1945, 1956: 286–90). The lower belt reaches up to 200 m above sea level. Although boulder fields predominate here, there are areas with a sufficient amount of fine soil, where tundras have developed with *Dryas octopetala* or *Salix polaris, Carex ensifolia* ssp. *arctisibirica, Hylocomium alaskanum* and *Tomenthypnum nitens*, and on surfaces with insufficient drainage sedge–moss and cottongrass–moss mires with peat layers. In the middle belt a lower belt of the oro-polar deserts (200–300 m.) is represented by fragments of tundra phytocoenoses with an impoverished composition and small patches of mires in very favorable conditions only. There are no *Dryas* tundras here. Boulder fields predominate, partly covered by an impoverished vegetation of polar desert associations with a layer of crustose lichens and variously formed fragments of a plant cover. The upper belt is represented by the upper belt of the oro-polar desert type above 300 m. in altitude. Here such phytocoenoses as tundras and mires with peat layers are totally lacking. The boulder fields and the cliffs are almost devoid of vegetation. On the ground, where some fine soil has collected, formations of polar-desert type have developed, and where there is enough water, there are open associations with a scattered growth of species such as *Phippsia algida, Saxifraga rivularis, Cetraria delisei* and individual tufts of *Deschampsia brevifolia*. Rarely, small mineral mires (without any peat) with scattered *Eriophorum scheuchzeri* are encountered.

Within this subprovince, three geobotanical districts can be distinguished: the Southern Novaya-Zemlya – Vaigach, the Western Novaya Zemlya, and the Eastern Novaya Zemlya districts.

The first of these, the *Southern Novaya Zemlya – Vaigach* district, embracing the rolling plains with low hills on southern Novaya Zemlya south of Propashchy Bay as well as the flat, northern part of Vaigach and its swampy, marine terraces and depressions, displays, typically, tundras with *Dryas octopetala* on the stony ground on the hill tops formed from bedrock. In the *Western Novaya Zemlya* district both the southern and the northern belts of the arctic tundras are represented, while in the *Eastern Novaya Zemlya* district, which reaches northwards from Cape Ratmanov, only the northern belt is found in spite of the fact that both districts are situated within the same latitudes. This can be explained by the effect of the very different climates on the Barents Sea side and the Kara Sea side of Novaya Zemlya. In addition, in the western associations there is admixture of more amphi-atlantic and European species than in the eastern ones, and in the eastern associations more Siberian species.

The Yamal–Gydan–Taimyr–Anabar subprovince of the arctic tundras. The northern part of Yamal, the Gydan Peninsula, Taimyr (except the area around Cape Chelyushkin which belongs to the polar deserts), and the narrow strip of coast from the Khatanga to the Olenek Inlet, as well as the northern part of Begichev Island, belong to this subprovince of the arctic tundras. In spite of the fact that in the territories situated farther south there is a distinctly expressed boundary between two distinct provinces of the subarctic tundras starting in the south at the right bank of Pyasina River and proceeding northeastwards, here the areas belonging to the arctic tundras do not display any provincial differences in the vegetation cover all the way from west to east, from Yamal as far as the northwest edge of the Olenek Inlet, which would justify a division either at a provincial or a subprovincial level. Instead the vegetation cover displays on similar types of substrate traits of uniformity in its basic characters; particularly in associations on zonal, mesic habitats with loamy soils one or more of the following species generally dominate: *Carex ensifolia* ssp. *arctisibirica, Dryas octopetala,* or *D. punctata, Aulacomnium turgidum, Hylocomium alaskanum, Tomenthypnum nitens* or, sometimes, *Ptilidium ciliare.* The role of *Carex ensifolia* ssp. *arctisibirica* starts to diminish at the edge of northeastern Taimyr Peninsula.

This subprovince is distinguished from the Novaya Zemlya–Vaigach subprovince by a number of characteristics. In zonal, mesic conditions on loamy substrates there are, side by side with the spotty tundras, also tundras with a closed or almost closed vegetation cover. Starting from the northern edge of Yamal, Mikhailichenko (1936) has described associations on loamy soils in zonal, mesic habitats with a vegetation coverage amounting to 95–98%, while on Novaya Zemlya on loamy, zonal, mesic habitats there is never a closed vegetation but only spotty tundras with a 40–70% coverage. Further, *Deschampsia brevifolia* s.l., which plays such an important role on Novaya Zemlya, is disappearing as one of the dominants in zonal, mesic associations here, although it still occurs as an admixture. The composition of the dominating mosses is also changing. Mosses such as *Aulacomnium turgidum* and, sometimes, *Ptilidium ciliare,* which play secondary roles on Novaya Zemlya, are here becoming as abundant as *Hylocomium alaskanum* and *Tomenthypnum nitens.* In addition, on loamy substrates in this subprovince, *Dryas* tundras are becoming frequent, while on Novaya Zemlya and Vaigach they are not met with on loam, but distributed only over stony grounds. An interesting difference is the change in the ecological amplitude of *Eriophorum angustifolium.* On Novaya Zemlya, this species relates only to mire phytocoenoses, mainly with limited drainage, and is never met with on zonal, mesic habitats, but on Yamal (Mikhailichenko, 1936: 89) and in the northwestern and northern Taimyr area *E. angustifolium* occurs as an admixture in the composition of the associations on zonal, mesic habitats, where *Dryas* and *Carex ensifolia* ssp. *arctisibirica* co-dominate. It is also of importance that polygonal mires, which were not found in Novaya Zemlya, do occur and have a wide distribution in the Yamal–Gydan–Taimyr–Anabar subprovince.

The flora assumes a distinctly expressed Siberian character due to the sharp decline in the circumpolar element. Amphi-atlantic species, which may be seen on Novaya Zemlya, occur here only exceptionally. Amphi-beringian species are still not met with, even if some of them, as e.g., *Salix alaxensis,* do penetrate from the more southerly situated areas belonging to the East Siberian province of the subarctic tundras (cf. Fig. 5).

Four districts can be distinguished within the subprovince: the North Yamal, the North Gydan–North Taimyr lowlands, the Byrranga Mountains, and the Khatanga-Anabar coastal plains districts.

The North Yamal District. Northern Yamal, built up from Quaternary marine sediments, is distinguished by a low relief with elevations not exceeding 30–50 m. Its particular types of vegetation are related to the predominance of sandy and loamy–sandy soils and the poorly developed lichen associations on these soils. This latter peculiarity has been explained by some investigators of Yamal (Andreyev, 1935, and others) as a result of overgrazing, because the arctic tundras have always been the summer pasture of reindeer. Consequently, although moss–lichen tundras with *Rhacomitrium lanuginosum, Cladonia uncialis, Alectoria ochroleuca* and other species do occur on sand, associations with dominating *Salix nummularia,* locally with *Dryas octopetala, Rhacomitrium lanuginosum* and a relatively variable amount of herbaceous species, are much more common.

On mesic habitats with loamy soils, small-hummocky sedge–moss tundras with a total coverage of 95–98% have developed, according to Mikhailichenko (1936: 89). Among the angiosperms *Carex ensifolia* ssp. *arctisibirica* dominates together with *Salix polaris, Luzula confusa,* a number of arctic grass species, and herbaceous species with an admixture of *Salix nummularia* and *Eriophorum angustifolium.* Dominating mosses are *Aulacomnium turgidum, Dicranum elongatum, Drepanocladus uncinatus, Polytrichum juniperinum,* and some *Ptilidium ciliare.* Lichens such as *Cladonia uncialis, Dactylina arctica, Cetraria crispa, C. delisei,* etc., are met with in small amounts. Mires, both polygonal and homogeneous herb–moss types as well as polygonal, swampy tundras, have been described from the northern parts of the subzone of arctic tundras by Mikhailichenko (1936: 91–4, 97–100) and from the southern part by Boch *et al.* (1971*a*). The most common components in the interpolygon troughs as well as in low-lying mires in this region are *Carex stans* and *Eriophorum angustifolium,* and in the north also *E. scheuchzeri,* and in the south *E. medium.* Both in the north and the south peatmosses such as *Sphagnum fimbriatum* and *S. balticum* are observed on the raised borders of the low center polygons in the polygonal mires. There is also some *Carex ensifolia* ssp. *arctisibirica.*

The North Gydan—North Taimyr Lowland district. The North Gydan Peninsula has an elevation of not more than 60 m, but in comparison with Yamal its relief is more varied. There are more hilly areas, and loamy soils are more widely distributed. As associations characteristic for the arctic tundras on zonal, mesic habitats with a loamy substrate Gorodkov (1944, 2:28) has described a tundra with a closed or almost closed coverage, where he includes in first place among the dominant angiosperms *Carex ensifolia* ssp. *arctisibirica, Salix polaris* and *Dryas punctata*. The mosses are *Aulacomnium turgidum, Hylocomium alaskanum, Tomenthypnum nitens* and *Ptilidium ciliare* and the lichens *Cladonia elongata,* etc. Gorodkov also reports on the wide distribution in this area of polygonal mires, where in the inter-polygonal troughs and in the central depressions of the low center polygons *Carex stans* (Gorodkov, 1944, 1:11) is the most common species, while on the raised borders of the low center polygons *Dicranum elongatum, Polytrichum alpestre, P. alpinum, Oncophorus wahlenbergii, Sphenolobus minutus* as well as peatmosses such as *Sphagnum fimbriatum* predominate. The lichens *Cladonia gracilis, C. mitis, C. rangiferina, Cetraria islandica* are also found there and some dwarfshrubs as *Salix polaris* and herbs like *Luzula confusa*.

The vegetation of northern Taimyr has been briefly described by Tikhomirov (1948*b*: 12–14). On the plains with elevations up to 25–30 m formed by Quaternary, mainly loamy, sediments (marine but locally of glacial origin) there appear some low flat hills, but also outcrops of bedrock (metamorphic shales, etc.) in the form of rocky ridges, 50–60, rarely 100 m tall. In fairly well drained localities with loamy soils and a moderate snow cover tundras have developed with predominating *Carex ensifolia* ssp. *arctisibirica, Dryas punctata, Hylocomium alaskanum* and *Tomenthypnum nitens*, sometimes together with *Salix arctica,* and in drier habitats with less snow cover a spotty tundra with *Dryas punctata*. In more moist and more snow-rich localities a tundra with a closed vegetation has formed, where *Hylocomium alaskanum, Tomenthypnum nitens* and also *Eriophorum angustifolium* are abundant. Here and there fragments of tussocks with *Eriophorum brachyantherum* (Tikhomirov, 1948*b*: 14) are found. On the outcrops of bedrock there are lichen and moss–lichen associations (*Cladonia rangiferina, C. mitis, Cetraria cucullata, Cornicularia divergens, Alectoria ochroleuca, Rhacomitrium lanuginosum,* etc.) and on the boulders grow *Umbilicaria proboscoidea, U. hyperborea, Cetraria hepatizon, Rhizocarpon geographicum,* and others.

A similar vegetation (Tikhomirov, 1956, 1957; Sdobnikov, 1959) has developed over the wide plains of northern Taimyr north of the Byrranga Mountains except for the far northern area around Cape Chelyushkin which belongs to the polar deserts (Matveyeva & Chernov, 1976). This lowland is formed from Quaternary glacial, fluvioglacial, and marine sediments and has locally some low hills and outcrops of bedrock. On the fairly well drained, moderately snow-covered, loamy substrates, a spotty tundra has developed, where *Carex ensifolia* ssp. *arctisibirica, Dryas punctata, Hylocomium alaskanum, Tomenthypnum nitens,* and *Aulacomnium turgidum* dominate, associated with *Festuca brachyphylla, Eriophorum angustifolium, Luzula confusa, Minuartia macrocarpa, Cassiope tetragona, Novosieversia glacialis, Salix reptans, Thamnolia vermicularis,* and other plants. On level, more moist and more snow-rich habitats a tundra with a closed cover is widely distributed. Together with mosses, such as *Hylocomium alaskanum, Aulacomnium turgidum, and Polytrichum alpestre,* the cottongrass *Eriophorum angustifolium* is met with, forming the basis for a closed sward in which also *Salix polaris, Polygonum viviparum* and other species are found. *Carex stans* is also encountered here. In localities where snow accumulates *Cassiope tetragona* associations are noted, and on outcrops of bedrock there are moss–lichen covers similar to those mentioned above. Homogeneous herb–moss and tetragonal mires have developed on thoroughly wet ground. In the troughs *Carex stans* and species of *Calliergon* and *Drepanocladus* are abundant. In the depressions and small pits there are cottongrass–moss associations with *Eriophorum scheuchzeri.* Peatmosses such as *Sphagnum squarrosum* and *S. subsecundum* occur.

In the area around Maria Pronchishcheva Bay in the far northeastern part of the Taimyr Peninsula there is, according to N.V. Matveyeva and O.I. Sumina (personal communication), a vegetation belonging to the northern belt of the arctic tundras represented on zonal, mesic habitats by spotty tundras with dominating *Salix polaris, Tomenthypnum nitens, Hylocomium alaskanum, Ditrichum flexicaule, Aulacomnium turgidum* and a considerable amount of *Luzula confusa* and *L. nivalis.*

In the arctic tundras of eastern Taimyr in the valley situated near the steeply breaking slope of the Byrranga Mountains there are at 75°10′ extrazonal, isolated small islands of *Salix lanata* thickets, growing 1.5 to 2 m tall (Dibner, 1961). This valley is protected on one side by the wall of the mountain and on the other by ridges of morainic hills. It is also very likely that föhn winds may occur here swooping down over the Byrranga Mountains and causing a locally warmer climate.

The Khatanga–Anabar coastal plains district. The narrow belt along the coast from Nordvik Bay eastwards to the Olenek Inlet belongs to the southern belt of the arctic tundras. Its vegetation is known from the publications by Sochava (1933*c*, 1934*a, b*). He emphasizes the absence here of any *Betula exilis* and hyparctic dwarfshrubs. Most typical for zonal, mesic habitats on loam are tundras with an almost continuous coverage, in which *Carex ensifolia* ssp. *arctisibirica* dominates together with mosses such as *Aulacomnium turgidum, Hylocomium alaskanum, Rhytidium rugosum* and some *Ptilidium ciliare, Polytrichum alpinum, Tomenthypnum nitens,* etc. and also *Salix reptans* and a number of other species among which there are a number of herbaceous ones. Spotty tundras with *Dryas punctata* or *Cassiope tetragona* as well as both *Dryas* and *Cassiope* are also present. Low-lying mires are widely distributed, where *Carex stans* predominates in the sward, and *Drepanocladus revolvens, D. sendtneri, Calliergon richardsonii,* etc. in the moss carpet. Tetragonal mires found in the river valleys and in sites of former lakes are also common here.

The East Siberian province of the arctic tundras

The transition to this province is indicated by the disappearance from the zonal, mesic habitats of associations with *Carex ensifolia* ssp. *arctisibirica.* Instead zonal, mesic associations appear, which are endemic to this province and where one of the important members is *Alopecurus alpinus* occupying a dominating position together with *Salix polaris* and mosses in the northern belt of the arctic tundras and being one of the abundant components in the southern belt. *Carex ensifolia* ssp. *arctisibirica* occurs in the composition of the flora as far as to the lower reaches of the Indigirka River and northwards all the way to Big Lyakovsky Island. But its aggressivity within the arctic tundras diminishes sharply, although in more southerly areas belonging to the subarctic tundras it continues to take part sometimes in the composition of the tussocky cottongrass tundras and it disappears just east of Indigirka River. Thus, the zones, where *Carex ensifolia* ssp. *arctisibirica* and *Alopecurus alpinus* (the latter typical of the high arctic over the major part of its area) are highly active, are displaced southwards in the area under discussion. I have already discussed the abnormal behaviour in various parts of their areas of some species such as, e.g. the role of *Eriophorum scheuchzeri* in subarctic mires in Greenland (cf., p. 80) etc. In the present case there is a tendency to expansion within associations of the species, which have a more northerly distribution as a consequence of the steadily diminishing snow cover and the severe winters due to the increasingly continental climate in Eastern Siberia (cf. pp. 51).

Three districts can be distinguished: the Sellyakh Inlet–Indigirka Delta, the Lower Alazeya Kolyma, and the Novosiberian Islands districts.

The Sellyakh Inlet–Indigirka delta district belongs to the southern belt of the arctic tundras and is distinguished from the Lower Alazeya – Kolyma district to the east of it by the presence in the flora of *Carex ensifolia* ssp. *arctisibirica* and the absence of *C. lugens*. Extremely scanty data on the vegetation have been published (E. F. Skvortsov, 1930; Shchelkunova, 1969; Andreyev & Nakhabtseva, 1974; Andreyev, 1975). There is some information on the flora thanks to the existence of herbarium specimens collected by various persons. E. F. Skvortsov, the topographer of the Vollosovich expedition, made an overland journey through this area in 1909, but in his detailed diary there are only dilettantic although vividly written observations on the nature of the landscape and its vegetation. Other authors have used aerial observations only or studies of aerophotographs, on the basis of which the boundaries of the vegetation subzones have been plotted (Andreyev, 1975). The district is, with the exception of the low heights on Cape Svyatoy Nos with its smooth summit at 400 m and a small, isolated height at Khaarastan, represented by a completely flat, low-lying plain full of mires and a multitude of shallow lakes. The gently sloping seashores with lagoons and large, silty low beaches and spits are typical. They are, according to the Skvortsov's diary (1930), overgrown with *Puccinellia phryganodes* and *Stellaria humifusa*. The low beaches and spits are especially wide between the Khrom Inlet and the Svyatoy Nos headland but also at Ebelyakh and Sellyakh Inlets, where the beach slopes so gradually that it may vary in width by as much as 15 km depending on whether there is an on- or off-shore wind.

In 1976 the area of the Svyatoy Nos headland was visited by I. N. Safronova. According to her (personal communication) a spotty tundra is typical of this area with *Alopecurus alpinus*, *Luzula confusa*, dwarf-shrubs like *Salix polaris*, and mosses associated with some *Salix reptans*. In better drained habitats tundras with *Dryas punctata* can be seen. Tussocky *Eriophorum vaginatum* associations are also met with.

The Lower Alazeya–Kolyma district. Information on the vegetation in this district may be found in data published by Perfil'yeva & Rykova (1975) from the lower reaches of the Chukoch'ya River. As stated by these authors the area is a plain with alternating low hills and depressions with lakes, often with 'cemetery mounds' resulting from the thermokarst process on the steep slopes. Along the sea there are marine terraces with a large number of small and large lakes. The authors place the vegetation of the area investigated within the southern belt of the arctic tundras on the basis of the following criteria: (1) complete dominance on zonal, mesic habitats of dwarfshrub tundras with prostrate and creeping willows (*Salix polaris, S. sphenophylla, S. reticulata, S. reptans*), arctic grasses (*Arctagrostis latifolia, Poa arctica, Alopecurus alpinus,* etc.), and mosses (*Aulacomnium turgidum, Hylocomium alaskanum, Ptilidium ciliare,* and others), but also of *Dryas–Salix* tundras (*D. punctata, Salix sphenophylla, S. reptans*) with arctic herbaceous species (*Carex lugens, Arctagrostis latifolia, Alopecurus alpinus,* etc.) and with a negligible quantity of mosses and lichens (Perfil'yeva & Rykova, 1975: 53–4); (2) the development in depressions of mires with a predominance of peatmosses on the raised edges of the low center polygons and *Carex stans* with an admixture of *Eriophorum angustifolium* and *Dupontia fisheri* in the central depressions and in the troughs as well as *Arctophila fulva* in the deepest ones; (3) the lack here of *Betula exilis* and hyparctic dwarfbushes other than *Vaccinium vitis-idaea* ssp. *minus;* and (4) the lack of willow thickets; *Salices* are only met with in the synusia of creeping *Salix pulchra.* From the subarctic zone tussocky tundras with *Eriophorum vaginatum,* associated with *Carex lugens, Salix pulchra, S. fuscescens, S. reticulata, Vaccinium vitis-idaea* ssp. *minus,* etc., penetrate into this district, taking on a specific arctic aspect and characterized by a slight (about 25%) formation of tussocks, the participation of arctic species and the lack of *Betula exilis* and hyparctic dwarfshrubs except for *Vaccinium vitis-idaea* ssp. *minus.* Perfil'yeva & Rykova also describe a number of associations observed here of herb – moss – willow and herb–*Dryas* tundra type, meadow-like communities, the glycophytic and halophytic marshes, chionophytic associations, and so on.

The *District of the Novosiberian Islands.* A number of papers have been published on the vegetation of the Novosiberian Islands (Gorodkov, 1956; Aleksandrova, 1958, 1960*a*, *b*, 1962, 1963, 1970; Aleksandrova & Zhadrinskaya, 1963; Kruchinin, 1963; Mikhailov, 1963*a*, *b*; Sisko, 1970; Sumina, 1975; etc.). This area is situated within the northern belt of the arctic tundras. However, Gorodkov has referred the tundras of Kotel'ny Island as well as those on Wrangel Island to the polar deserts, but his description of the zonal, mesic associations (Gorodkov, 1956: 71, 74, 78–86) indicate that these are no doubt arctic tundras close to those which have been described from Big Lyakovsky Island. On Kotel'ny Island a polar desert has developed only as a vertical belt on the high plateau in the interior part of the island.

The major parts of the Novosiberian Islands are occupied by rolling plains with low hills developed to an important extent in the paleodelta areas of the major rivers existing during the maximal emergence of the continental shelf. These hills reach an elevation of 30–50 m and consist of loose, Quaternary sediments, mainly loam or loam mixed with small stone fragments. Most hold ice-bodies of long-lasting iceveins (Romanovsky, 1960; and others). Thermokarst is widely distributed, just as are vegetation formations relating to it (Aleksandrova, 1963: 22–6; Sumina, 1975). There are in inland areas of the Kotel'ny Island, Big Lyakhovsky, and Stolbovy Islands some isolated heights with flat summits and nival terraces on their slopes. These heights are the remnants of an ancient plateau. On Kotel'ny Island the plateau, usually 170–180 m high, locally reaching 374 m, occupies the major part of the interior area of the island. On Big Lyakhovsky Island there are some isolated heights up to 311 m, and on Stolbovy Island they reach 220 m.

The climate is severe and continental. An anticyclonal régime predominates, as a consequence of which the amount of precipitation is low, on Kotel'ny Island 131 mm, on Big Lyakhovsky 140 mm per year. There are signs of a relatively dry climatic period. The plants receive their moisture not so much from atmospheric precipitation during the summer months as from the water produced by the melting ice from the annually developing ice-wedges in the cracks of the spotty tundras. The ice-wedges are formed by snow accumulating in the cracks and from meltwater filling the cracks during the spring before the temperature has started to rise in general. Water is also obtained from the melting of ancient buried giant ice-bodies of longlasting ice-veins, which mainly nourish the river net of the island.

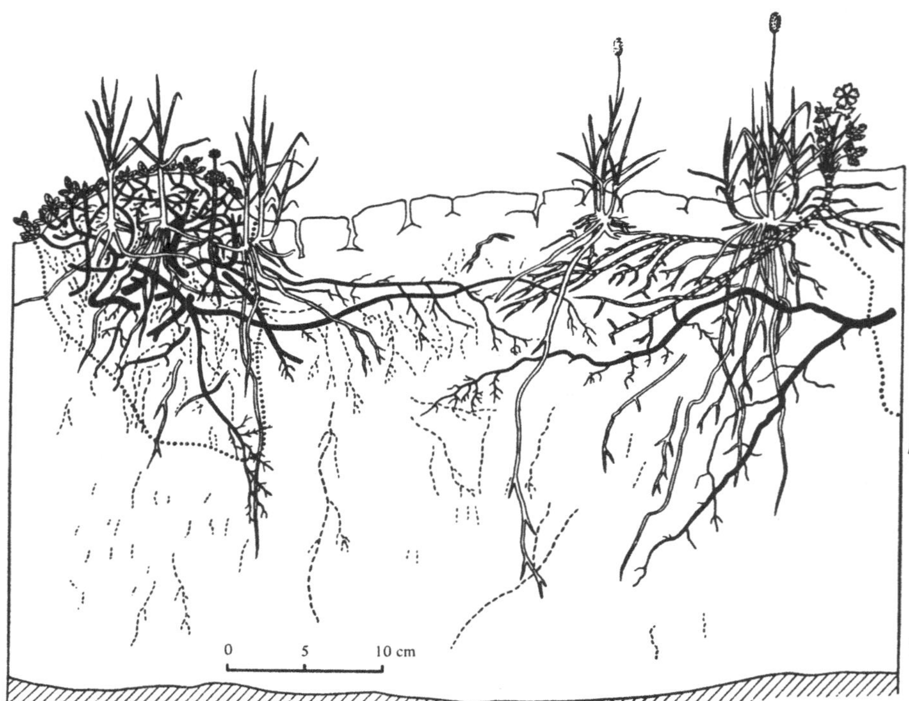

Fig. 19. The underground structure of an arctic tundra phytocoenosis. (Big Lyakhovsky Island; Aleksandrova, 1962.) 1, underground organs of *Salix polaris*; 2, *Alopecurus alpinus*; 3, *Draba pseudopilosa*; 4, *Potentilla hyparctica*; 5 unidentified roots; 6, limit of clusters full of roots; 7, limit of permafrost in August.

On zonal, mesic habitats with loam and under optimal snow conditions as demonstrated in studies made by Gorodkov (1956: 73–86) on Kotel'ny Island, by Mikhailov (1963*a*, *b*: 127) on Faddeyevsky Island, and by myself (Aleksandrova, 1963) on Big Lyakhovsky Island, herb – dwarfshrub – moss (*Hylocomium alaskanum, Drepanocladus uncinatus, Polytrichum alpinum*, etc.) tundras have developed with a coverage of 50–65%. As a rule *Salix polaris* dominates. Its underground organs are buried in the moss carpet (Fig. 19) and its leaves, according to data by Gorodkov and Aleksandrova, cover on an average 15%, in isolated cases up to 30%, of all the area occupied by the coenosis as such, but up to 60% of the actual area covered by the vegetation mat. As co-dominants (coverage 15%) there often appear *Alopecurus alpinus* (in this area being an especially aggressive, eutrophic species) and sometimes *Luzula confusa*. There is also an abundance of herbaceous species such as *Potentilla hyparctica, Ranunculus sulphureus, Papaver radicatum* s.l. (*P. polare*), *P. lapponicum, Cerastium bialynickii*, and many others, which give, however, less coverage than *Salix polaris* and the monocotelydons. But the number of species is considerable: lists from 11 localities published by Gorodkov include 22 herbaceous dicotyledon species. Eight observations by myself of zonal, mesic phytocoenoses from Big Lyakhovsky Island show 31 species of herbaceous dicotyledons. There are rather few lichens (*Cetraria crispa, Stereocaulon alpinum*, etc.).

In localities with a slight snow cover zonal, spotty tundras alternate with strongly denuded, polygonal moss–herb tundras and sometimes grass – moss or a willow–moss tundras where a cushion-like (*Potentilla hyparctica, Papaver lapponicum*, etc.) species or tufts of *Deschampsia brevifolia, Festuca brachyphylla*, etc. occur. The sward formed, including the mosses which are fewer here than the angiosperms, grows along the cracks around the bare-soil polygons without closed vegetation (some individual plants grow there). The vegetation occupies about 15 to 40% of the surface. The abundance of poppies, up to 40 flowers/m^2 in certain cases, does noticeably liven up the landscape when they flower.

Dryas tundras with *D. punctata* are met with only under more favorable conditions of wind and solar exposure; then the coverage may amount to 80%. Also, in similar, favorable conditions, tundras with low-growing *Salix reptans* are found.

Over the wide, flat areas with poor drainage and stagnating moisture enormous tetragonal mires have developed, their peat layers reaching 15 cm in thickness. In the central parts of the low-centre polygons with stagnant moisture *Dupontia fisheri, Eriophorum angustifolium, E. medium, Drepanocladus revolvens, D. vernicosus, Calliergon sarmentosum* and other species predominate. On the raised borders of the polygons *Polytrichum alpestre, Aulacomnium palustre, Dicranum elongatum* and a few angiosperms such as *Saxifraga foliolosa, Chryso-*

splenium alternifolium and others occur. Some *Sphagnum fimbriatum, S. squarrosum, S. girgensohnii, S. contortum*, and *S. obtusum* are also found there. In the troughs along frostcracks where the present ice-veins are developing more hydrophilous vegetation, with *Arctophila fulva* and sometimes *Carex stans*, occurs. In the shallow, rather narrow depressions, stretching along the gentle slopes mires with *Eriophorum angustifolium* are widely distributed. Often small thermokarst swamplets with *Eriophorum scheuchzeri* and *Alopecurus alpinus* are observed. Mires with *Carex stans*, forming a peat layer up to 22 cm thick, are rare and are found mainly on floodplain terraces along the major rivers on Big Lyakhovsky Island.

Vertical belts are found on the low heights in the southern and northern parts of Big Lyakhovsky Island (Aleksandrova, 1963: 29–33). Three belts can be distinguished: (1) a lower belt with herb–moss tundras on skeletal–loamy soils; sometimes small hanging peat mires with *Eriophorum brachyantherum* occur; (2) an upper tundra belt with an impoverished mountain tundra on loamy – skeletal and coarse, stony soils; and (3) a belt of alpine, polar deserts above 300 m, on patterned, stony soils and boulder fields. According to descriptions by Gorodkov (1956), a similar type of vegetation is found also on Kotel'ny Island, but the belt of oro-polar deserts starts there at a lower elevation and occupies more area.

The flora of vascular plants is poor on the Novosiberian Islands (about 135 species) and 'incomplete': it contains no leguminous species, and there are no pollinators (bees). On Big Lyakhovsky Island no *Saxifraga oppositifolia* has been found. There are no endemics.

The main body of species consists of moderately arctic, circumpolar and Siberian species. High-arctic species make up a rather important part; the hyparctic species are few, although some, like *Eriophorum angustifolium*, are quite aggressive.

The Wrangel Island — West American province of the arctic tundras

Wrangel Island has a vegetation on zonal, mesic habitats represented by associations with dominating *Carex lugens* associated with mosses such as *Tomenthypnum nitens, Aulacomnium turgidum* and others and some *Salix polaris*. According to Hultén this willow should be *Salix polaris* ssp. *pseudopolaris*, but in the opinion of A. K. Skvortsov (1966:

49) *S. pseudopolaris* Flod. (*S. polaris* ssp. *pseudopolaris* (Flod.) Hult.) does not exceed the limits of variability for *S. polaris* and hardly 'deserves separation even as a subspecies'. There are also tundras with *Dryas integrifolia*, a species with a mainly American area.

Wrangel Island is united into one province with the area around the Point Barrow in Alaska and the southwestern islands of the Canadian Arctic Archipelago, where within the limits of the arctic tundras there occur communities with a specific composition like those mentioned above.

In spite of its small size Wrangel Island can be distinguished as a subprovince of its own. The other subprovince unites the Point Barrow area with the southwestern part of the Canadian Arctic Archipelago, which is situated within the limits of the arctic tundra.

The Wrangel Island subprovince of the arctic tundras

Two-thirds of Wrangel Island is occupied by mountains with elevations up to 1096 m, to the south and north of which spread plains. To the north there is a broad, flat plain, the Academy Tundra, to the south narrow parts of a plain dissected by mountains reaching all the way to the sea.

Usually the climate of Wrangel Island is described in atlases, text-books, and various papers as having a mean temperature of 2.4 °C in July. This is measured at Roger's Bay on the southeastern low-lying coast of the island and reflects a local, oceanic climate, characterized by much fog, considerable precipitation, and low summer temperatur-es. But in the central part of the island, due to the heating of the air over the land surface and especially thanks to the effect of föhn winds, there are often air temperatures in the river valleys and the intermon-tane depressions which are considerably higher, as well as a much drier climate. Thus, in 1951, Borodin (in Svatkov, 1961: 12) measured an average July temperature of 8 °C at the foot of Mt Inkala and 10 °C in the valleys of the Soviniy and Khrustal'niy Rivers. Because of these peculiarities of the climate in the central parts of the island, there are areas with an extra-zonal vegetation characteristic of the northern belt of the subarctic tundras, i.e., with thickets of *Salix lanata* ssp. *richard-sonii* growing 1 m tall.

Information on the vegetation of Wrangel Island can be found in a number of works (Gorodkov, 1943, 1958*a*, *b*; Svatkov, 1958, 1961,

1970; Petrovsky, 1967; Kitsing *et al.*, 1974; Yurtsev, 1974*b*; etc.). In his papers, especially in the detailed paper *Soils and Vegetation Cover on Wrangel Island*, Gorodkov (1958*b*) has furnished invaluable information on the vegetation of this island in the form of excellent descriptions and complete lists of all groups of plants, lucid discussions on the structure of the vegetation, its habitats at various elevations above sea level, its succession, and its floristic composition, etc. When using this paper it is, however, necessary to keep in mind that this author is using a particular terminology; he calls the tundras 'polar deserts', the snow layer vegetation with predominately chionophilic lichens 'nival meadows', etc.

In the southern and central parts of Wrangel Island the zonal, mesic vegetation belongs to the southern belt of the arctic tundras. It has been described by Gorodkov from the southern plains at an elevation of 20 m (Gorodkov, 1958*b*: 28–32) and by V. V. Petrovsky (1967: 340) from the central part of the island on gentle slopes and terraces. There are sedge–moss tundras with dominating *Carex lugens, Tomenthypnum nitens, Aulacomnium turgidum, Hylocomium alaskanum* and a considerable abundance of *Alopecurus alpinus, Deschampsia borealis, Salix polaris*, and other plants. Sometimes prostrate *Salix pulchra* and *S. reptans* take part in the communities. The lichens are represented by *Thamnolia vermicularis, Stereocaulon alpinum, Cetraria islandica*, and others. *Dryas* tundras are widely distributed. *Dryas punctata* occurs mainly on acid substrates, *D. octopetala* on both basic and neutral ones, and the American species *D. integrifolia* only on basic soils (Gorodkov, 1958*b*: 33–4; Petrovsky, 1967: 337). Together with these species *Salix rotundifolia* co-dominates in fairly moist conditions. *Saxifraga oppositifolia* is met with in large quantities, and also *Parrya nudicaulis, Deschampsia borealis, Lloydia serotina, Carex misandra, Saxifraga serpyllifolia*, and others occur. Among the lichens *Thamnolia vermicularis* is the characteristic one. Associations of *Dryas integrifolia* var. *canescens* are observed on gently sloping, warm and dry hill sides formed by basic soils. The *Dryas* communities have a 65–70% coverage, while 20–25% of the surface is occupied by patches of basic blocks colonized by a number of single individuals of species such as *Carex rupestris, Oxytropis middendorffii, Saxifraga oppositifolia, Carex hepburnii*, etc. *Cetraria cucullata* and *C. nivalis* are among the lichens met with.

In addition to forming communities where they dominate, *Dryas punctata* and *D. octopetala* are often associates in many other plant communities on the island. Petrovsky (1967: 334) has also described chionophytic associations with *Cassiope tetragona* in depressions at a few localities, sometimes at the lower parts of slopes.

Kobresia associations relate to well drained habitats on slopes, where *Kobresia bellardii* dominates together with the co-dominant sedges *Carex rupestris* and *C. hepburnii* in the cryomesophytic, meadow-like communities (Petrovsky, 1967: 339–40). In addition Yurtsev (1974*b*) has described tundra–steppe associations with *Carex obtusata* and *Pulsatilla multifida* on the dry parts of south facing slopes.

In the central part of the island, in the wide valleys and the intermontane hollows in localities under the influence of föhn winds, extra-zonal islands of northern subarctic tundras have developed. Sometimes they occupy large surfaces (Petrovsky, 1967: 333) covered by willow brush – (*Salix lanata* ssp. *richardsonii, S. glauca,* etc.), dwarf-shrubs – (*Dryas punctata, D. octopetala, Salix rotundifolia*), graminoid – (*Carex lugens, Arctagrostis latifolia*, etc.), moss – (*Tomenthypnum nitens, Hylocomium alaskanum, Aulacomnium turgidum*, etc.), tundras; dense flood-plain willow thickets of *Salix lanata* ssp. *richardsonii*, growing 60–70 cm and locally up to 1 m tall, are also met with.

The mires, both in the mountainous areas and on the southern plains (Gorodkov, 1958*b*: 40–5; Petrovsky, 1967: 340), are represented mainly by sedge (*Carex stans*)–moss (*Calliergon sarmentosum*, etc.) communities with some peatmosses (*Sphagnum contortum, S. subsecundum, S. squarrosum*, etc.). The peat layer has been measured up to 15 cm thick (Gorodkov, 1958*b*: 44). In conditions where the run-off is limited, a sward of *Eriophorum angustifolium* has developed associated with mosses such as *Aulacomnium palustre, Drepanocladus revolvens, Tomethypnum nitens*, etc.

In the mountainous areas large surfaces are occupied by boulder fields (Gorodkov, 1958*b*: 8–13; Petrovsky, 1967: 338), which are distinguished by vegetation covers developed on and between the boulders. These vegetation formations consist of *Lecidea, Rhizocarpon*, and *Umbilicaria* species, *Parmelia omphalodes*, *Cornicularia divergens*, *Alectoria ochroleuca*, *A. pubescens, Cetraria nivalis, Sphaerophorus globosus, Cladonia mitis* and other lichens and mosses such as *Rhacomitrium lanuginosum, Andreaea papillosa, Schistidium* *gracile, Dicranoweisia crispula*, etc. as well as a considerable variety of angiosperms growing within the limits of the lower and middle mountain belts (*Luzula confusa, L. nivalis, Oxytropis tschuktschorum, Minuartia macrocarpa, Androsace ochotensis, Artemisia furcata, A. glomerata, Saxifraga serpyllifolia, Poa pseudoabbreviata, Salix phlebophylla*, etc.).

As shown by Svatkov (1970: 472) the vertical belts are well developed; this can also be seen from the data published by Gorodkov (1958*b*). The lower belt, up to 250 m, has a vegetation which is richer than that of the zonal, mesic habitats on the southern plains. The favorably acting temperature inversion occurs due to the effect of the föhn winds. The middle belt, 250–400 m, locally up to 600 m, is an area of impoverished arctic tundras. Descriptions from these elevations are furnished by Gorodkov (1958*b*: 11, 15, 21–23). The upper belt, above 600 m, is occupied by oro-polar deserts, where the role of the flowering plants has decreased sharply in the dominating boulder fields. According to observations made by Gorodkov (1958*b*: 9–10) angiosperms are rarely seen above 700 m; on the high peak of Mt Berry (1000 m) the granite boulders are completely covered by lichens and some *Rhacomitrium lanuginosum*, but no other mosses. There are no angiosperms either.

The northern plains, the Academy Tundra, belong to the northern belt of the arctic tundras. According to observations made by Chevykalov (in Svatkov, 1970: 460–3) the air temperature in the area of Cape Ushakov is lower and it is windier than at Roger's Bay. The snow starts to fall earlier and lasts longer than on the southern coast. Gorodkov (1958*b*: 26–7) described how the zonal, mesic associations are represented by spotty tundras with predominating *Deschampsia borealis, Salix polaris, Alopecurus alpinus, Polytrichum alpinum, Oncophorus wahlenbergii, Aulacomnium turgidum*; the crustose lichens have a coverage of up to 30%, mainly of *Ochrolechia frigida, Psoroma hypnorum*, etc. The fruticose and tubular ones, represented by *Thamnolia vermicularis* and *Stereocaulon alpinum*, among others, cover about 5%. The Academy Tundra lowland is very swampy (Gorodkov, 1958*b*: 40–9; Svatkov, 1970: 470).

Having escaped complete glaciation during the Ice Age, Wrangel Island became part of the extensive Bering land area at the time of the maximum emergence of the continental shelf. As shown by Yurtsev

(1974*a*: 106), supported by information in native and foreign literature, the mountains on Wrangel Island were at this time united by low ridges, dissected only by narrow canyons east of the present Geral'd Island, with Brooks Range in Alaska, and spread out over a single continental landmass including northern Chukotka and the Kolyma River basin. This made it possible for the island flora to become saturated with Beringian and Angaran species. At the same time periodic isolation of the island created conditions for the development of endemism. As a result there is an unusually rich flora here for such a high latitude and such a small island; according to information from Petrovsky, the number of vascular plant species exceeds 320.

The particular characteristics of the flora of Wrangel Island are emphasized by its comparatively low content of circumpolar species, 45%, while at the same time on, e.g. the Novosiberian Islands, these amount to 63%. Only 12% are species of Siberian and Eurasiatic origin not met with in America. In contrast 34% are species met with both in Asia and America (not including circumpolar taxa); here belong species with Siberian–American areas as well as amphi-beringian and beringian species and also a small number of species with wide amphiatlantic areas, the western part of which extends all the way to Alaska and Chukotka (e.g. *Campanula uniflora*). About 3% are endemic or subendemic, *Poa wrangelica, Potentilla wrangelii, Oxytropis wrangelii*, etc. Most significant is the presence on the island of a number of species, amounting to less than 3% of its flora, which outside of Wrangel Island are found in America or America and Greenland, but are not found in Asia, *Poa hartzii, Erigeron compositus, Chrysosplenium rosendahlii*, and others. A part of these are not even reported from Alaska, but are found only on Wrangel Island, in the Canadian Arctic Archipelago and on Greenland.

The subprovince of Point Barrow and the southwestern islands of the Canadian Arctic Archipelago

This area, which includes Point Barrow and the Banks and Victoria Islands with the exception of their southernmost parts, where the vegetation belongs to the subarctic tundras, is united into one province together with Wrangel Island in spite of a number of differences, because the vegetation here is closer to the one on Wrangel Island than to that on the rest of the Canadian Arctic Archipelago. Thus, in this territory, the presence of associations with *Salix polaris* (*Salix polaris* spp. *pseudopolaris*, according to Hultén) is typical, while not found on the other islands in the archipelago. In the vegetation cover there are amphi-beringian components, such as *Salix phlebophylla, Cardamine digitata, Carex lugens*, and others, which do not reach the eastern islands in the archipelago, where instead amphi-atlantic species such as *Carex bigelowii, Salix herbacea, Cerastium alpinum*, etc. occur, which are not found on the western islands.

The *Point Barrow Area*, distinguished at the level of a district, is represented by a low-lying, swampy plain of its own consisting of marine alluvial terraces with many lakes. Judging from the data available (Hanson, 1951, 1953; Wiggins, 1951; Britton, 1957, 1966; Drew *et al.*, 1958; Wiggins & Thomas, 1962; Brown, 1968; Johnson, 1969; Dennis & Johnson, 1970; Andreyev, 1976; etc.), the vegetation here is quite uniform. A considerable part of the area is occupied by polygonal mires. The main components of the sward formed in the shallow depressions are *Carex stans, Eriophorum angustifolium* and *Dupontia fisheri*. The mosses are represented by taxa of genera such as *Calliergon* and *Drepanocladus*. There is a thick growth around lakes of *Arctophila fulva, Carex stans, Eriophorum scheuchzeri, E. angustifolium, Dupontia fisheri* and *Alopecurus alpinus*. Between the shallow depressions with lakes and mires a swampy tundra is situated on the only slightly elevated parts of the terrain. According to observations by Andreyev (1976), in addition to the polygonal mires on the low marine terraces tundras with a thin turf and mosses such as *Aulacomnium turgidum, Pogonatum* sp., and others dominate. In the closed cover here low-growing *Eriophorum angustifolium*, 10–12 cm tall, *Poa arctica*, and *Alopecurus alpinus* predominate; specimens of *Salix pulchra* are also found. Britton (1966: 37) has pointed out the occurrence in the swampy tundras of *Eriophorum vaginatum*, which is scattered among other species and does not form here the tussock community typical of the more southerly situated subarctic tundras. On the low marine beaches at Point Barrow *Puccinellia phryganodes* and *Carex ursina* grow, and on the pebbly slopes down towards the sea there are *Mertensia maritima, Honkenya peploides*, and *Leymus mollis*. On the pebbly terraces at a higher elevation *Salix polaris, Cerastium beeringianum, Cochlearia arctica, Saxifraga caespitosa, S. rivularis* s.l., and *Festuca brachyphylla* are found (Wiggins & Thomas, 1962: 16), as well as *Dryas integrifolia, Salix phlebophylla, Saxifraga oppositifolia, S. flagellaris, Astragalus umbellatus, A. alpinus, Papaver radicatum* s.l., *Pedicularis lanata, Cardamine bellidifolia* and other species. Ground-hugging *Salix pulchra* (Wiggins & Thomas, 1962: 17) is less common. *Poa arctica, P. alpigena, Arctagrostis latifolia* etc., thrive around fox dens.

The flora of vascular plants in the Point Barrow area amounts to about 135 species. It has a well expressed arctic character. Moderately arctic species predominate together with some high-arctic ones, *Phippsia algida, Alopecurus alpinus, Carex ursina, Ranunculus sabinei, Papaver radicatum* s.l., *Cardamine bellidifolia, Saxifraga oppositifolia, S. platysepala,* etc., and some hyparctic ones, *Eriophorum angustifolium, Salix pulchra, S. lanata* ssp. *richardsonii, Valeriana capitata, Nardosmia frigida,* etc. Circumpolar species constitute about 55%, species with Beringian areas about 16% of the flora.

The *Banks and Victoria Islands*, with the exception of their southernmost parts which belong to the subregion of the subarctic tundras, can also be distinguished at the level of a district. We are obliged to Porsild (1955) for a brief account on their vegetation. The central part of Banks Island is occupied by a low plateau, not above 300 m above sea level with a few lakes and mires. The coastal lowland on the western side of the island is also to a great extent very swampy. In the northeastern and northwestern parts of Victoria Island there are also lowlands, rarely exceeding 90 m in altitude, with many lakes; in the central part of the island the terrain is crossed by heights at some points reaching 500–600 m elevation.

Moist hummocky tundras and mires, mostly polygonal, occupy large areas of the island lowlands. On the hummocks grow *Arctagrostis latifolia, Carex atrofusca, C. misandra, C. membranacea, C. stans, Salix arctica, S. reticulata, Hierochloë pauciflora,* etc.; *Salix lanata* ssp. *richardsonii* is also met with. In the mires a sward has been formed consisting mainly of *Dupontia fisheri, Carex stans, Eriophorum scheuchzeri* and *Arctophila fulva*; the mosses are not mentioned. The aquatic vegetation is represented by *Arctophila fulva, Hippuris vulgaris, Ranunculus gmelinii, R. hyperboreus* and *Pleuropogon sabinei.* On loamy or stony-loamy soils *Dryas*–herb tundras are typical on better-drained habitats. Common species are *Dryas integrifolia* and *Potentilla rubricaulis. Oxytropis arctica, Festuca baffinensis, Carex scirpoidea* and a number of other herbaceous species occur together with them. On stony habitats and on rocky cliffs, favorably exposed, there are also various combinations of *Dryas integrifolia, Festuca baffinensis, Poa abbreviata, P. glauca, Kobresia bellardii, Carex rupestris, Luzula confusa, Salix arctica, Polygonum viviparum, Papaver radicatum* s.l., *Lesquerella arctica, Saxifraga oppositifolia, S. tricuspidata, Potentilla rubricaulis, Oxytropis arctica, O. arctobia, Pedicularis capitata, Erigeron compositus*, and many other species. The composition of mosses is not reported. *Salix polaris* predominates in areas with a deeper snow cover in association with *Alopecurus alpinus, Cerastium regelii, Minuartia rossii, Gastrolychnis apetala, Saxifraga hirculus, S. cernua, Potentilla hyparctica* and others, mosses not mentioned.

On marine spits with clayey soils *Puccinellia andersonii, P. phryganodes, Carex maritima, C. ursina, Stellaria humifusa* and *Cochlearia arctica* grow in protected wetlands. A rich and variable vegetation has developed on ledges of birdcliffs (Porsild, 1955: 64).

The Northeastern Canadian–North Greenland province of the arctic tundras

The eastern islands in the Canadian Arctic Archipelago and the part of Greenland above latitude 68–70 °N, belonging to the subregion of the arctic tundras, consist of mountainous country with considerable elevations, in Greenland reaching up to 2940 m, on Axel Heiberg Island up to 2100 m, on Ellesmere Land and on Devon Island above 1000 m, on Baffin Island up to 2591 m, and on Bylot Island to 1890 m. The land is partly covered by glaciers and ice caps. The deep fjords cutting into the land areas are always very typical. Basic soils are widely distributed. The strong aridity of the climate is one of its essential characteristics, a cryo-aridity, associated with the effect of the mighty Greenland high pressure maintained not only during the winter but also during the summer months. Owing to this, the amount of precipitation is very low and decreases rapidly towards north. While the precipitation is 236 mm at Myggbukta, central E. Greenland, it is only 210 mm at Upernivik, W. Greenland, 131 mm at Backe Peninsula,. Ellesmere Land, about 80 mm at Inglefield on northwest Greenland, and in Peary Land on northeast Greenland only about 25 mm per year. The aridity of the climate is sharply enhanced by föhn winds.

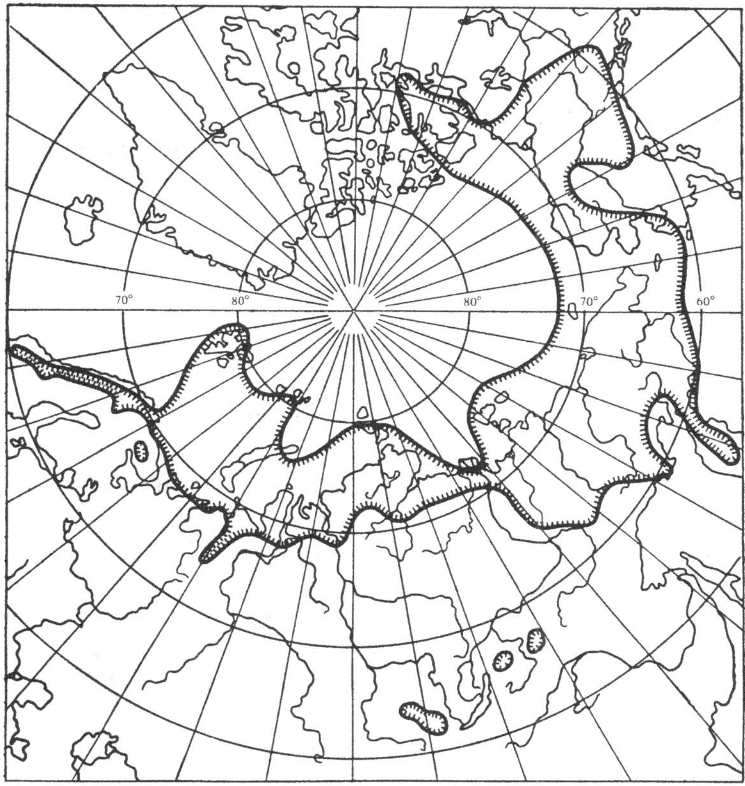

Fig. 20. The area of *Salix polaris* Wahlenb. s.l. According to A. K. Skvortsov, 1968; Yurtsev, 1968*b*; Hultén, 1968; and *The Alpine Flora of the Stannovoye Foothills*, cf. Malyshev, 1972.

The vegetation of this province is quite distinct. First and foremost there is no *Salix polaris* (Fig. 20), the most typical component of all other zonal, mesic habitats in the subregion of the arctic tundras and its main dominant species. Then there is the extraordinarily expanded ecological amplitude of *S. arctica*, which has a disjunct area in Eurasia (A. K. Skvortsov, 1968: 135), but does not play any prominent role in the composition of the vegetation cover. In this province it has become a very active, eurytopic species reaching very high latitudes. Here its stems and branches are completely buried in the moss cushion with only leaves and catkins appearing above it (Savile, 1961: 923; Tedrow *et al.*, 1968: 17). However, farther south and under slightly improved conditions this polymorphic and vegetatively stratified willow develops into a creeping or upright but low small shrub met with in very variable habitats and growing on substrates from boulder fields and snow-free localities to mires and deeply snow-covered patches (Holmen, 1957: 86, etc.), although it never attains a dominant position. In localities, which may be considered analogous to zonal, mesic habitats in this mountainous country, two species dominate: *Dryas integrifolia* and *Cassiope tetragona*. They form *Dryas* (mainly on habitats with a poor snow cover), *Dryas–Cassiope*, and *Cassiope*–tundras. The latter is particularly characteristic of the province under discussion.

During the Ice Age this province was never completely glaciated, but did house refugia for the flora. The present floristic complex, in spite of being definitely dominated by circumpolar and nearly circumpolar species, has a specific character which is emphasized by the presence of such endemics as *Puccinellia rosenkrantzii, P. poacea, Braya intermedia* and *Taraxacum arctogenum*, and subendemics such as *Puccinellia groenlandica, P. bruggemannii, Draba allenii, Braya thorild-wulffii, Potentilla rubella*, and *Antennaria glabrata*. There is also a large quantity of co-differentiating species, eastern (*Carex bigelowii* s. str., *Salix herbacea,* etc.) as well as western (*Dryas integrifolia,* etc.).

Two subprovinces can be distinguished within the province: the Northeastern Canadian–northwestern Greenlandic, and the northeastern Greenlandic subprovinces with Peary Land included in the latter. To the former belong some islands in the Parry Archipelago on the basis of such guiding characters as the lack of *Salix polaris* in their vegetation cover, although there are also other differences which will be discussed below.

The Northeastern Canadian–Northwestern Greenlandic subprovince of the arctic tundras

This subprovince covers northwestern Greenland and those islands in the Canadian Arctic Archipelago which are situated within the limits of the arctic tundras, with the exception of Banks and Victoria Islands. It is distinguished from the other subprovinces by the presence of associations in which *Carex membranacea* appears as one of the dominants. This is a Chukotka–American species of sedge with a growth-type similar to that of *C. bigelowii*, but more hygrophilic and not found in eastern Greenland. In the vegetation cover there are also other species, not reaching to northeastern Greenland, *Hierochloë pauciflora*, *Salix reticulata*, *Stellaria monantha*, *Androsace septentrionalis*, *Pedicularis sudetica*, *Senecio congestus*, etc.

There are two districts: the Parry Archipelago and that of the eastern islands in the Canadian Arctic Archipelago together with northwestern Greenland.

The Parry Archipelago district. This district includes the islands within the northern belt of the arctic tundras, i.e. the Melville, Bathurst, Prince William, etc. Islands, situated west of the Axel Heiberg and Devon Islands. These islands adjoin towards west and south the so-called 'barren wedge' of Beschel (1969: 879) and are partly included in it. As a rule they have a relatively hilly relief with low elevations. As a consequence of this general character, the vegetation cover differs from that developed among the high mountains of the eastern islands and in northern Greenland. But a number of characteristics, among which are the lack of associations with *Salix polaris* and the distinctly increasing activity of *S. arctica*, make it possible to place these islands in the subprovince under discussion.

According to available information (Porsild, 1955, 1957*a*,*b*; Savile, 1961; Tedrow *et al*., 1968; Beschel, 1969; Babb & Bliss, 1974*a*, *b*; Price *et al*., 1974; Thannheiser, 1975; etc.), it is possible to conclude that the vegetation of these islands belongs to the northern belt of the arctic tundras, except for that of Meighen Island and the southern parts of the Ellef Ringnes, Borden, Brock, and Prince Patrick Islands, which lie within the limits of the polar deserts. The belt of southern arctic tundras occurs in this sector of the archipelago on the Banks and Victoria Islands. Among the sources available to me there are careful but very brief descriptions of the vegetation on the southern part of Prince Patrick Island, situated in the northern part of this district, in a paper by Tedrow, Bruggemann & Walton (1968: 15–22). The authors report on the presence on sloping hillsides of a patchy cover, in which occur species such as *Salix arctica*, *Dryas integrifolia*, *Poa alpigena*, *P. arctica*, *Alopecurus alpinus*, *Arctagrostis latifolia*, *Deschampsia brevifolia*, *Puccinellia angustata*, *Festuca brachyphylla*, *Luzula confusa*, *L. nivalis*, *Juncus biglumis*, *Oxyria digyna*, *Polygonum viviparum*,

Stellaria ciliatosepala, *Cerastium arcticum*, *Ranunculus nivalis*, *R. sulphureus*, *Papaver radicatum* s.l., and others. The stems of *Salix arctica* are pressed to the ground and partly buried in the moss cover or the soil, and only the leaves and the catkins project above it. In wind-exposed areas there are polygonal or stony–polygonal tundras rich in lichens, but mosses and a small amount of angiosperms occur only in the depressions along the cracks or grow in the shelter of large boulders. Among the angiosperms are *Poa abbreviata*, *Festuca brachyphylla*, *Alopecurus alpinus*, *Luzula confusa*, *L. nivalis*, *Salix arctica, Papaver radicatum* s.l., *Cerastium beringianum*, *Saxifraga oppositifolia*, *Oxyria digyna*, and *Potentilla vahliana.*

On slopes covered by boulders there are individual specimens of *Saxifraga oppositifolia*, *Carex misandra*, *Cardamine bellidifolia*, *Draba macrocarpa (D. bellii)*, *Salix arctica*, *Alopecurus alpinus*, *Puccinellia angustata*, *P. bruggemannii*, *Luzula confusa*, *Cerastium beringianum*, *Crepis nana*, *Dryas integrifolia*, *Papaver radicatum* s.l., *Oxyria digyna*, *Potentilla rubricaulis*, *P. vahliana*, *Oxytropis arctica*, *Erigeron eriocephalus*, etc. In graminoid mires grow *Dupontia fisheri*, *Eriophorum scheuchzeri*, *E. triste*, *Carex stans*, *Arctophila fulva*, and others, and on low beaches and spits with muddy soil *Puccinellia phryganodes* and *Stellaria humifusa*. Marine salt-marshes with *Puccinellia phryganodes* have been described from Cornwallis Island by Thannheiser (1975).

The district of the Eastern Islands of the Canadian Arctic Archipelago and Northwestern Greenland. The vegetation of this district, situated within the southern belt of the arctic tundras, can be characterized according to the description by Polunin (1948: 83–144) of the northern and central parts of Baffin Island. *Salix glauca* ssp. *callicarpaea*, *S. lanata* ssp. *richardsonii* (although not forming thickets) and fragments of associations with *Vaccinium uliginosum* ssp. *microphyllum* are found there. The latter species does also take part in the formation of the small-hummocky tundras developed here. A small-hummocky tundra with a closed vegetation cover occupies, locally, a major part of the lowland habitats (Polunin, 1948: 84–7, 105–7). As co-dominants appear *Carex membranacea*, *Salix arctica*, *Cassiope tetragona*, *Eriophorum angustifolium*, and *Aulacomnium turgidum*. On the small hummocks grow *Carex misandra*, *Vaccinium uliginosum* ssp. *micro-*

phyllum, *Dryas integrifolia*, *Salix reticulata*, *Dicranum groenlandicum*, *Ditrichum flexicaule*, *Cetraria crispa*, *Cladonia mitis*, *Parmelia omphalodes*, *Psoroma hypnorum*, and others. Between the small hummocks there is much *Arctagrostis latifolia*, *Eriophorum angustifolium*, *Polygonum viviparum*, etc.; *Carex stans* also occurs. Among the mosses mentioned by Polunin are *Tomenthypnum nitens*, *Ptilidium ciliare*, and *Sphagnum warnsdorffii*. *Salix lanata* ssp. *richardsonii*, *S. arctophila*, and *S. herbacea* are also met with. The first of these species forms flat patches, 20–25 cm in height and 1–2 m in diameter. The quantity of *Carex stans* and the peatmosses, a few species only, increases in the very wet variant of this kind of tundra. Graminoid, mossy mires have a sward of *Carex stans*, *Eriophorum angustifolium*, *Arctagrostis latifolia*, etc.

On drier slopes in protected situations a *Cassiope*-tundra has developed. There *Cassiope tetragona* predominates, associated with *Vaccinium uliginosum* ssp. *microphyllum*, *Carex rupestris*, *Salix arctica*, *Dryas integrifolia*, etc. In the moss cover of *Dicranum groenlandicum*, *Hylocomium alaskanum*, *Rhacomitrium lanuginosum*, etc. there are many liverworts such as *Gymnomitrium coralloides*. Lichens such as *Cetraria nivalis*, etc. are frequent. In the outer zone around the edges of snow banks, where the snow is moderately deep and does not last very long, associations have developed with *Cassiope tetragona* associated with *Luzula nivalis*, *Salix herbacea*, *S. arctica*, *Carex bigelowii*, and others, mosses such as *Rhacomitrium lanuginosum*, *Aulacomnium turgidum*, and lichens such as *Cetraria crispa*, *Cladonia mitis*, *Sphaerophorus globosus*, etc. The next zone, where the snow is deeper, is occupied by *Salix herbacea*, *Luzula confusa*, *Oxyria digyna*, etc., associated with mosses. Then follows a zone with *Phippsia algida*, *Luzula confusa*, *Saxifraga rivularis*, and *Bryum* species. Finally, under the innermost part of the snow bank, there are only tiny patches of *Bryum* spp., *Andreaea crassinervia*, and algae (Polunin, 1948: 109–11). Occupying the major part of the area on the summits of stony ridges and mounds impoverished associations dominate with much *Saxifraga oppositifolia* and some *Salix arctica*, *Carex hepburnii*, *Festuca brachyphylla*, *Luzula confusa*, *Dryas integrifolia*, *Rhacomitrium lanuginosum*, *Polytrichum piliferum*, *Umbilicaria* spp., *Cetraria nivalis*, *Cornicularia divergens*, *Thamnolia vermicularis*, etc.

The northern belt of the arctic tundras is seen on Ellesmere Land,

Axel Heiberg, Devon and a number of other islands as well as on the northern edge of Baffin Island and in corresponding parts of western Greenland. A short account of the vegetation on the northern part of Hayes peninsula, northwest Greenland, by Tedrow (1970) is available and a number of papers have been published on the vegetation of Ellesmere Land (Hart, 1880; Polunin, 1948; Savile, 1964; Brassard & Beschel, 1968), of Axel Heiberg Island (Porsild, 1955; Beschel, 1961, 1963*a*, *b*, 1970), on Devon Island (Polunin, 1948; Bliss, 1971, 1972; Muc, 1972, 1973, 1977; Svoboda, 1972, 1973, 1977; Bliss & Kerik, 1973; Pakarinen & Vitt, 1973; Richardson & Finegan, 1973, 1977; Babb & Bliss, 1974*a*, *b*; Pakarinen, 1974; Bliss, Kerik & Peterson, 1977; Muc & Bliss, 1977; Vitt & Pakarinen, 1977), on Bylot Island (Drury, 1962), and on the northern edge of Baffin Island (Polunin, 1948).

In territories where there are level habitats, usually with skeletal soils or patterned ground, or in habitats analogous to the zonal, mesic ones, *Dryas* tundras with *D. integrifolia* is the most typical in the low-montane belt, where there is a light snow cover, and *Cassiope* tundras on the slopes which are better protected by a snow cover. In the *Dryas* tundras there are *Salix arctica*, *Saxifraga oppositifolia*, *Polygonum viviparum*, *Carex rupestris*, *C. hepburnii*, *Kobresia bellardii*, *Carex misandra*, *Pedicularis hirsuta*, *Tortula ruralis*, *Ditrichum flexicaule*, *Thamnolia vermicularis*, and *Cetraria nivalis*, in the *Cassiope* tundras such species as *Salix arctica*, *S. reticulata*, *Carex misandra*, *C. bigelowii*, *Dryas integrifolia*, *Vaccinium uliginosum* ssp. *microphyllum*, *Equisetum variegatum*, *Rhacomitrium lanuginosum*, *Hylocomium alaskanum*, *Dicranum scoparium*, *Ptilidium ciliare*, *Blepharostoma trichophyllum*, *Cetraria cucullata*, and *Stereocaulon alpinum*, etc.

Where they occur, the mires are represented by sedge (*Carex stans*)–, cottongrass–sedge (*Carex stans*, *Eriophorum triste*)–, and cottongrass (*E. triste*)–moss associations. The sward is composed of *Arctagrostis latifolia*, *Eriophorum scheuchzeri*, *Dupontia fisheri*, *Carex membranacea*, *Hierochloë pauciflora*, and *Pedicularis sudetica*; the mosses *Calliergon giganticum*, *Drepanocladus brevifolius*, *D. revolvens*, *Cinclidium arcticum*, and *Meesia triquetra* predominate. The mires on the northern shore of Devon Island have been studied in detail at the International Biological Project camp under the direction of L. C. Bliss (Muc, 1972, 1973, 1977; Pakarinen & Vitt, 1973; Pakarinen, 1974; Vitt & Pakarinen, 1977). The vegetation on the beach ridges was also studied there (Svoboda, 1972, 1973, 1977). Twenty-six different surfaces of these ridges of various age and altitudes were distinguished; the change in the vegetation and the phytomass was found to relate to the age of the terraces, which were from 3283 to 9500 years old. The width of the pebbly ridges varied between 20 and 150 m. On their tops closed formations with predominating *Saxifraga oppositifolia*, *Carex hepburnii*, and *Salix arctica* are the most common; on the slopes *Dryas integrifolia*, *Saxifraga oppositifolia*, *Salix arctica*, and *Carex misandra* dominate, and in low places *Carex misandra*, *Dryas integrifolia*, *Salix arctica* and *Cassiope tetragona*.

With the increase in cryo-aridity, aggravated by the effect of föhn winds, cryo-xerophilous associations are often found on dry habitats, where the snow disappears early; there occur arctic–alpine species such as *Carex hepburnii*, *Kobresia bellardii*, or *Carex rupestris* with an admixture of *Poa hartzii*, *Festuca hyperborea*, *Puccinellia bruggemannii*, *P. angustata*, *Braya purpurascens*, *B. thorild-wulffii*, *Armeria maritima* s.l., and *Caloplaca bracteata*, *Toninia coeruleonigricans*, etc. ('dry steppe', Beschel, 1963*b*).

Where the vertical belts are not affected by inversions, the vegetation forms, when ascending the mountains above 350 m (Svoboda, 1972: 150), a belt of oro-polar deserts, dominated by cryptogams and scattered angiosperms, *Saxifraga oppositifolia*, *S. cernua*, *S. caespitosa*, *Cerastium alpinum*, *Minuartia rubella*, *Papaver radicatum* s.l., etc., mainly in the shelter of large boulders.

Temperature inversion (Porsild, 1955; Holmen, 1957; Beschel, 1963*a*) is often connected with local air currents of föhn winds at this elevation, and sometimes with cold winds penetrating from the sea along the foothills of the mountains and blowing only in the lower stratum of the atmosphere. This phenomenon was encountered by Beschel (1963*a*: 205) on Axel Heiberg Island. He described an inversion occurring on 7 August 1962, when at the foot of the mountains at sea level there was a northwesterly wind and a temperature of 4.6 °C, while at an altitude of 360 m it was calm and the temperature was 10 °C.

In the subnival–alpine belt situated higher up the only flowering plants left are *Phippsia algida* and *Saxifraga rivularis* (Beschel, 1961). In the nival belt on the nunataks there are only crustose lichens such as *Acarospora chlorophana*, *Lecidea dicksonii*, and foliose ones such as *Umbilicaria leiocarpa* and *U. decussata* (Beschel, 1961: 183).

The Northeastern Greenlandic subprovince of the arctic tundras

This subprovince is distinguished first and foremost by the extreme aridity of its climate (cf. above, p. 120) reaching its climax in Peary Land. The aridity is aggravated by the effect of föhn winds in the interior parts of the major fjords and only slightly eased in the narrow coastal belt. The development of the vegetation depends almost entirely on the supply of meltwater from snow and ice. Here a number of unique associations are met with as, e.g., those developed on the so called 'earth glaciers' (Sørensen, 1937: 121) or solifluction lobes near glaciated areas. Besides *Dryas integrifolia*, *Dryas punctata* occurs (Porsild, 1957*a*) which has a differentiating role for this subprovince in comparison with the former one. The occurrence of Atlantic (European–Greenlandic) species in the composition of the associations is typical (*Carex parallela*, etc.). The amount of the European element in the flora here is, however, strongly reduced in comparison with that in southern Greenland. *Braya intermedia* and *Saxifraga nathorstii* are endemic to the subprovince.

Within the subprovince two districts can be distinguished: Northeastern Greenland, and Peary Land.

The district of Northeastern Greenland. The landscape of this district is dominated by open associations on stony, often basic, habitats, the poverty of which is aggravated by the aridity, as has been shown by a number of investigators (Warming, 1889, 1928; Hartz & Kruuse, 1911; Lundager, 1912; Seidenfaden, 1931; Sørensen, 1933, 1935, 1937, 1941, 1945; Gelting, 1934; Seidenfaden & Sørensen, 1937; Oosting, 1948; Schwarzenbach, 1961; Raup, 1965, 1969; etc.). Besides mosses such as *Schistidium gracile*, *Rhacomitrium lanuginosum*, etc., and lichens such as *Cetraria nivalis*, *Alectoria ochroleuca*, *A. nigricans*, etc., there are only isolated individuals of *Saxifraga oppositifolia*, *Poa abbreviata*, *Carex hepburnii*, and rare *Erigeron compositus*, *Dryas punctata*, *Draba subcapitata*, *Papaver radicatum* s.l., *Saxifraga caespitosa*, *Minuartia rubella*, *Arenaria pseudofrigida*, *Poa glauca*, *Festuca brachyphylla*, *Draba cinerea*, and *Lesquerella arctica*.

In habitats which may be considered analogous to the zonal, mesic ones, *Cassiope* tundras are widely distributed giving a colorful aspect to the landscape because of their brownish tones.

Cassiope tetragona belongs to the group of species which in various parts of the Arctic display interesting changes in ecology and level of activity. In the Anabar basin (Sochava, 1934*b*: 281) and farther to the east, e.g. in the lower part of the Lena basin (Shamurin, 1966: 47), this species, found also in zonal, mesic habitats, forms well developed chionophytic associations, where there is a combination of deep snow and loose, skeletal soils provided with good drainage. According to Polunin (1948: 109–11), *Cassiope tetragona* is moderately chionophilic on Baffin Island, but shows none of this character in Greenland. Thus, in subarctic Greenland, there is no mention of *Cassiope tetragona* in any of the detailed descriptions of snow layer vegetation by Böcher (1933*b*, 1954, 1963*b*). There the species is part of the zonal dwarfshrub tundras on habitats which are moderately snow-covered. Its role in the vegetation cover increases both in a northerly direction and towards higher altitudes. Finally, in the arctic tundras of northeastern Greenland, it becomes the basic dominant in localities analogous to zonal, mesic habitats. It is interesting to note that *Cassiope tetragona* does not form nival associations in the mountain tundras on the Kola Peninsula nor in the Polar Urals.

In the district of Northeastern Greenland in *Cassiope*-tundras a closed cover is formed by *Cassiope tetragona* associated with *Salix arctica*, *Carex bigelowii*, *Poa arctica*, *Hierochloë alpina*, *Luzula confusa*, *L. nivalis*, *Polygonum viviparum*, *Papaver radicatum* s.l., *Pedicularis hirsuta*, and other plants. Mosses such as *Aulacomnium turgidum*, *Dicranum fuscescens*, *Hylocomium alaskanum*, *Rhacomitrium lanuginosum*, *Ptilidium ciliare*, etc. have been noted, and lichens such as *Stereocaulon alpinum*, *Dactylina arctica*, *Cetraria islandica*, *C. nivalis*, and *Nephroma arcticum*.

In the interior parts of the fjords on habitats where in more southerly areas there would be shrub thickets, i.e. in localities protected from wind and kept moderately wet all summer by water flowing downslope, dwarfshrub tundra communities are formed with dominant *Vaccinium uliginosum* ssp. *microphyllum* (Sørensen, 1937: 21). In addition to this blueberry, in the closed cover *Salix arctica*, *Rhododendron lapponicum* occur and in the southern part of the district also *Empetrum hermaphroditum*. There is usually an admixture of *Cerastium alpinum*, *Polygonum viviparum*, *Poa glauca*, *P. arctica*, *Hierochloë alpina*, *Carex rupestris*, *Kobresia bellardii*, *Saxifraga caespitosa*, *S. nivalis*, rarely of *Draba glabella*, *Cystopteris fragilis* and *Tofieldia coccinea*, and in drier places of *Dryas punctata*.

Small patches of cryo-xeromesophytic meadow-like communities are quite typical on slopes in these cryo-arid areas, moistened during spring by meltwater but completely desiccated during summer. Here a closed cover is formed mainly by *Kobresia bellardii* associated with *Carex hepburnii*, *C. rupestris*, *Dryas integrifolia*, *Saxifraga oppositifolia*, *Minuartia rubella*, *Draba cinerea*, and *Papaver radicatum* s.l. On slopes with a southern exposure and a moderate snow cover in habitats where some water flows downslope all summer small patches of herbaceous–mesophytic meadow-like communities are found. They are very colorful in composition and sometimes mixed with mosses. Typical components are *Taraxacum arcticum*, *Ranunculus nivalis*, *R. sulphureus*, *Trisetum spicatum*; *Silene acaulis*, *Oxyria digyna* and *Gastrolychnis apetala* are abundant, as well as *Saxifraga cernua*, *Salix arctica*, *Polygonum viviparum*, *Pedicularis hirsuta*, *Cerastium alpinum*, *Stellaria edwardsii*, *Minuartia rubella*, *Saxifraga oppositifolia*, *Potentilla hyparctica*, *Carex misandra*, *Luzula confusa*, *L. nivalis*, and *Juncus biglumis*. Where the soils are acidic, *Salix herbacea*, *Draba lactea*, and *Carex bigelowii* are met with.

Special associations have developed on the socalled 'earth glaciers' (Sørensen, 1937: 121) or solifluction lobes. These are formed by a deluvium of fine soil, slowly slipping downhill and spreading out as an even, slightly sloping surface. In the interior, drier areas, their motion is very slow and there terraced solifluction lobes are formed, the frontal edges of which are pushed up in an abrupt berm above the main ground to a height of 0.5 to 0.75 m. On the surface of these solifluction lobes representatives mainly of grasses and herbs will develop: *Arctagrostis latifolia*, *Alopecurus alpinus*, *Poa arctica*, *Carex bigelowii*, *C. misandra*, *Eriophorum triste*, *Luzula confusa*, *L. nivalis*, and *Juncus biglumis*. Where the substrate is strongly basic, *Carex atrofusca*, *Kobresia simpliciuscula*, *Juncus triglumis*, *Carex maritima* and sometimes

Eriophorum callitrix are found. *Carex glacialis* is typical for the southern part of the district. Also *Gastrolychnis apetala*, *Polygonum viviparum*, *Potentilla hyparctica*, *Stellaria longipes*, *Draba lactea*, *Salix arctica* and sometimes *Equisetum arvense* and *E. variegatum* occur there. *Pedicularis flammea* and *Eutrema edwardsii* are common on basic soils. On the outer edge of a stable solifluction lobe an especially luxuriant graminoid vegetation may develop. In it dominate *Poa arctica*, *P. glauca*, *Hierochloë alpina*, *Festuca brachyphylla*, *Carex bigelowii*, *C. misandra*, *C. hepburnii*, *C. rupestris*, *C. capillaris*, *Kobresia bellardii* and *Luzula confusa*. *Polygonum viviparum*, *Stellaria longipes*, *Pedicularis flammea*, *Draba alpina*, *D. fladnizensis*, *Armeria maritima* as well as *Salix arctica*, *Dryas punctata* and sometimes also *Rhododendron lapponicum* are also met with here.

In localities where the snow melts late open associations are found composed of scattered specimens of *Phippsia algida*, *Sagina intermedia*, *Saxifraga foliolosa*, *S. cernua*, *S. tenuis*, *Cardamine bellidifolia*, *Luzula confusa*, *Pedicularis hirsuta*, *Draba lactea*, and *D. micropetala*. The latter two species are lacking on definitely basic soils. *Saxifraga rivularis*, *Oxygraphis glacialis* and *Cochlearia groenlandica* are rare occurrences.

Chamaenerium latifolium is very typical on the sandy and pebbly terraces along rivers and their deltas, which are flooded by high water during the spring. The species often dominates physiognomically. Together with it occur *Saxifraga nathorstii*, *Polemonium boreale*, *Poa alpigena* ssp. *colpodea*, *Carex maritima*, *Equisetum arvense*, and *E. variegatum*, but also *Salix arctica*, *Polygonum viviparum*, *Oxyria digyna*, *Stellaria edwardsii*, *Draba lactea*, *Gastrolychnis apetala* and sometimes *Cerastium regelii*.

Because the coast of Greenland is to a large extent formed by steep cliffs plunging straight down into the sea, sedimentary marine terraces covered by loam are usually found only at the mouths of rivers, where they form their natural banks. Although these localities are protected by surrounding mountains from severe deflation, snow does not accumulate there, and during the summer these flat plains are so severely dried out that the vegetation is represented by only open associations of scattered individuals distributed far and wide. *Puccinellia angustata*, *Poa hartzii*, *Potentilla pulchella*, *Braya purpurascens* and *B. thorild-wulffii* are most typical here. Sometimes *Poa abbreviata*,

P. glauca, *Draba subcapitata*, *D. cinerea*, *D. macrocarpa* (*D. bellii*) and *Gastrolychnis triflora* are observed, occasionally also *Potentilla nivea*. The dry, pebbly ridges, separating the lagoons from the sea, are almost devoid of vegetation. *Potentilla pulchella* appears to be the most widespread species. The silty low beaches and spits, stretching mainly along the edges of the deltas, are overgrown with *Puccinellia phryganodes* and *Stellaria humifusa*. Also *Carex ursina* occurs here, and at a higher level *C. subspathacea*, *Cochlearia groenlandica*, *Sagina intermedia*, and *Saxifraga rivularis*.

Because of the mountainous relief, but also due to the aridity of the climate, mires occupy only small areas. As a rule *Eriophorum scheuchzeri* and *E. triste*, rarely *Carex stans*, associated with mosses such as *Calliergon sarmentosum* and *Drepanocladus revolvens*, dominate. Sometimes *Arctagrostis latifolia*, *Carex atrofusca*, etc. help form the sward. Around small waterholes on the shore *Eriophorum scheuchzeri* is abundant, and in the water occurs *Pleuropogon sabinei*, and sometimes also *Ranunculus hyperboreus*. In the southern parts of the district there is, in addition, *Hippuris vulgaris*. In localities constantly irrigated by melting snow-water and subject to solifluction, chionophytic mineral mires have developed with relatively large tussocks, formed by *Deschampsia brevifolia* and *Carex misandra*. Also *Cerastium regelii*, *Saxifraga foliolosa*, *S. cernua*, *Phippsia algida*, *Juncus biglumis*, *Alopecurus alpinus*, and *Oxyria digyna*, *Luzula nivalis*, etc. are common here. Where the solifluction is very pronounced, the vegetation is represented by scattered patches of a dense turf, 'floating' in the wet loam. On these 'islands' *Arctagrostis latifolia*, *Eriophorum triste*, *E. scheuchzeri*, *Carex misandra*, *Puccinellia vahliana* and *Poa alpigena* ssp. *colpodea* are predominant. *Saxifraga foliolosa* and *Salix arctica* occur and, in lesser quantities, also *Carex maritima* and *Equisetum arvense*.

The Peary Land District. Previously (Aleksandrova, 1951) I placed Peary Land with the polar deserts, as Korotkevich (1967) now does. However, the data published by Holmen (1957), who studied the flora and vegetation of this area during four summers and one winter, furnish convincing evidence that this area belongs to the northern belt of the arctic tundras. This notion is supported also by the data of other investigators (Ostenfeld, 1925, 1927; Holmen, 1955, 1957; Fristrup,

1952; Fredskild, 1966). The individuality of the area is connected with its extensive cryo-aridity. In spite of the fact that the elevation of southern Peary Land reaches 1100 m, there is a complete absence of glaciers due to the extreme scarcity of precipitation. In part this appears to be a consequence of the air currents of dry föhn winds often reaching a force of strong gales. The total yearly precipitation amounts to a mere 25 mm (Holmen, 1957), the major part of which falls during winter, but during summer it comes in the form of either snow or freezing rain, which never soaks the ground to a depth of more than 1 cm. During winter hurricane force winds blow snow from a considerable distance and pile it up against slopes as enormous snow-drifts. Meltwater coming from this snow during the summer provides the only water for the plants. The interior area of Peary Land is especially dry; due to this 97–98% of the surface is completely lacking vegetation (Holmen, 1957: 138), but because the July temperature is relatively high, 6.0–6.4 °C, a tundra or mire vegetation has developed where the soil is able to hold some water.

According to data published by Holmen, *Dryas* tundras have developed in habitats more or less close to zonal, mesic conditions. In slightly moist places with a moderate snow cover *Dryas integrifolia* forms a cushion vegetation together with *Carex misandra* and *Oxyria digyna*. In somewhat drier, moderately snow-covered localities it grows together with *Carex hepburnii*, but in very dry localities with a poor snow cover it forms associations with *Kobresia bellardii*. *Cassiope* tundras are met with in patches, the surface of which rarely exceeds a few square meters. The vegetation is closed; together with *Cassiope tetragona* grow *Salix arctica*, *Carex misandra*, *Saxifraga oppositifolia*, *Carex hepburnii*, *Poa arctica*, and *Silene acaulis*, and mosses such as *Hypnum revolutum*, *Aulacomnium turgidum*, *Orthothecium killiasii*, and *Tortula ruralis*. In the very dry areas of the interior *Cassiopeta* are lacking.

On the boulder fields, except for in the extremely dry, interior area, patches of *Rhacomitrium lanuginosum* occur mixed with *Dicranoweisia crispula* and species of *Cetraria*, *Alectoria* and *Usnea* as well as a small number of angiosperms, *Potentilla hyparctica*, *Luzula confusa*, etc. More often there are scattered individuals of *Potentilla hyparctica*, *Luzula confusa*, sometimes *Poa glauca*, *Trisetum spicatum*, *Carex rupestris* and *Saxifraga oppositifolia* among the boulders, mixed with a few mosses, *Orthothecium killiasii*, etc., and lichens, *Cetraria nivalis*, and others.

On the slopes where there are hollows between the boulders on ledges and in similar situations, there are patchy fragments of various herb and herb–grass vegetation; on dry soils there are various combinations of *Lesquerella arctica*, *Poa abbreviata*, *Gastrolychnis triflora*, *Draba cinerea*, *Dryas integrifolia*, *Saxifraga oppositifolia*, and *Papaver radicatum* s.l., etc.; on more moist soils, of *Braya thorild-wulffii*, *Ranunculus sulphureus*, *Draba lactea*, *D. oblongata*, *Carex rupestris*, *Festuca hyperborea*, *Braya purpurascens*, *Carex misandra*, *Oxyria digyna*, *Taraxacum pumilum*, etc. In the southernmost parts of Peary Land it is possible to find fragments of associations dominated by *Vaccinium uliginosum* ssp. *microphyllum*. Where the snow melts late, associations are formed of *Oxyria digyna*, *Saxifraga oppositifolia*, *S. nivalis*, *Draba oblongata*, *D. macrocarpa*, *Ranunculus sulphureus*, *Luzula nivalis* and *Alopecurus alpinus*, and on pebbly-sandy riverbeds open complexes occur of *Puccinellia vahliana*, *Braya purpurascens*, *Draba macrocarpa*, *Alopecurus alpinus*, *Deschampsia brevifolia*, *Poa hartzii*, *Gastrolychnis apetala*, *Silene acaulis*, and *Chamaenerium latifolium*, etc. *Phippsia algida*, *Puccinellia andersonii* and *P. angustata* occur on marine, low, silty beaches and spits and pebbly terraces.

Mires with *Carex stans* and mosses such as *Drepanocladus brevifolius* are met with in the shallow depressions in the valleys, which are kept constantly moist by meltwater from the big snowdrifts all summer long. It is paradoxical that these occupy the largest surface in the interior, extremely dry area. Up to 2–3% of the surface covered by vegetation consists of mires, which are the basic source of food for the musk-oxen (Holmen, 1957: 109). *Carex stans* forms a dense, almost closed sward, sometimes together with *Eriophorum triste*. Along brooks and in the very wet hollows with running water but also on alluvial river terraces there are also mires, sometimes occupying relatively large surfaces and dominated by *Eriophorum scheuchzeri* and *Drepanocladus brevifolius*. Where the water is stagnant, *Eriophorum triste* is often the dominant species. It grows around lakes and in depressions together with *Carex stans*, *Ranunculus sulphureus*, *Polygonum viviparum* and other species. Along brooks there are often small, mossy, mineral mires with *Bryum obtusifolium*, *Drepanocladus revolvens*, *Tomenthypnum nitens*, *Meesia triquetra* and *Aulacomnium turgidum* associated with some *Cerastium regelii* and *Alopecurus alpinus*. *Pleuropogon sabinei* and *Ranunculus hyperboreus* grow often in the water along beaches and in shallow lakes; at the water's edge *Cerastium regelii* can be found.

The vertical belts of the vegetation are affected by the temperature inversions in Peary Land. The Danish expedition, of which Holmen was a member, studied the vertical belts in southern Peary Land on the slopes of a plateau formed by rows of terraces up to an elevation of 1100 m. The lower belt was found to have an extremely poor vegetation with hardly any plant associations, except for small patches of *Dryas integrifolia* and *Carex hepburnii* in localities kept moist by melting snow water. No *Cassiope tetragona* was noted there. However, at an elevation of 400 m fragments of *Cassiope* tundras appear, and on the boulder fields *Potentilla hyparctica*, *Poa glauca*, *Trisetum spicatum* and *Luzula confusa* occur in association with mosses. There are very small patches of an association consisting of *Carex stans* and *Eriophorum triste*, as well as fragments of a cover with *Dryas* associated with *Kobresia bellardii*. Above 700 m the vegetation is distinctly impoverished and belongs in fact to the oro-polar deserts, where only individual specimens of plants hide among the boulders.

The Spitsbergen autonomous district of the arctic tundras

Except for the island of Nordaustlandet, which is situated in the southern belt of the polar deserts, Spitsbergen belongs to the northern belt of the arctic tundras. I have earlier united Spitsbergen with northern Greenland into one province (Aleksandrova, 1971*b*) on the basis of some common characteristics in the vegetation, particularly the wide distribution of *Cassiope* tundras. Summerhayes & Elton (1923: 248) have also commented on the similarity in physiognomy between the vegetation of Spitsbergen and northeastern Greenland, while Polunin (1945: 88) has pointed out its similarity with Ellesmere Land. However, a number of guiding traits, on which I am basing this book, cause me to revise my former ideas in respect to the geobotanical position of Spitsbergen. The presence on Spitsbergen of *Salix polaris* speaks against uniting it with Greenland. This species is not only widely distributed, but dominates together with mosses in the zonal, mesic phytocoenoses on loamy habitats. The lack of *Salix arctica* on Spitsbergen is also important, while in arctic Greenland it is the typical high-arctic species. There is no *Dryas integrifolia* on Spitsbergen; it is completely exchanged for *D. octopetala*, which forms widely distributed tundras here. These islands are also characterized by a damp, cryo-humid climate in contrast to the extreme cryo-aridity of northern Greenland.

On the other hand, it is impossible to accept the ideas of Eurola (1971*b*: 103), who unites Spitsbergen with Novaya Zemlya. On the latter islands there are none of the *Cassiope tetragona* communities which are so typical of Spitsbergen; on Novaya Zemlya *Deschampsia brevifolia* s.l. is an aggressive basic species and one of the dominants in the zonal associations, while on Spitsbergen it is not very conspicuous. The complete lack of Siberian species in the composition of the flora of Spitsbergen, which give to Novaya Zemlya the particular 'Siberian coloration', is also important. Thus we find on Spitsbergen the amphi-atlantic species *Carex bigelowii* ssp. *bigelowii* instead of *C. ensifolia* ssp. *arctisibirica* on Novaya Zemlya, etc. Therefore I now separate Spitsbergen on the basis of the characteristics enumerated above as an autonomous geobotanical district.

Spitsbergen is the best known area in the Arctic from a geobotanical point of view (Walton, 1922; Summerhayes & Elton, 1923, 1928; Michelmore, 1934; Scholander, 1934; Dobbs, 1939; Acock, 1940; Triloff, 1943; Polunin, 1945; Hadač, 1946, 1960; Rønning, 1963, 1964, 1965, 1969, 1970; Lid, 1967; Hofmann, 1968; Eurola, 1968; 1971*a*, *b*; Kuc, 1969; Philippi, 1973; etc.). This mountainous country with elevations up to 1717 m (Mt Newton) is more than 50% covered by glaciers. The mountains reach all the way to the sea, sometimes the cliffs rise straight up from the ocean, but often a narrow belt of pebbly or sandy–pebbly terraces have developed at their feet. There are hills or rolling plains only in a few areas. Fjords cut deeply into the land on the western and northern sides of the island of West Spitsbergen. The western coast is washed by the Gulf stream, which together with föhn winds makes possible the development of a relatively rich flora and vegetation in relation to the high latitude in the Isefjord area. Extrazonal 'islands' of vegetation are found, here representing the southern arctic tundras. At the same time, the major part of Spitsbergen belongs to the northern belt of the arctic tundras. Föhn winds have a warming effect, which favors the vegetation in all the areas of the Arctic. The drying effect of these winds, where the climate is already dry, e.g. in northern Greenland, can lead to a strong desiccation of the vegetation, but on Spitsbergen with its humid climate, frequent fogs and low clouds, the winds have a favorable influence on the vegetation.

Arctic and arctic–alpine species with circumpolar or almost circumpolar areas dominate in the flora of Spitsbergen. There were apparently no Ice Age refugia and endemism is weakly expressed (cf. *Puccinellia svalbardensis*; Rønning, 1970). The amphi-atlantic element is abundantly represented; together with species of European distribution, this element constitutes 18% of the flora.

Because of the predominantly mountainous relief and stony ground, there are few zonal, mesic habitats, although the latter are found and not too rare, e.g. on the Reindyrsflua (Reindeer Peninsula) on the northern tip of the West Spitsbergen Island. This area is represented by a slightly undulating plain, the elevation of which in its central part does not exceed 60 m. The bedrock is covered by loose sediments, mostly bouldery loam. From here Summerhayes & Elton (1928: 215–20) have described a willow (*Salix polaris*)–herb–moss tundra. Variants of the zonal associations with dominating *Tomenthypnum nitens*, *Drepanocladus uncinatus*, and *Salix polaris*, frequently also with *Saxifraga oppositifolia*, and not rarely with an admixture of *Dryas octopetala* have been described by a number of investigators (Michelmore, 1934: 161; Hadač, 1946: 155–7; Rønning, 1965: 23–7; Eurola, 1968: 26–7; Hofmann, 1968: 18; Philippi, 1973: 53–8); with the exception of the first of the authors enumerated, all have published detailed tables containing lists of the angiosperms as well as the cryptogams.

Dryas tundras with *D. octopetala* are very characteristic for Spitsbergen. These occupy the stonier and drier habitats in comparison with the willow–herb–moss tundras and are considerably more widespread, except in some areas especially affected by the fog and cooled by the icebergs drifting along the eastern shores of the archipelago. The coverage by *Dryas* associations varies greatly, but closed ones are also found, the socalled '*Dryas* heath' according to some authors. Detailed lists of the species composing the *Dryas* tundras have been published (Summerhayes & Elton, 1928: 228; Hadač, 1946: 158–60; Rønning, 1965; Eurola, 1968: 16–17, 22–3). Rønning, using the Braun-Blanquet methods, distinguishes four associations included under the alliance *Dryadion* Du Rietz 1942. In addition to *D. octopetala*, which has a dominant position or shares it with mosses such as *Tomenthypnum nitens*, *Hypnum bambergeri*, *Drepanocladus uncinatus*, and *Oncophorus wahlenbergii,* there are abundant *Salix polaris*, *Saxifraga oppositifolia*, and sometimes *Cassiope tetragona*; in drier habitats *Carex rupestris* participates as well as such lichens as *Stereocaulon alpinum*, *Cetraria islandica*, and *Ochrolechia frigida*.

The *Cassiope* tundras are mainly distributed in the interior parts of the major fjords. In some areas they cover large surfaces on the lower parts of the mountains and occur on old, raised, pebbly beach terraces. At an elevation of 60–90 m they usually give way to *Rhacomitrium* tundras. Brief descriptions of *Cassiope* tundras are found in a number of works (Summerhayes & Elton, 1923; Polunin, 1945; etc.); in other publications, they are given a complete, floristic evaluation in the form of detailed lists (Summerhayes & Elton, 1928: 233–4; Hadač, 1946: 160–2; Rønning, 1965: 28–33; Lid, 1967: 461 – the *Cassiope tetragona* synedrium; Eurola, 1968: 22–3). Hadač distinguished the association *Cassiopetum tetragonae spitsbergense* Hč. The coverage of *Cassiope tetragona* amounts to 70–80%, according to the publications by Hadač and Eurola. It is accompanied by *Salix polaris*, *Luzula confusa*, *Dryas octopetala*, *Saxifraga oppositifolia*, and other species. In the moss tier the most common species are *Hylocomium alaskanum*, *Tomenthypnum nitens*, *Drepanocladus uncinatus*, as well as *Rhacomitrium canescens*, *Aulacomnium turgidum*, *Polytrichum alpinum*, and sometimes also *Ptilidium ciliare*. Lichens, *Stereocaulon alpinum*, *Cetraria islandica*, *Ochrolechia frigida*, *Rinodina turfacea*, *Psoroma hypnorum*, and so on, play a subordinate role. In the area of Wijde Fjord there are small patches of *Cassiope tetragona* with an admixture of *Empetrum hermaphroditum* and some small fragments of a dense brush of the latter species alone.

The associations described above occupy only a small part of the surface of Spitsbergen, because of the predominance of mountainous relief and rocky ground.

On patterned ground with stone nets and stone rings the stones are more or less covered by *Rhizocarpon geographicum*, *Lecidea dicksonii*, *Lecanora polytropa*, *Andreaea alpestris* and other epilithic species. Between the stones and on the loamy–skeletal soils in the center of the stone rings, scattered or forming groups of plants,

Rhacomitrium lanuginosum, *R. canescens, Drepanocladus uncinatus*, *Ditrichum flexicaule, Cetraria nivalis*, *Stereocaulon alpinum*, *Saxifraga oppositifolia*, *Papaver dahlianum* (*P. radicatum* p.p.), *Cerastium alpinum*, *Saxifraga caespitosa*, *Luzula confusa*, *Pedicularis hirsuta*, *Silene acaulis, Salix polaris* and, occasionally, *Dryas octopetala* are found, and, in the inner parts of the fjords, also small quantities of *Cassiope tetragona.*

The boulder fields in the areas, constantly affected by fog, are sometimes blackened by a thick coat of lichens, such as *Umbilicaria proboscoidea*, *U. arctica*, *U. erosa*, *U. hyperborea*, *Cetraria hypatizon*, *Parmelia omphalodes*, *Alectoria pubescens*, *Haematoma ventosum*, and others. The crannies between the boulders are colonized by *Rhacomitrium lanuginosum*, *Hylocomium alaskanum*, *Sphaerophorus globosus*, *Stereocaulon alpinum*, *Cladonia pyxidata*, *Cetraria islandica*, *Luzula confusa*, *Saxifraga caespitosa*, *S. nivalis*, and *S. oppositifolia*. In the drier areas at the inner parts of the fjords crustose lichens predominate and the angiosperm composition is richer; *Cystopteris fragilis* is found there. There are many moving screes almost devoid of vegetation (Eurola, 1971*a*), but those relatively stable and with a favorable snow cover are frequently covered by a blanket of *Rhacomitrium lanuginosum* associated with some other mosses, lichens, and a small number of angiosperms, such as *Luzula confusa*, *Salix polaris*, *Saxifraga oppositifolia*, *S. caespitosa*, *S. cernua*, *S. nivalis*, *Cerastium alpinum*, *Papaver radicatum* s.l. (*P. dahlianum*), *Draba subcapitata*, *Potentilla hyparctica*, and others (Summerhayes & Elton, 1928: 231–2; Hadač, 1946: 145–6, – *Rhacomitrietum lanuginosi spitsbergense* Hč; Eurola, 1968: 16–17, 1971*a*: 19; etc.).

Because of the mountainous character of the country mires do not occupy a very large area of Spitsbergen. However, Eurola (1971*b*) has described one such from the Belsun Bay area covering about 20–25 km^2 in a river valley and belonging to the type of mires with flat-topped peat mounds so typical of an oceanic, arctic climate. In respect to their genesis the mounds seem similar to the ones on Novaya Zemlya, but are here represented by a more high-latitude, impoverished variant. Against a background of depressions the mounds themselves reach 30–50 cm in height, and owe the origin of their growth to lenses of ice under the 8–18 cm thick peat layer. Alongside the just developed mound old ones are falling apart, because of the melting of the ice, in turn due to the decomposition of the peat on the top of the mounds. Eurola (1971*b*: 96) has furnished a descriptive illustration in his Table 11. In the depressions between the mounds there are such mosses as *Calliergon sarmentosum* and *Drepanocladus exannulatus*, and graminoids such as *Dupontia fisheri* with an admixture of *Eriophorum scheuchzeri*. Around the edges of the younger mounds peatmosses predominate, mainly *Sphagnum subfulvum*, with less *S. fimbriatum*, *S. balticum*, and *S. squarrosum*. *Drepanocladus uncinatus* together with some *Aulacomnium turgidum* cover the tops. On the old mounds dying mosses predominate, while the living ones make up only about 20% of the cover (*Aulacomnium turgidum*, *Dicranum angustatum*, *Drepanocladus uncinatus*, etc.). Some angiosperms, single individuals of *Cochlearia arctica* and *Ranunculus spitsbergensis*, occur rarely. This type of mire occurs very rarely only in some inner parts of the big fjords. Another type is common: the small mineral mire without peat. It is met with along brooks, frequently on a stony substrate, at small waterholes, along the bottom of well-watered gullies, on various parts of slopes moistened by water from melting snow, etc. Such mires have been described by many authors (Summerhayes & Elton, 1923: 248, 253–4, 1928: 225–6, 237; Michelmore, 1934: 161; Dobbs, 1939: 140; Lid, 1967: 486; Hofmann, 1968: 31–2; Eurola, 1971*b*: 92: etc.). Here a small-hummocky sward of mosses has formed: *Calliergon sarmentosum* and other species of *Calliergon*, *Drepanocladus brevifolius*, *D. revolvens*, *Campylium stellatum*, *Orthothecium chryseum*, *Bryum obtusifolium* and other species of *Bryum*, *Polytrichum alpinum* (on top of the small hummocks), *Drepanocladus uncinatus* (also on top of the small hummocks), etc. There are many liverworts: *Orthocaulis quadrilobus*, *Blepharostoma trichophyllum*, *Cephaloziella rubella*, *Tritomaria quinquedentata*, and others. Lichens such as *Peltigera aphthosa*, *P. malacea*, and others are sometimes met with. The coverage by vascular plants amounts to 30–50% according to Hadač and Eurola. The most common are *Alopecurus alpinus*, *Eriophorum scheuchzeri*, *Salix polaris* (on top of the small hummocks), *Ranunculus sulphureus*, and also *Cardamine pratensis* ssp. *angustifolia*, *Juncus biglumis*, *Dupontia fisheri*, *Saxifraga rivularis*, *S. cernua*, *S. foliolosa*, *Stellaria crassipes*, *Cerastium regelii*, *Cochlearia arctica*, *C. groenlandica*, *Chrysosplenium tetrandrum*, *Equisetum arvense*, *E. variegatum*, and others. On the bottom of very wet gullies, where the vegetation period is shortened due to long-lasting snow, mossy mineral mires have developed, where the angiosperms are almost lacking (Summerhayes & Elton, 1923: 248, etc.).

On wet, low silty beaches and spits halophytic arctic wetlands (marshes) have developed. There are associations of *Puccinellia phryganodes* and *Stellaria humifusa* together with algae such as *Enteromorpha* sp. and *Nostoc* sp., but also communities of *Carex subspathacea* together with *Stellaria humifusa, Bryum ventricosum, B. nitidulum, B. pallescens, Distichium inclinatum*, etc. Farther from the beach they change into more developed associations with participation of some facultative halophytes, such as *Dupontia fisheri*, etc. (Walton, 1922; Dobbs, 1939; Hadač, 1946: 141–2; etc.).

It should also be mentioned that the vegetation on pebbly, marine terraces and their succession (Walton 1922; Summerhayes & Elton, 1923, 1928; Dobbs, 1939; etc.) as well as the vegetation on birdcliffs (Summerhayes & Elton, 1928: 238–47, etc.) on Spitsbergen have been described in detail. Expecially interesting is the direct observation of the course of succession, studied and mapped by Dobbs in 1936 (Dobbs, 1939) 15 years after the same surfaces had been studied and mapped by Walton (1922).

The vertical belts. According to data published by a number of authors (Summerhayes & Elton, 1923: 263–4, 1928: 231; Michelmore, 1934: 165; Hadač, 1946: 132–8; Sunding, 1966; Eurola, 1968: 7, 43, etc.), it is possible to distinguish some vertical belts on Spitsbergen. The belt of impoverished arctic tundras is found at an elevation from 60–120 to 100–200–250 m. The lowest limits are in the coastal areas and on slopes with northern exposures and the upper ones in the inner parts of fjords and on south-facing slopes. In this belt, the scattering of the open associations, so prominent in this landscape, is even more extensive and the composition of the associations is impoverished. At lower altitudes, where there are *Cassiope* tundras, these are intermixed with *Rhacomitrieta*. The lower belt of the oro-polar deserts reaches from 100–250 up to 500 m elevation. Tundra associations are rare, found mainly under favorable conditions (Eurola, 1968: 43); open vegetation predominates. The surface becomes increasingly more denuded, in places almost devoid of vegetation. The flora is extremely poor.

According to Hadač, who studied one of the areas in Isefjord, the number of vascular plant species is diminished by more than half (Hadač, 1946: 134–7 – the 'hill flora'). The upper belt of the oro-polar deserts is situated above 500 m. Tundra associations and even fragments thereof are completely lacking, boulder fields predominate with single, rare individuals of the flowering plants, high-arctic species only, and crustose and foliose epilithic lichens. The growth of the lichens is enhanced by the prevalence of fog at this altitude. The occurrence of *Schistidium gracile* among the mosses is typical. In respect to angiosperms, *Phippsia algida* and *Saxifraga cernua* have been found in the mountains, as high up as 1250 m (Sunding, 1966).

The Bear Island with its cold, maritime climate belongs also to the Spitsbergen district. There the average temperature for the warmest months is 4.5 °C, for the coldest −12 °C. The northern part of the island is a stony plain with shales and sandstones; the southern part is hilly with lakes and some heights reaching 600 m. The flora of the island is extremely poor and 'incomplete'. Most notable is the almost complete lack of any representative for Cyperaceae. The early explorers (Summerhayes & Elton, 1923; Hanssen & Holmboe, 1925) on the whole did not observe any sedges or cottongrasses; later ones (Engelskjön & Schweitzer, 1970) have found *Carex subspathacea* and *C. tripartita* (*C. lachenalii*).

In conditions more similar to zonal, mesic ones herb (*Saxifraga oppositifolia, S. caespitosa*, etc.) – willow (*Salix polaris*) – moss (*Drepanocladus uncinatus*, etc.) tundras have developed; sometimes also *Salix herbacea* and *S. reticulata* are found there. In habitats with much snow accumulation there is a blanket consisting almost exclusively of *Drepanocladus uncinatus.* There are also peculiar mires with an absolute predominance of mosses, e.g. *Calliergon sarmentosum, Drepanocladus brevifolius, D. fluitans, D. vernicosus, Meesia triquetra, Campylium stellatum*, and only a few representatives of herbs, such as *Saxifraga rivularis, S. hirculus, Ranunculus sulphureus.* Most of the island surface is occupied by stony habitats, where the boulders are covered by lichens, *Umbilicaria* spp., *Lecanora* spp., *Lecidea* spp., *Rhizocarpon* spp., and mosses, *Schistidium apocarpum, Dicranoweisia crispula,* and others. Between the boulders patches have developed of a moss–lichen sward with such species as *Rhacomitrium lanuginosum, Polytrichum alpinum, Tortula ruralis, Dicranum bonjeanii, Hylocomium alaskanum, Drepanocladus uncinatus, Pohlia cruda, Ptilidium ciliare, Lophozia* spp., *Cetraria islandica, C. delisei, Cladonia mitis, C. pyxidata, C. gracilis, Sphaerophorus globusus, Stereocaulon paschale, Parmelia omphalodes, Peltigera canina, Ochrolechia frigida*, and others. The composition of angiosperm species is poor, some *Saxifraga oppositifolia, Papaver radicatum* s.l., *Oxyria digyna, Salix polaris* and a few others. All of them occur as single, isolated specimens (Summerhayes & Elton, 1923: 220, 225).

3

The geobotanical regions of the Arctic: the region of the arctic polar deserts

At high latitudes in both the northern and the southern hemispheres a particular type of zonal landscape is found, where the major area of the land as well as the sea around it is covered by ice all the year. The annual radiation balance does not exceed 10–20 kcal/cm^2 and the mean temperature of the warmest month of the year, as a rule, does not go above 2 °C. In areas free from snow and ice the vegetation is scanty, and in places there is none at all; where there is vegetation it is represented only by formations of cryptogams and the angiosperms play an insignificant role or are entirely lacking. As a term for this zonal landscape Passarge (1921) suggested the designation 'cold deserts' for both the Arctic and the Antarctic. Berg (1928) preferred the term 'the glacial zone' for the Arctic zone. Later other terms have been applied, such as 'the snow-bound zone' (Gorodkov, 1935*b*), 'the high-arctic, nival region' (Leskov, 1947), 'the zone of the polar deserts' (Korotkevich, 1958*a*, etc.), 'the region of the polar deserts' (Aleksandrova, 1971*b*).

Gorodkov (1935*b*: 127) named the type of vegetation which is typical of 'the snow-bound zone' in the Arctic, 'an arctic desert'. In 'the snow-bound zone' he included 'only part of the islands in the North Polar Sea; the northern island of Novaya Zemlya, the Franz Joseph's Land Archipelago, Severnaya Zemlya, a part of the Novisiberian Islands, and definitely, the Geral'd Island'. Later, Gorodkov extended his concept of the polar deserts and expressed his opinion that they reached farther south than the area he had originally defined as 'the snow-bound zone'. Thus he distinguished the following three zonal types: (1) in the southernmost part, dwarfshrub – moss 'arctic deserts' with abundant or relatively variable dwarfshrubs (e.g. *Salix arctica*,

S. polaris and *Dryas punctata*); (2) the impoverished dwarfshrub – moss 'arctic deserts' (with *Salix polaris* and *Saxifraga oppositifolia* as the typical dwarfshrubs); (3) in the Far North, herb–moss arctic deserts with an extremely poor angiosperm flora (Sochava & Gorodkov, 1956: 80). The notion here of *Saxifraga oppositifolia* as a dwarfshrub seems irrational, as this species is herbaceous (Warming, 1912; Polozova, 1976). It is, however, only the northernmost of these types, the 'herb–moss arctic deserts with an extremely poor angiosperm flora', which deserves to be regarded as a special type of vegetation different from the tundras. It should actually rather be called a 'herb–moss–lichen polar desert'. The other two cannot be distinguished from the arctic tundras (Aleksandrova, 1957).

My own comparative studies of the structure of the vegetation formations in the Arctic (Aleksandrova, 1962, 1969*b*, 1971*a*, 1977) and data from investigations by other authors (Perfil'yev, 1928; Korotkevich, 1958*a*; Matveyeva & Chernov, 1976) allow me to list the following typical characteristics of the zonal type of vegetation in the polar deserts as it has developed on zonal, mesic habitats, that is, sufficiently well drained habitats with a fine soil and a moderate snow cover, melting off relatively early:

(1) The predominance of cryptogamous components over angiosperms in respect to number of species, coverage, and phytomass. According to available data (see Table 1) the phytomass of the living parts of the cryptograms above as well as below ground compared with that of the angiosperms amounts to 16–19% in the arctic tundras, to 8–23% in the subarctic ones, but to 78% in the zonal, mesic habitats of the polar deserts.

(2) The dominating role of the lichens. In contrast to the zonal tundra phytocoenoses, where the mosses predominate among the cryptogams, the lichens have assumed first place in the polar deserts in respect to both coverage and phytomass (Korotkevich, 1958*a*; Shamurin *et al.*, 1975: 18); the crustose lichen synusiae appear as characteristic.

(3) The particular species composition. The angiosperms in the zonal vegetation are represented by only high-arctic hyperhekistothermal herbaceous plants, e.g., *Phippsia algida, Drabae, Saxifragae, Papaver*, etc. Hyperhekistothermal means plants able

to grow in July temperatures below 2 °C. The dwarf-shrubs occur in the region of the polar deserts (e.g. *Salix polaris*) but are never a part of the zonal communities. Moss species common in the tundras, such as *Hylocomium alaskanum* and others, are here only found as admixtures, and the one most frequently occurring is *Ditrichum flexicaule*, often associated with *Polytrichum alpinum*, etc. Small mosses, such as species of *Bryum, Pohlia*, and *Myurella* are very typical. Among cushion-forming lichens species of *Cetraria* and *Stereocaulon* (*S. rivulorum*, etc.) predominate. There are especially many crustose lichens, e.g. *Ochrolechia, Pertusaria, Toninia, Collema*, etc. The hepatics play an important role (*Cephaloziella arctica*, etc.).

(4) The particular growth form of the plants. The individual specimens of the angiosperms are very small ('miniaturization of the living beings' Matveyeva & Chernov, 1976). Cushion-forming plants predominate, e.g. *Papaver, Drabae*, etc. or those forming small tufts, e.g. *Phippsia algida, Poa abbreviata*, etc. The 'mobility' of the vegetative parts of the plants so typical of the tundras is almost completely lacking here. Peduncles have become extremely short; for instance, *Saxifraga caespitosa* has sessile flowers buried in the leaf cushion. However, the poppy peduncles are not reduced in length but are adpressed to the surface of the leaf cushion, straightening up only slightly at the time of flowering.

(5) The complete loss of a formative role in the association of the angiosperms. In spite of the abundance of mosses and their great ecological importance in the tundras the angiosperms appear as the main association formants in the zonal, mesic phytocoenoses. Norin (1976) has demonstrated experimentally the great formative role of the dwarfshrubs in comparison with that of the mosses in the spotty tundras at Sivaya Maska (the Komi ASSR). In the tundra there the angiosperms furnish the main part of the phytomass (cf. Table 1). Not only do their underground organs form a dense mat in the surface layer of the soil, but they penetrate into the soil also under the patches of bare ground (Fig. 19; see also Aleksandrova, 1962, 1971*a*). In contrast, the angiosperms in the zonal, mesic vegetation in the polar deserts grow as separate individuals and the lichens and the mosses are the main formants of the phytocoenoses. It is important to note that there are no closed root systems here (Fig. 21). This characteristic, on the diagnostic importance of which Perfil'yev (1928: 71) was the first to

Table 1. *Phytomass of the living parts of plants in the polar deserts and the tundras under zonal, mesic and near-zonal, mesic conditions. In each case, the figure above the line is air dried material (g/m²) and that below the line is %.*

Plant association	Total	Above ground mass			Underground Mass (Vascular plants)	Total mass		Literature source
		Total	Mosses and lichens	Vascular plants		Mosses and lichens	Vascular plants	
Moss–lichen–polygonal polar desert; Franz Joseph's Land	$\frac{158}{100}$	$\frac{129}{81}$	$\frac{123}{78}$	$\frac{6}{3}$	$\frac{29}{19}$	$\frac{123}{78}$	$\frac{35}{22}$	Aleksandrova, 1969*b*
Graminoid–willow (*Salix polaris*)–moss northern, arctic tundra; Big Lyakhovsky Island	$\frac{696}{100}$	$\frac{185}{27}$	$\frac{114}{16}$	$\frac{71}{11}$	$\frac{511}{73}$	$\frac{114}{16}$	$\frac{582}{84}$	Aleksandrova, 1958
Moss–herbaceous plant–polygonal northern arctic tundra; Big Lyakhovsky Island	$\frac{379}{100}$	$\frac{119}{31}$	$\frac{70}{19}$	$\frac{49}{12}$	$\frac{260}{69}$	$\frac{70}{19}$	$\frac{309}{81}$	Aleksandrova, 1958
Sedge (*Carex ensifolia* ssp. *arctisibirica*)–dwarfshrub–moss middle subarctic (typical) tundra; E. Taimyr	$\frac{3422}{100}$	$\frac{1081}{31}$	$\frac{604}{17}$	$\frac{477}{14}$	$\frac{2341}{69}$	$\frac{604}{17}$	$\frac{2818}{83}$	Ignatenko *et al.*, 1973
Sedge (*Carex globularis*)–lichen–moss–dwarfshrub subarctic, southern subarctic tundra Koryakskaya Zemlya	$\frac{2878}{100}$	$\frac{491}{17}$	$\frac{241}{8}$	$\frac{250}{9}$	$\frac{2387}{83}$	$\frac{241}{8}$	$\frac{2637}{92}$	Vikhireva-Vasil'kova *et al.*, 1964
Willow–birch–moss southern subarctic tundra; Komi ASSR.	$\frac{2370}{100}$	$\frac{668}{28}$	$\frac{360}{15}$	$\frac{308}{13}$	$\frac{1702}{72}$	$\frac{360}{15}$	$\frac{2010}{85}$	Shamurin, 1970
Dwarf-birch–*Pleurozium*–*Polytrichum*–tundra in the forest tundra; Komi ASSR.	$\frac{5858}{100}$	$\frac{2142}{36}$	$\frac{1325}{22}$	$\frac{817}{14}$	$\frac{3716}{64}$	$\frac{1325}{23}$	$\frac{4523}{77}$	Rakhmanina, 1971

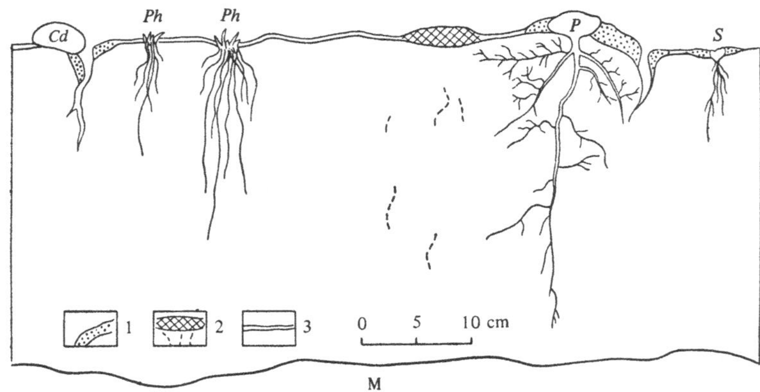

Fig. 21. The underground structure of zonal, mesic phytocoenoses in the polar deserts on the island of Alexandra Land, Franz Joseph's I and (diagram by the author). 1, lichen (*Cetraria delisei*, etc.) – moss (*Ditrichum flexicaule*, etc.) – sward along the cracks; 2, remnants of disintegrating *Cerastium regelii*; 3, crust of crustose lichens (*Pertusaria, Ochrolechia, Collema*, etc.) with an admixture of liverworts (*Cephaloziella arctica*, etc.). *P.*, *Papaver polare*; *C.d.*, *Cetraria delisei*; *S.*, *Saxifraga hyperborea*; *Ph.*, *Phippsia algida*; M., permafrost layer in August.

comment, is always essential from the point of view of phytocoenology, because it indicates a disruption of the phytocoenotic interdependence between the angiosperms in the type of vegetation under discussion.

(6) The complete lack of separate layers. Although layer formation is strongly reduced in the arctic tundras, there are at least two micro-layers in the zonal, mesic associations (Fig. 19), the herbaceous one, 10–15 cm high, and the moss tier. In the polar deserts, however, there are no such separate layers at all. The angiosperms are scattered in the cover formed by the mosses and the lichens and hardly reach above it (Fig. 21).

(7) The particular horizontal structure. The zonal vegetation developed on fine soil has a polygonal structure. The cracks are here formed by desiccation and consolidated by cryogenic processes. In addition to other traits they differ from the polygonal tundras in the small size of the polygons, 20 × 30, 30 × 40, 40 × 50 cm, separated by the narrow cracks. Strips of vegetation consisting of mosses, lichens both fruticose and crustose, and a few angiosperms relate to the edges of the cracks (Fig. 21). The polygonal surface itself is only more or less

covered by a thin crust, about 0.5 cm thick, composed mainly of crustose lichens.

(8) The basic role of the crustose lichens, the liverworts, and the blue-green algae. Although these groups of plants play a role in the composition of the associations in the arctic tundras, their role in the polar deserts has become fundamentally important. Crustose lichens form the typical synusia of the vegetation type under discussion. They are the basic material in the vegetation crust mentioned above and often have the largest projected coverage compared to other groups of plants. They conclude a cycle of micro-succession; the tufts of the higher plants are never long-lived here, they soon die off and are overgrown with mosses, which in turn are completely covered by a layer of crustose lichens. As in the tundras, the liverworts are associated with the moss sward, but here, in addition, they form black crusts on the surface of some substrates for which their growth form, anatomy, and other characteristics are particularly suited (Zhukova, 1973*a*, *b*). Similar black crusts are also formed by lichens (*Collema* spp., etc.), the algal components of which seem to consist of blue-green algae. These black crusts, forming patches measuring from 1 to 3–5 cm in diameter and alternating with the white, yellowish, and grey patches of the crustose lichens belonging to the genera *Ochrolechia, Pertusaria, Toninia*, etc., make this type of vegetation particularly colorful and very typical for the zonal associations of the polar deserts. Microscopic algae also take part in the formation of this cover. Here, as in all the soil microflora, the blue-green algae predominate, while in the arctic tundras the green algae are most important (Novichkova-Ivanova, 1972). The small, spherical, elliptical, or plate-like clumps of *Nostoc* are often associated with the mosses.

(9) The basic type of soil formation. In contrast to those of the tundras the soils here have not developed a humus horizon; humus is present only in 'pits' or 'pockets'. In addition, an almost neutral pH appears typical for soils, and the high saturation of absorbant soil complexes is fundamental together with other characteristics (Mikhailov, 1960; Mikhailov & Govorukha, 1962; Tedrow, 1966, 1968, etc.).

(10) The basic composition of the soil fauna. As shown by investigations in the Cape Chelyushkin area by Chernov (in Matveyeva & Chernov, 1976), many components of living beings which inhabit the

soil and are typical for the tundras, including the arctic ones, are absent from the soil fauna here. The dominant groups of invertebrates, Nematoda, Enchytraeidae, Collembola, chironomid larvae, and so on, are strongly reduced in respect to number of species. Some species are distinguished by very particular morphology and very small forms predominate ('miniaturized animals'). Specific characters have trophic interrelations caused by the particular composition and structure of the vegetation formations in the polar deserts.

The characteristics listed above confirm that at the high latitude of the periglacial Arctic a particular type of vegetation is found, which is different from that of the tundras. The area of zonal, mesic habitats over which this vegetation is distributed may be considered as the special geobotanical region of the arctic polar deserts.

An important characteristic for the region of the polar deserts as a particular geobotanical area is the lack of mires with peat as well as the absence of *Carex stans* – the most typical component of the subregion of the arctic tundras. The boundary of the polar deserts in the Arctic is therefore situated north of the area of *Carex stans*. In conditions where there is a surplus of moisture in the ground, there are high-arctic, mineral mires without peat, with mosses, *Orthothecium chryseum, Campylium* spp., *Bryum* spp., and others, and without or with only a few separate individuals of flowering plants.

The moss–lichen associations with a relatively high degree of coverage as well as the swampy vegetation occupy only a very small part of the surface. Open formations predominate as do habitats completely lacking in plants, where the snow persists past the critical date mentioned. The limestone areas are also completely without life.

The angiosperm flora in the region of the polar deserts is distinguished by strongly negative characteristics; many families typical of the tundras are completely absent, and as a result there is a qualitative difference in comparison with the floristic complexes of the tundra region.

The *Sphagna* are completely absent from the moss flora. Some moss species commonly distributed on the skeletal soils are, on dry soils *Rhacomitrium lanuginosum, Andreaea* spp., and others, and on moist soils *Onchophorus wahlenbergii*, etc. An especially important role is played by those species which in the tundras are met with just as admixtures or have established themselves together with crustose lichens and small liverworts, such as *Gymnomitrium coralloides*, in habitats of decaying turf around the edges of bare spots, on the edges of cracks, and so on, e.g. *Ditrichum flexicaule, Distichium capillaceum, Polytrichum alpinum, Pohlia cruda, Encalypta rhabdocarpa, Myurella julacea, Psilopilum cavifolium*, and others. There are also many active species of *Bryum*.

In the lichen flora – the richest one in this region in respect to number of species – the genus *Cladonia* is almost completely lacking, the only common species being the cup-like ones, mainly of *C. pyxidata*. The presence of *Neuropogon sulphureus* (*Usnea sulphurea*) is typical. This taxon is, like a number of other bi-polar species of lichens, often found in the Antarctic. There are particularly many crustose species of lichens here.

In my opinion the region of the polar deserts is considerably narrower and more fragmentarily expressed in the Arctic than was suggested by Gorodkov in his later works. The southern boundary coincides closely with the July isotherm for 2° C. The region agrees almost completely with Young's 'zone 1' (1971) and with Yurtsev's 'subzone of the periglacial, high-arctic tundras' (in Rebristaya & Yurtsev, 1973). Two belts can be distinguished within the region of the polar deserts, the northern and the southern ones (Fig. 2).

To the *northern belt* belong Franz Joseph's Land, the northernmost part of the Severnaya Zemlya archipelago (the Pioneer, Komsomolets, and other islands), the De Long Islands, Meighen Island and a number of small islands in the northern part of the Canadian Arctic Archipelago, as well as a number of small islands at high latitudes in the North Polar Sea. The angiosperm flora consists of a small number of species, fewer than 50, mainly high-arctic and to a minor degree moderately arctic ones. In respect to the nature of the life-forms and the structure of the vegetation this belt is analogous to the northern ('warmer') subregion of the antarctic polar deserts.

To the *southern belt* of the polar deserts in the Arctic belong the northernmost tip of Novaya Zemlya, the southern part of the Severnaya Zemlya archipelago, the area around Cape Chelyuskin, the group of islands included in the northern parts of the 'barren wedge' of the Canadian Arctic Archipelago, the narrow coastal belt of northern Ellesmere Land and Greenland, the Nordaustlandet of Spitsbergen and the islands in King Charles Land. The number of vascular plants varies. There are, for instance, only 44 in the area of Cape Chelyushkin, 61 on Severnaya Zemlya. The largest number of vascular species, 67, has been observed on the island of Nordaustlandet. Extra-regional plant associations typical of the tundras appear within these areas in localities with favorable conditions. These are either isolated phytocoenoses or the disjunct occurrence of an entire extra-zonal area with a plant cover reminiscent of the vegetation in the northern belt of the arctic tundras, as e.g. on the southern islands of Severnaya Zemlya, where the place in the flora of moderately arctic species is expanded and the activity of the dwarfshrubs – in Eurasia of *Salix polaris*, in the Canadian North of *S. arctica* – is increased. This belt is, thus, a transition zone to the northern belt of the arctic tundras.

The provinces in the region of the arctic polar deserts

While the distinction of province within the tundra region is based on the distribution of plant associations with dominating or characteristic hyparctic or hyparctic–alpine shrubs and dwarfshrubs in the subarctic tundras, or arctic and arctic–alpine dwarfshrubs in the arctic tundras, the diagnostic characteristics in the polar deserts have shifted to the cryptogamous components. The flowering plants have lost their diagnostic value not only because of their always unimportant participation in plant associations, but also in relation to the great uniformity of the circumpolar flora.

A comparison between my own data from Franz Joseph's Land and the data from the area around Cape Chelyuskin, published by Matveyeva (in Matveyeva & Chernov, 1976) reveals that the most conspicuous difference between the vegetation complexes, which in general are very similar in both areas, is displayed in the composition and development of the layer which is formed mainly by the crustose lichens and appears to be the most characteristic phytocoenotic component of the polar desert vegetation. On Franz Joseph's Land with its strongly oceanic climate species of *Pertusaria* and *Ochrolechia* predominate in the lichen crust of the zonal, mesic associations. A combination of white and lightly yellowish patches of these lichens form the 'background' synusia; among them black spots are scattered, formed by a crust of liverworts and lichens with blue-green algal components. Around Cape Chelyushkin, where the climate has taken on the traits of a greater continentality, *Toninia lobulata* is of basic importance, as shown by Matveyeva (in Matveyeva & Chernov, 1976: 300). It forms a crust of greyish color with black apothecia. However, the species of *Pertusaria* play an insignificant role here. There is reason to believe that the lichen crust on the islands of Severnaya Zemlya is of a similar composition. It is noteworthy that none of the species belonging to the genus *Toninia* has ever been included in our collections from Alexandra Land or even been mentioned from any other island of Franz Joseph's Land (Lynge, 1931). The change in the specific composition of the characteristic synusia of crustose lichens is well correlated with a number of additional characteristics in the vegetation.

It is, however, difficult to decide on a division of this region into geobotanical areas at present, because of the very slight knowledge we have of the plant associations in the polar deserts. General descriptions – and even those are few – do not often reflect on the compositional structure of the associations, since the majority of the investigators pay attention only to the higher plants, and – at most – also to mosses and fruticose lichens, but the localities covered by crustose lichens are usually described as 'barren'. The following excerpt from Matveyeva (in Matveyeva & Chernov, 1976: 303) may serve as a good illustration: '. . . the fragments of such polar deserts should only formally be called "barren ground". As a matter of fact, they are always active and filled with communities of typical coenotic relationships. Their surface is usually covered by algae [when somewhat moist – V.A.] or by a lichen crust'.

Preliminarily, the distinction of the following provinces in the region of the polar deserts is suggested: the Barents, the Siberian, and the Canadian provinces.

The Barents province of the arctic polar deserts

Areas distinguished by an oceanic, cryo-humid climate are united within this province: Franz Joseph's Land, the northern edge of Novaya Zemlya, and the Nordaustlandet of the Spitsbergen Archipelago. According to data available from Franz Joseph's Land (Aleksandrova, 1969*b*, 1977) species of *Pertusaria* (*P. glomerata*, *P. octomela*, *P. freyii*, etc.) and *Ochrolechia* (*O. frigida*, *O.* cfr. *tartarea*) dominate in the lichen crust and play the main role in the formation of the delicate vegetation cover on zonal, mesic habitats. Some amphi-atlantic species take part in the composition of the angiosperm flora (*Cerastium arcticum* on Franz Joseph's Land and *C. alpinum* as well as *Draba norvegica* on the island of Nordaustlandet, Spitsbergen).

Franz Joseph's Land. The archipelago of Franz Joseph's Land, consisting of 186 islands, 85% of the surface of which is covered by glaciers, belongs to the northern belt of the arctic polar deserts. Thanks to the intense cyclonic activity the characteristics of an oceanic, arctic climate are strongly expressed. The moist and very cold summers are distinguished by an extremely variable weather with frequent fogs, considerable cloudiness and high relative humidity. The mean July temperature at different places in the archipelago oscillates between −1.2° at Tikhaya Bay and 1.6 °C on Hayes Island; the August temperature varies similarly between 0.6 °C and 0.4 °C. The precipitation falls mainly in the form of snow. The average thickness of the snow cover near sea level is 40–60 cm (Govorukha, 1970: 339). On elevated ground the snow blows away during winter, but collects as thick drifts in depressions and below slopes, where permanent snow fields are formed. The snow masses melt slowly, which leads to a shortening of the vegetative period to which also the coolness of the air and the soil contribute as well as the fact that the ground is soaked with meltwater. The frequently falling but scanty summer precipitation adds to this moisture.

Information on the flora and vegetation is found in a number of publications (Palibin, 1903; Lynge, 1931; Hanssen & Lid, 1932; Størmer, 1940; Mikhailov & Govorukha, 1962; Norvichkova-Ivanova, 1963; Govorukha, 1968, 1970; Aleksandrova, 1969*b*, 1977; Ladyzhenskaya & Zhukova, 1971, 1972; Zhukova, 1973*a*, *b*; Tolmachev & Shukhtina, 1974, etc.). The results of my own investigations made during 1959 from 30 May to 3 September on Alexandra Land will serve as the basic material on the plant formations, their composition and structure.

Polygonal polar deserts appear here as the zonal, mesic plant associations developed on loamy ground under optimal snow conditions. The snow depth reaches 10–25 cm and melts off by 15 June. Along the cracks bordering the polygons, which measure 15×20 to 30×40 cm, there are narrow strips, 1–12 cm wide, of a lichen–moss cover where *Ditrichum flexicaule*, *Polytrichum alpinum*, *Cetraria delisei*, and

Stereocaulon rivulorum dominate. The surface of the polygons is almost completely covered by a colorful crust of crustose lichens. Here a mosaic of patches, 1–5 cm in diameter, is formed consisting of whitish *Pertusaria octomela* and *P. freyii*, pale yellow *Ochrolechia frigida* and bright white *P. glomerata* and *Ochrolechia* cfr. *tartarea*, the latter rare. Against this background there are scattered black spots formed by *Collema* cfr. *ceranicum* as well as black crusts of liverworts, a mixture of *Cephaloziella arctica*, *Gymnomitrium concinnatum*, *Tritomaria scitula*, and others. Small patches of bare ground occupy 10–15% of the entire surface. Odd specimens of very small-size angiosperms are scattered among the sporophytes. *Phippsia algida* dominates together with some *Papaver radicatum* s.l. (*P. polare*), *Saxifraga hyperborea*, *Draba oblongata*, *Cerastium alpinum*, *C. regelii* and a few other species. In general the coverage by the angiosperms amounts to around 3%. In Fig. 21 the open root systems and the lack of tiers are illustrated (cf. also Fig. 19, where a vertical section through a phytocoenosis in the arctic tundra is shown). The coverage by crustose lichens amounts to about 75%.

In habitats where the snow disappears at a similar time but the ground consists of fine soil mixed with stones, the vegetation develops better, because the stones, completely or partly buried under the surface of the fine soil, heat up to 1–1.5 °C above the temperature of the loam during the daylight hours of the warm season. *Cetraria cucullata* is abundant in such places (coverage up to 4%). There are also more angiosperms here, covering another 4% of the surface, and the species variety is increased. Their root systems are, however, still not closed. The coverage by mosses amounts to 20%, that of fruticose lichens up to 15% and that of crustose ones up to 55%.

Optimal snow conditions are, however, found in not more than 3–4% of the island territory free of glaciers. Where the snow disappears later, the plant associations are impoverished due to the shortened vegetative period, but in habitats where the snow is blown away by the winds the vegetation may, in spite of the severe conditions during the winter, cover up to 85–90% of the surface, mainly with lichens. Angiosperms are almost completely lacking. Such areas as these are found on the summits of low hills, the edges of terraces and the upper parts of old beach ridges. In places where boulders, blocks and pebbles are piled up and almost no snow accumulates during winter but the vegetative period starts during the first days of June, the surface of the stones is covered by *Umbilicaria proboscidea*, *U. artica*, *Alectoria pubescens*, *Neuropogon sulphureus*, epilithic, crustose lichens of the genera *Lecidea*, *Lecanora*, *Rhizocarpon* and others, but also by the black moss *Andreaea rupestris.* In the gaps between the stones fragments of a moss–lichen sward has developed, formed by *Alectoria ochroleuca*, *A. nigricans*, *Cornicularia divergens*, *Sphaerophorus globosus*, *Cetraria nivalis*, *C. crispa*, *Rhacomitrium lanuginosum*, *Ditrichum flexicaule*, *Gymnomitrium coralloides*, etc. Angiosperms are usually absent, although sometimes single individuals of *Cerastium arcticum*, *Draba oblongata*, and *D. subcapitata* may be found. When, under similar conditions, there is a floor of flat basalt blocks, the plant associations are also similar, but the coverage does not exceed 25%. In habitats with a poor snow cover, where the ground consists of a mixture of sand, gravel, and pebbles, e.g., on the summits of old, sandy–pebbly beach ridges, the vegetation can cover from 30 to 90% of the surface and consists almost exclusively of crustose lichens such as *Ochrolechia frigida*, *Lecanora campestris*, *Pertusaria* spp. and others. They create a bright background of grey and white patches, on which are scattered the small black patches of lichens with blue-green phycobionta as well as black crusts of liverworts. The fruticose lichens covering up to 5% are represented mainly by *Stereocaulon vesuvianum* var. *depressum.* Small mosses, *Encalypta alpina, E. rhabdocarpa*, etc., cover 1% of the surface. The odd angiosperm is encountered, such as *Poa abbreviata*, *Papaver radicatum* s.l., *Cerastium arcticum*, occupying less than 1% of the surface.

In habitats where the snow is deeper than under optimal conditions, the deeper the snow the more impoverished the vegetation because the later the snow disappears the shorter is the vegetation period. The paucity of the vegetation already becomes apparent where the snow is from 25 to 40 cm deep and starts to melt off only around 1 July. Here carpets of *Stereocaulon rivulorum* are typical on skeletal or sandy–pebbly ground covering in some cases up to 40% of the surface. On moist, silty loams on the primary beach terraces in habitats becoming free from snow at the time mentioned we observed a polygonal pattern, similar to that of the aspect described above, but here the mosses were only occasionally found along the cracks in the form of isolated strips of *Hygrohypnum polare* and a few small, scattered groups of tiny stalks belonging to *Psilopilum cavifolium*. A black film covering up to 90% of the surface dominates. It is formed mainly of liverworts, *Cephaloziella arctica*, *Lophozia alpestris*, *L. excisa*, *L. grandiretis*, *Scapania calcicola* and *S. globulifera* and black, mucilagineous lichens together with some small patches of crustose lichens such as *Lecanora campestris* and *Caloplaca subolivacea* and small cushions of *Stereocaulon rivulorum* as well as primary thalli of *Cladonia pyxidata*. Here and there sterile tufts (5–15 cm wide and 3–5 cm tall) of *Phippsia algida* and single, tiny specimens of *Saxifraga cernua* and *S. hyperborea* are scattered.

The carpets and cushions of *Stereocaulon rivulorum* are also found in habitats free from snow only by 10 July (about 15% of the territory), on sandy–skeletal or sandy–pebbly substrates. Here, however, they cover only 1–20% of the surface. In moist, silty habitats with polygons and cracks there are only fragments of the black film mentioned above covering 10–30% of the ground. Under such long-lasting snow conditions small mires have developed – small patches along the brooks with a thin cover of mosses such as *Orthothecium chryseum*, *Campylium stellatum*, *Bryum rutilans* and others, and a liverwort typical of this habitat *Blepharostoma trichophyllum*. Single, very small and sterile tufts of *Phippsia algida* and sometimes an odd little plant of *Saxifraga cernua* and *Cochlearia groenlandica* are met with here.

By 15 July the snow melts off another 35% of the territory. Here it may have been more than 50 cm but not more than 1 m deep. In such areas only single individuals of plants occur, of very poor coverage amounting to only fractions of one percent. Here and there very small, sterile individuals of *Stereocaulon rivulorum*, *Cetraria delisei*, *Pohlia obtusifolia*, *Polytrichum alpinum*, *Phippsia algida*, *Saxifraga hyperborea*, *S. cernua* and rarely *Draba oblongata* and *D. micropetala* are encountered.

In places occupying about 25% of the territory the snow is above 1 m deep and does not melt before 20 July. There is a complete lack of even the smallest sign of vegetation.

Only 47 species of angiosperms have been found on all of Franz Joseph's Land and there are no vascular cryptogams either. The list of vascular plants (according to Engler's system) ends abruptly with the family Rosaceae, which is represented only by *Potentilla hyparctica*.

Salix polaris and *S. arctica* have been found, but only as single specimens at a few points on the archipelago.

The northern edge of Novaya Zemlya. In this area of the southern belt of the Arctic polar deserts the major part of the land is covered by a glacial shield reaching up to 1000 m in altitude. The fjords are filled with ice from its discharging glacier tongues. Ice barriers are typical at the coast. The land free from ice is a narrow strip or disappears to nothing in front of the ice, but from the area of Cape Zhelaniya it widens out along the northern edge of the island. The average temperature for June is 1.2 °C, for July 1.7 °C, for August 2.1 °C, and for September −0.1 °C.

No flora on the vascular plants of this area has been published. A list furnished by Tolmachev (1936) for the northern part of Novaya Zemlya includes data which apply also to the area north of latitude 75°N. However, his data from the more southerly parts of the area belong to the northern belt of the arctic tundras.

Very little is known of the vegetation on the northern edge of Novaya Zemlya, only some very general information. Thus, for instance, Tolmachev writes (1936: 167): '...the vegetation cover is everywhere very poor and in cases even the concept of plant association is inapplicable, so that we have reason to speak rather of only individual plants in the vast space of uninhabited land'. A. I. Zubkov made a 60 km long journey in the area of Cape Zhelaniya during 1932–3. Unfortunately, the material he collected has not been published. In a short preliminary note it is stated only that 'the ice-free land on the northeast coast of the North Island ... carries a very poor vegetation developed in the form of individual specimens, and fragments of phytocoenoses are encountered only in a few spots... The lack of a snow cover in one locality and the extreme accumulation of snow in another leads in both cases to very unfavorable conditions for the development of any vegetation... On the summits of frontal moraines, at the foot of south-facing slopes, and where the snow is almost always blown away by the strong winds, it is possible to find among the boulders single specimens of *Saxifraga oppositifolia* having settled in the micro-relief with a shelter on its northern side... In habitats where the snow accumulates the vegetation has the character of small fragments of phytocoenoses. However, due to the late snow-melt the angiosperms are very insignificant in numbers. Here mainly lichens are found associated with odd specimens of flowering plants, e.g. *Cerastium regelii*, *Poa alpigena* and *Oxyria digyna*. On very wet slopes of various exposure and in shallow depressions stony-polygonal formations have developed, where single individuals of *Saxifraga oppositifolia*, *Cerastium regelii* and *Draba alpina* grow on the polygons, and at the edges of the stone rings there is *Cetraria delisei*, crustose lichens, and mosses of the genus *Drepanocladus*. In wet places, along rills of meltwater and along the banks of pools there are a few patches with tussocks of *Deschampsia borealis* or fragments of a graminoid-moss mire-like phytocoenoses.' (Zubkov, 1934).

Nordaustlandet, Spitsbergen. The islands of Nordaustlandet, Storøya, Kvitøya, and the group called King Charles' Land, are situated within the southern belt of the arctic deserts. Information on the vegetation of Nordaustlandet, 77% of which is covered by a glacial dome, is found in a number of papers (Summerhayes & Elton, 1928: 201–12; Scholander, 1934; Neilson, 1968). In the flora of the vascular plants, numbering 67 species, there is a single species belonging to the cryptogams, i.e., *Lycopodium selago* ssp. *arcticum*. It has been found in a single locality in the inner end of Murchison Fjord on a south-facing slope. This was why Young (1971) placed this island in his '2nd zone'. Under favorable conditions there are sometimes fragments of a tundra vegetation here with some admixture of *Salix polaris*, but in spite of that the vegetation has in general a polar desert character.

Stony and patterned ground predominates, sandy substrates are rarely encountered. Zonal, mesic habitats with loam have not been mentioned by the authors. On the stony ground lichen associations have developed on the surface of the boulders, and between them mosses are scattered. In addition to the crustose lichens *Lecidea dicksonii, Rhizocarpon geographicum, Xanthoria parietina, Buellia disciformis, Lecanora polytropa,* and so on; black, foliose and fruticose lichens like *Umbilicaria cylindrica*, *U. erosa*, *U. hyperborea*, *U. proboscidea*, *Cetraria hepatizon*, *Hypogymnia apicola*, *Parmelia minuscula*, *Alectoria pubescens* are abundant, *Stereocaulon fastigiatum* is common. There is also the black, epilithic moss *Andreaea papillosa*. Between the stones associated with the mosses such as *Schistidium gracile*, *Dicranoweisia crispula*, *Bartramia ithyphylla*,

Polytrichum alpinum, *Drepanocladus uncinatus*, *Distichum capillaceum*, *Pohlia cruda*, grow such lichens as *Alectoria nigricans*, *Cetraria crispa*, *C. delisei*, *C. nivalis*, *Cladonia elongata*, *Parmelia omphalodes*, *Physcia muscigena*, *Solorina bispora*, *Sphaerophorus fragilis*, *Stereocaulon rivulorum*, and *Thamnolia vermicularis*. Angiosperms are found only in insignificant amounts and, as a rule, as isolated individuals. They belong to *Saxifraga cernua*, *S. caespitosa*, *Papaver radicatum* (*P. dahlianum*), *Cochlearia groenlandica*, *Draba macrocarpa* (*D. bellii*), *D. oblongata*, *D. subcapitata*, *Cerastium alpinum* and *Phippsia algida* (Summerhayes & Elton, 1928, 202–3; Scholander, 1934: 111–12).

The development of open formations with *Luzula confusa* and *L. nivalis* is typical on sandy localities on Nordaustlandet, e.g. in Lady Franklin's Fjord. The formations consist of scattered tufts and do not form closed associations. Some mosses and lichens grow in the shelter of the tufts, e.g. *Lobaria linita, Nephroma expallidum, Cladonia bellidiflora, Solorina crocea* and *Peltigera rufescens*, and on small stones *Umbilicaria deusta*. There are also some species, which are typical of stony habitats. In addition, *Oxyria digyna, Draba lactea, Sagina intermedia, Saxifraga flagellaris, Cardamine bellidifolia* and *Salix polaris* occur (Scholander, 1934:112–13).

Small, mossy mires are met with along brooks and at the edges of small water-holes (Summerhayes & Elton, 1928: 205; Scholander, 1934: 115). The mosses *Bryum crispulum*, *Drepanocladus brevifolius*, *Calliergon sarmentosum*, *Campylium stellatum*, *C. zemliae*, *Drepanocladus uncinatus*, and others, form a slightly hummocky thin carpet. Most typical among the angiosperms, which grow scattered among the mosses, are *Saxifraga rivularis* and in the wettest localities *Ranunculus hyperboreus*. It is also possible to come across some of the following species: *Ranunculus sulphureus*, *Saxifraga foliolosa*, *S. caespitosa*, *S. nivalis*, *Alopecurus alpinus*, *Poa alpigena*, *Cardamine pratensis* ssp. *angustifolia* (sterile rosettes), *Cerastium regelii*, *Juncus biglumis*, *Phippsia algida*, *Dupontia fisheri*, *Eriophorum scheuchzeri* (rare), *Puccinellia vahliana*, and ×*Pucciphippsia vacillans*.

Sometimes on south-facing slopes there are extrazonal fragments of a tundra vegetation with some *Salix polaris*. Scholander (1934: 115–17) writes that very rarely one may come across localities, where '...the angiosperms stand side by side and may really be said to be associated. The only place exclusive of the birdcliffs and other places with manured soils, where this was observed in Nordaustlandet was the "rich" locality on the north side of Snaddvika,* where a small patch of some few square meters was covered with a loose mat consisting mainly of *Carex rupestris* and mixed with *Papaver radicatum*, *Potentilla hyparctica*, *Cerastium alpinum*, *Oxyria digyna*, *Polygonum viviparum*, *Salix polaris*, *Poa rigens*, and in between, some tufts of *Carex misandra* and *C. hepburnii*' (Scholander, 1934: 117). The best developed vegetation is found on habitats fertilized by birds on birdcliffs as well as on the islands used as nesting habitats by birds.

The Siberian province of the arctic polar deserts

Severnaya Zemlya, the area around Cape Chelyuskin, and a number of small islands at high latitude in the Polar Sea belong to the Siberian province of the polar deserts. Although during the autumn and winter seasons low pressures penetrate to the Severnaya Zemlya area from the Kara Sea and give rise to a snow cover thinner than that on Franz Joseph's Land but still reaching 30–40 cm in thickness, a high pressure régime reigns during spring and summer, when the climate takes on the traits of moderate continentality. The cryo-humidity which distinguishes Franz Joseph's Land is not so pronounced here; in contrast according to data presented by Korotkevich (1958*a*) the plants regularly suffer from a lack of water during summertime.

As already mentioned a change in the components of the characteristic synusiae of the zonal, mesic communities forms the basis for a differentiation of this area from the Barents province. While on Franz Joseph's Land species of *Pertusaria* and *Ochrolechia* dominate, the predominant position around Cape Chelyushkin is taken by *Toninia lobulata* (Matveyeva & Chernov, 1976), which has never been found on Franz Joseph's Land. Siberian and Sibero-West American species enter into the composition of the angiosperm flora, e.g. *Saxifraga serpyllifolia*, *Novosieversia glacialis* and *Androsace triflora*, beside the absolutely predominant circumpolar species. The amphi-atlantic species met with in the Barents province are absent here.

* Translator's note. The inner end of Murchison Fjord.

Within this province the De Long Islands and the northern islands of the Severnaya Zemlya archipelago belong to the northern belt of the polar deserts.

The De Long Islands. These islands belonging to the Novosiberian Archipelago are also known as the 'Small Islands' (Malye Ostrova). They consist of the five islands of Bennet, Henrietta, Jeanetta, Zhokhov, and Vil'kitsky. About half of their surface is covered by glaciers. According to data recorded over several years the July temperature on Henrietta Island is 0.7 °C. On Bennet Island the average July temperature in 1956 was 0.2 °C according to data from Kartushin (1963). Twenty species of flowering plants have been found on Bennet Island (Tolmachev, 1959). However, as stated by Kartushin (1963), although the air temperature at 2 m above ground is unfavorable, the temperature at the soil surface is considerably higher, which creates possibilities for the development of vegetation, although this is widely scattered. It consists of crustose and other lichens, mosses and very few angiosperms. The flowering plants grow as single specimens or in small groups. Sometimes in the most favourable conditions *Salix polaris* grows in the dense moss cushions; only once a single specimen of *Dryas punctata* has been discovered.

The northern islands in the Severnaya Zemlya archipelago. Korotkevich (1958*a*) has observed a total of 17 species of flowering plants, exclusively of the high-arctic element of the flora, on the Pioneer and Komsomolets Islands as well as on some of the small islands, where the major part of the surface is occupied by glacier domes. Korotkevich (1958*a*: 656–57) describes the vegetation of the Middle Island as characteristic of the northern belt of the polar deserts within the limits of the Severnaya Zemlya. Here three main kinds of vegetation have developed: (1) Polygonal arctic deserts on positive relief consisting of loams with a small quantity of pebbles and large stones. The vegetation is represented by separate cushions and small-size plants, mainly lichens, mostly the crustose ones. The angiosperms met with are *Saxifraga oppositifolia*, *S. cernua*, and *Papaver radicatum* (*P. polare*); other species are very rare. (2) Fissured arctic deserts in localities with finer soils and smaller rocks in rather dry habitats. The small polygons,

20–30 cm in diameter, are separated by cracks. The angiosperms are mainly isolated specimens of *Cerastium regelii* and *Phippsia algida.* Relatively large patches are occupied by a black coating of liverworts and other components common to the soil surface growth in the polar deserts. (3) Small-hummocky arctic deserts developed in depressed, moist habitats composed of loams. The small hummocks consist mainly of various mosses with a predominance of *Bryum* species and occasional tufts of *Deschampsia borealis.* A large portion is occupied by *Phippsia algida* sometimes forming small tufts of its own. There are also soil-surface algae, *Nostoc commune* and others, and black crusts of algae, liverworts and crustose lichens with blue-green algal components. The coverage by plants in these formations, when typical, amounts to 5–10%, rarely to as much as 15%.

The largest space is nearly devoid of vegetation. Under the most favorable conditions, formations where plants cover up to 50% may be observed, and the major part consists of a crust of algae, liverworts and lichens. The fruticose lichens are *Cetraria delisei* and *C. cucullata.* There are mosses such as *Distichium capillaceum* and *Ditrichum flexicaule* and angiosperms like *Saxifraga oppositifolia.* Also *Cerastium regelii*, *Draba macrocarpa*, *D. micropetala*, *Papaver radicatum* s.l. (*P. polare*), *Phippsia algida*, *Saxifraga caespitosa*, *S. cernua* and *Stellaria edwardsii* occur here.

On the Golomyanny Island in the Sedow Archipelago (the westernmost one of the islands) the vegetation is still more impoverished. Over wide areas only small isolated plant specimens are met with. *Phippsia algida* is the angiosperm most often encountered, and locally there is *Cerastium regelii. Stellaria edwardsii*, *Saxifraga caespitosa* and *S. cernua* are rare, and there are single individuals of *Papaver radicatum* s.l. (*P. polare*) and *Puccinellia angustata.* Mosses, *Ditrichum flexicaule* and a few others, are rare and poorly developed. There are hardly any fruticose lichens, only *Thamnolia vermicularis* as well as a few specimens of *Dufourea madreporiformis* and *Cetraria delisei.* In some places it is possible to find patches of crustose lichens and of a black film of algae, liverworts and lichens (Korotkevich, 1958*a*: 656–7).

The southern part of the Severnaya Zemlya archipelago and the area around Cape Chelyuskin belong to the southern belt of the Siberian province of the arctic polar desert.

The southern islands of the Severnaya Zemlya archipelago. The flora and vegetation on the Bol'shevik and the October Revolution Islands have been described by Korotkevich (1958*a*). In addition Semenov (1970) and Safronova (1975) have furnished information. The number of vascular plant species here amounts to 61. On the October Revolution Island glaciers cover 56% of the land area, on the Bol'shevik Island 29%. The air temperature during the summer months is below 2 °C, but as a result of föhn winds (Arngold', 1929 in Semenov, 1970) a considerably warmer, local climate occurs under certain conditions. Apparently this is also why in some areas described by Korotkevich (1958*a*) and Safronova (1975) there is a vegetation reminiscent of the northern arctic tundras against a background of arctic deserts. An additional condition, which ought to be favorable for the development of plant associations on bedrock, is, e.g. the presence of limestone, but according to the observations made by Safronova there is in general no sign of any vegetation cover there.

The zonal vegetation has been described by Korotkevich from the October Revolution Island where there are heights from 3 to 5 m up to 50 m tall in the western part, forming a flat to undulating relief. On level, rather well drained surfaces with fine soil polygonal deserts have developed. The vegetation covers about 65% of the surface, 30% of which consists of a black and grey colored crust of crustose lichens and the other usual components of similar soil-surface crusts in the polar deserts. About 25% of the cover consists of mosses, *Hylocomium alaskanum*, *Distichium capillaceum* and *Ditrichum flexicaule*, about 10% of lichens, *Cetraria cucullata*, *C. delisei*, etc., and about 15% of herbaceous plants, mostly *Alopecurus alpinus* and *Papaver radicatum* s.l. and a small quantity of *Phippsia algida*, *Deschampsia borealis*, *Cerastium regelii*, *Saxifraga cernua*, *S. oppositifolia*, *Stellaria edwardsii* and single specimens of *Draba oblongata* and *Ranunculus sulphureus*.

On loamy-skeletal, patterned grounds, where the stones are concentrated along the cracks bordering the polygons and forming stone nets, the vegetation has developed along the stone stripes in the form of small strips and patches covering about 20% of the general surface; 10% consist of portions of a crust with a predominance of crustose lichens, and another 10% are dominated by *Cetraria delisei* and a few mosses, *Distichium capillaceum*, *Hypnum bambergeri*, etc. The angiosperms grow scattered often with *Saxifraga oppositifolia* as the predominant species. In lesser quantity occur *Cerastium bialynickii*, *Deschampsia borealis*, *Eritrichium villosum*, *Minuartia rubella*, *Papaver radicatum* s.l. (*P. polare*), *Phippsia algida*, *Saxifraga caespitosa*, *S. nivalis* and *Stellaria edwardsii*.

Small arctic mineral mires supplied with water from melting snow (Korotkevich, 1958*a*) have a slightly uneven surface with a weakly developed moss carpet of *Orthothecium chryseum*, etc. Small tussocks are formed by *Deschampsia borealis* and to some extent by *Alopecurus alpinus* and *Phippsia algida*. Other angiosperms, *Cerastium regelii*, *Juncus biglumis*, *Saxifraga cernua*, etc., are rare and scattered about as single specimens. In drier habitats a black crust of blue-green algae, liverworts, and lichens with blue-green algal components has developed. The polygonal polar deserts on moist loam are covered by this type of black film occupying from 65–70% up to 90–95% of the surface. Here are also found mosses, *Orthothecium chryseum*, *Distichium capillaceum* and others, as well as a small quantity of lichens, *Cetraria delisei* and *Sphaerophorus globosus*, and angiosperms, mainly *Deschampsia borealis* (Korotkevich, 1958*a*: 647–50).

An extra-zonal vegetation of tundra type developed under the most favorable conditions mentioned also by Safronova (1975) has been described from a number of localities on the island by Korotkevich (1958*a*: 647–9, 654, 656). Here in the best developed phytocoenoses the vegetation cover may occupy 80%, the mosses covering up to 55%, and the angiosperms about 50%. The latter consist of about 20 species of herbaceous plants, of which *Salix polaris* is relatively common.

The Cape Chelyuskin district. This area is represented by a rather low plain of its own. Only in its eastern part is there a ridge with elevations up to 300 m. The July temperature is 1.5 °C and gale-force winds are often blowing. Until the last couple of years nothing was known about the vegetation of this area, except that there were plant associations of a generally high-arctic nature together with some moderately arctic species. This information was published in a list of the vascular plants by Tikhomirov (1948*a*). He used both his own collections and data from other authors (Kjellman, 1883; Birulya, 1902, etc.); the list contains 44 taxa. Today we have at our disposal information on both the composition and the structure of the vegetation thanks to the extensive team work by Matveyeva & Chernov (1976). In addition to the botanical data collected by Matveyeva, Chernov has studied the soil fauna, which, as shown above, clearly indicates the zonal characters distinguishing the polar deserts from the arctic tundras (see pp. 150–1).

As stated by Matveyeva (in Matveyeva & Chernov, 1976: 310), polygonal lichen–moss associations appear here as the zonal ones with a coverage from 5–10% to 60–70%. The mosses (*Ditrichum flexicaule*, *Aulacomnium turgidum*, *Rhacomitrium lanuginosum*, *Orthothecium chryseum* and *Oncophorus wahlenbergii*) and the lichens (*Cetraria delisei*, *C. crispa*, *C. cucullata*, *Thamnolia vermicularis*, and others) mainly relating to the cracks, form stripes 1 to 2–3 cm wide. *Saxifraga cernua*, *S. oppositifolia*, *Cerastium regelii* and *Draba oblongata* are mentioned among the angiosperms. Here and there on the polygons various amounts of a crust has developed consisting of crustose lichens, *Toninia lobulata*, etc., algae and liverworts, and a few tufts of *Phippsia algida.* On south-facing slopes larger closed patches of mosses are noted associated with a somewhat larger amount of herbaceous species (*Papaver* spp., *Ranunculus sulphureus*, *Saxifraga caespitosa*, *S. platysepala*). According to Matveyeva & Chernov there are no mires, but on well-watered ground polygonal deserts have developed, on the surface of which there is often a thin, black film of mucilagineous lichens with blue-green phycobionta of the families Collemataceae and Pannariaceae, algae and liverworts. There are also some small tufts of *Phippsia algida.*

The Canadian province of the arctic polar deserts

The Canadian province consists of a group of islands included in the northern part of the 'barren wedge' by Beschel (1969: 875) in the Canadian Arctic Archipelago, i.e. the Meighen, Amund Ringness, Ellef Ringness, Borden, Lougheed, and Bathurst Islands, and the northern part of Prince Patrick Island. These lowland islands, the elevation of which nowhere exceeds 200 m, represent a striking contrast to the mountainous islands situated farther east, Axel Heiberg Island, Ellesmere Land, and others. The major part of the province lies within the limits of the Sverdrup syncline, where the most ancient rockbase reaches a depth of 3000 m and is covered by layers of mesozoic sedimentary rocks. There is no glaciation at present because of the lack of precipitation, except for a low glacier dome on Meighen

Island, the most northerly of these islands at latitute 80°N. Together with some of the small islands it belongs to the northern belt of the polar deserts. The rest of the islands belong to the southern belt.

All their investigators have been surprised by the poverty of the flora of these islands, their barrenness, the poor development of vegetation and the extreme dwarfing of the plants, while at the same time on the islands to the east, Axel Heiberg, Ellesmere Land and Greenland, the area of vegetation and the flora are considerably richer not only at the same but even at higher latitudes (e.g. Peary Land). Savile (1961) believes that the total number of angiosperms on these islands does not exceed 60. According to Beschel (1969), this area is situated north of the 'isotaxe 80'. The flora is distinguished by the absence of *Salix polaris* and the presence of the high-arctic Canadian–Greenlandic endemic species *Braya thorild-wulffii.*

The vegetation of this group of islands is extremely poor. As stated by Savile (1961: 917) there are at present no distinct plant associations. The most variable habitats and the best developed vegetation are found in the area of Cape Isachsen on Ellef Ringness Island, but the larger part of the surface here is almost devoid of vegetation. Not only is the low temperature a hindrance for the development of a vegetation, but the lack of moisture is as well; the annual precipitation amounts to around 75 mm according to Porsild (1955). Because of this, the fragments of plant communities which include angiosperms are met with mainly in such habitats, where a greater or smaller amount of the moisture comes from melting snow. Under such conditions and on bouldery slopes grow species such as *Poa abbreviata*, *Festuca brachyphylla*, *Oxyria digyna*, *Ranunculus nivalis*, *R. sulphureus*, *Draba* spp., *Saxifraga nivalis*, *S. tenuis*, *S. flagellaris*, and *S. oppositifolia.* On loamy substrates there are groups with *Alopecurus alpinus*, *Phippsia concinna*, *Cerastium arcticum*, *Ranunculus sabinei*, *Papaver radicatum* s.l., *Draba oblongata*, *D. subcapitata*, *Saxifraga nivalis*, *S. cernua* and *Potentilla hyparctica.* The species composition of the mosses and lichens is not known. Forming small patches in depressions, where water collects and flows down all the summer from melting snow, there are graminoid–mossy, high-arctic mineral mires without peat. The mosses predominate (Savile does not furnish the species composition) and among them it is possible to come across such species as *Dupontia fisheri*, *Eriophorum scheuchzeri*, *Juncus big-*

lumis and *Ranunculus hyperboreus.* Where it is much drier, *Alopecurus alpinus* and *Saxifraga foliolosa* are found. On the other islands the vegetation is even more impoverished (Savile, 1961).

The extreme poverty of the flora can be blamed both on the present climate and on historical events. There is reason to believe that the flora on these flat islands was destroyed during different periods of the Ice Age and the subsequent marine transgressions. At such times the surface of the water stood 76 m above the present level. In part, as suggested by Savile (1961: 932), the destruction might be blamed on the accumulation of firn fields. The small and not very high nunataks were unable to serve as refugia and the re-colonization by the flora must have taken place in a not-too-distant past mainly coming from Axel Heiberg and Ellesmere Land. However, the poverty of the vegetation, the miniature size of the plants, their small surface cover and partial sterility are connected with the present, severe climate. Of special importance is the predominance of the low, dense cloud cover and the mainly northerly and northwesterly winds, which blow in from the ocean with a force frequently reaching 30 m/s. The cooling effect on the soil from these winds is that much stronger, because as stated by Savile (quoting data from Crary (in Savile, 1961)) the waters of the Polar Sea are especially cold in this area. At Spitsbergen the amount of heat in the surface layer of the waters amounts to 120 kcal, at the North Pole the figure is 25 kcal, but a minimum of 12 kcal is found close to the edge of the continental shelf near the northern Queen Elizabeth Islands.

In addition to the islands mentioned above the northwest edges of Devon and Axel Heiberg Islands belong evidently also to the southern belt of the polar deserts just as do the northern extremes of Ellesmere and Peary Lands, which lie outside the northern boundary of the area of *Carex stans.* The authors who have collected data on the vegetation of these areas (Holmen, 1957; Beschel, 1963*a*, *b*; Bliss, 1971, 1972, etc.) have suggested that the impoverished flora and vegetation in the coastal belt facing north and northwest can also be blamed on the destructive effects of the cold winds blowing in from the ocean.

4

Division of the Antarctic into geobotanical areas

The boundary of the Antarctic is in a wide sense usually defined by a line drawn along the northern edge of the antarctic convergence (cf. Korotkevich, 1970: 53). This is a line along the border between the more northerly, definitely warmer surface waters of the oceans and the more southerly, colder ones (Fig. 22). In this way the Antarctic continent as well as the islands in the South Polar Sea lying south of the antarctic convergence line will be included, as well as those close to this line such as Macquarie, Crozet, Prince Edward Islands and others.

The geobotanical regions of the Antarctic

If the same principles are adopted for the division of the Antarctic into geobotanical areas as were used for the Arctic, that is those based on a complex of characteristics of which the guiding ones are composition (the combination of ecobiomorphs and geographical elements of the flora) and structure (the character of the mosaic, the degree of closedness of the root system, etc.) of the zonal, mesic associations, it is evident that the Antarctic can be divided into two circumpolar, geobotanical regions. The first of these, which is aptly named the region of the subantarctic cushion plants, includes the islands situated north of the antarctic divergence line (essentially at the northern limit of where ice-floes and icebergs are found all year) and reaches to the boundary of the Antarctic itself approximately along the line of the subantarctic divergence. The second one is the region of the antarctic polar deserts. It includes the Antarctic continent and the islands lying south of the line of the antarctic divergence.

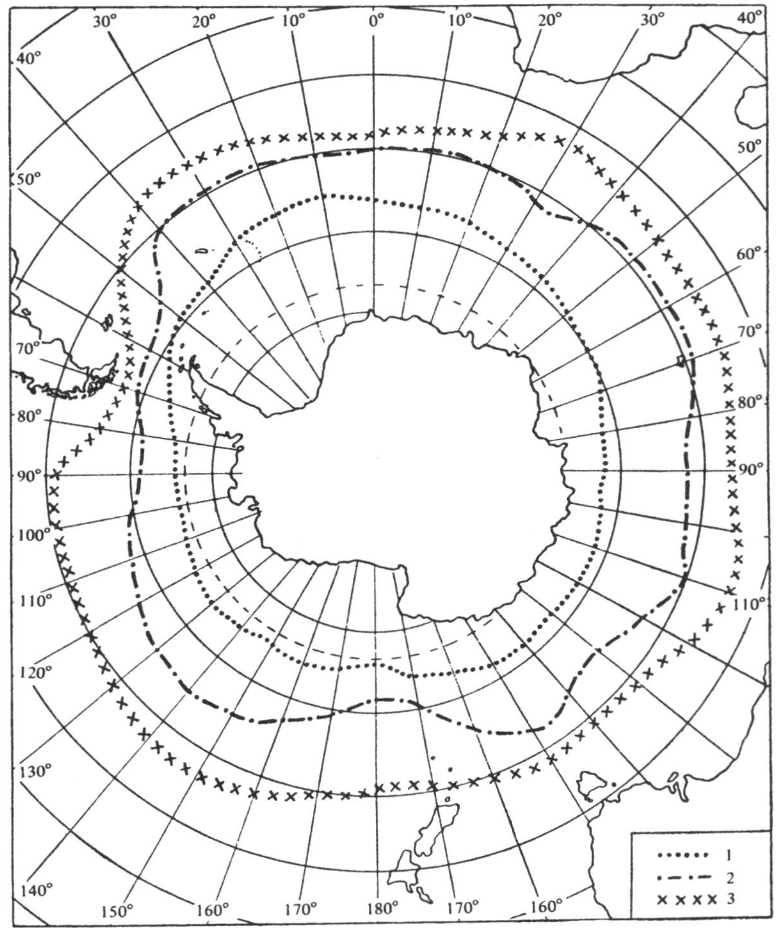

Fig. 22. The position of the basic, frontal systems in the Antarctic. 1, Extension of floating ice-floes during October – November; 2, the antarctic convergence; 3, the subantarctic divergence. According to *The Atlas of the Antarctic*, I, 1966.

The region of the subantarctic cushion plants

In this region, the 'subantarctic zone' in the sense of Greene (1964*a*), these islands are included: South Georgia, Kerguelen, Emerald, Macquarie, Prince Edward, as well as the Crozet archipelagos, Heard

Island and some others. Their flora and vegetation have been described in a number of papers (Skottsberg, 1912*b*, 1960; Taylor, 1955; Greene & Greene, 1963; Greene, 1964*b*; Huntley, 1967; Lindsay, 1973, 1975; Walton, 1973; Walton & Smith, 1973; Greene & Walton, 1975; cf. also reviews by Vul'f, 1944: 502–31, and Kats, 1971: 246–9). The region extends around the pole, its northern border coinciding approximately with the line of the subantarctic divergence its southern one with the northern extension of year round ice-floes and icebergs. With a few exceptions it approaches what Kats distinguished as 'the subantarctic province of peatbogs with cushion plants' (1966: 71–2).

The climate of the islands is always distinctly oceanic. With some exceptions the annual mean temperature is above 0 °C. Only those farthest south, South Georgia and Heard Island, have a period of frost during winter, but the average temperature of the coldest month does not fall below −4 °C. On the islands of Macquarie, Kerguelen and others, there is no period of frost and the mean temperature for all months is above 0 °C. The summer is very cold, and the temperature oscillation through the months is extremely small. Thus, on Macquarie Island the mean temperature for January and February is 7 °C, for July 4 °C. The average, annual amount of precipitation exceeds 1000 mm on most of the islands. Not only is permafrost lacking here, but the soils hardly ever freeze or freeze just briefly on the coldest days but never to any significant depth. As stated by Kats according to data from Troll (Troll, 1960 in Kats, 1966), the soils on, for instance, Kerguelen Island never freeze down to more than 5 cm below the surface. Because of this the patterned and cryogenic soils so typical of the Arctic are not found here and the subsequent cryogenic mosaic of the vegetation cover is lacking. Under the conditions of a constantly cold and wet climate a specific flora with a high degree of endemism has developed on the subantarctic islands. Herbaceous plants dominate, their typical life-forms being tall cushions and thick mats as well as tussocks of gigantic size formed by species such as *Poa foliosa*, *P. flabellata*, *Acaena magellanica* (*A. adscendens*), *Azorella selago*, etc. The plant cushions are also chiefly peat-formers in the bogs (Kats, 1971: 249). Mosses play a subordinate role.

Neither the type of landscape nor the nature of the vegetation can be used for identification with the tundras of the northern hemisphere.

Thus, I do not share the opinion of Korotkevich (1967, 1972: 23–5, etc.) on this problem. The tundra type of vegetation, as I have described it above, is represented by co-dominant associations with characteristic synusiae of semi-prostrate shrubs and dwarfshrubs, where mosses play a significant role in the ground tier; mosaic in connection with cryogenic nano-relief appears as its constant characteristic. This type of association is not found on the subantarctic islands. Here in zonal, mesic habitats a special type of vegetation has developed in the form of communities of hekistothermal herbaceous plant cushions belonging to the typical life-form of giant plant cushions, giant tussocks and mat-forms. Especially different is the large portion in the flora of ferns and the complete absence of shrubs and dwarfshrubs. A recent discovery of *Empetrum rubrum* on South Georgia has been judged to be an introduction (Smith, 1973; Walton & Smith, 1973). Even if mosses such as *Drepanocladus uncinatus*, *Tortula robusta*, *Chorisodontium aciphyllum*, *Rhynchostegium brachypterygium*, *Rhacomitrium lanuginosum* and others do take part in the composition of a number of associations (Huntley, 1967, etc.), their role is insignificant in the majority of cases. The typical wetland vegetation on the subantarctic islands is, as already mentioned, mainly composed of plant cushions, which appear to be the main peat-formers. Peatmosses are completely lacking on these islands but do occur at more northerly latitudes (Kats, 1971: 246–9).

According to the available data (Greene & Walton, 1975) 26 species of vascular plants have been counted on the islands of South Georgia, 21 on Prince Edward Island, 25 on Marion Island,* 28 on the Crozet Islands, 29 on Kerguelen, 8 on Heard Island, and 36 on Macquarie Island. In total the number of vascular plant species on these islands amounts to 70. In addition a large number of introduced plants have colonized them (cf. Walton & Smith, 1973; Greene & Walton, 1975). One of the signs of extreme specialization of the flora is its exceptionally high proportion of vascular cryptogams, 17 species (24%), belonging to the families Lycopodiaceae, Ophioglossaceae, Hymenophyllaceae and Polypodiaceae. In comparison, the richness of Polypodiaceae (12 species) is matched only by that of the family Gramineae (13 species). On Prince Edward Island, the cryptogams amount to 33% of the vascular plants (7 out of 21).

* Translator's note. Marion Island belongs to Prince Edward Archipelago.

While in the northern hemisphere we do not find any region analogous to the subantarctic cushion-plant region, the vegetation developed in the 'Antarctic zone' (sensu Greene, 1964*a*) and reaching south from the northern limit of constant presence of ice-floes and icebergs in the sea fully agree in essential traits with the vegetation type of the polar deserts in the northern hemisphere, insofar as their ecobiomorphs and structures are concerned.

The region of the antarctic polar deserts

The continent of Antarctica and the islands constantly surrounded by ice-floes (Fig. 23) are included in the region of the antarctic polar deserts. Glaciers occupy 99% of the land area, the major part belonging to the Antarctic Iceshield.

Along the northern limit for year-round extension of ice-floes and icebergs there is a very distinct climatic and botanical–geographical boundary between the above described subantarctic islands and the islands situated farther south belonging to the region of the antarctic polar deserts. These are the South Sandwich, South Shetland, South Orkney and Palmer Archipelagos, etc., where all angiosperms except for *Deschampsia antarctica* and *Colobanthus quitensis* (*C. crassifolius*) have disappeared from the flora. Both of these species, or either of them, are also met with in the northwestern parts of the Antarctic Peninsula and on its coastal islands. It should be mentioned that this small number of angiosperms cannot be blamed on ecological causes, because the conditions existing here are no worse than those on, e.g. Franz Joseph's Land. The reason for the scarcity of angiosperms is apparently the extreme isolation. In the Arctic the polar deserts are situated within the limits of the shallow continental shelf and had in the past dispersal contacts with the tundra and mountain areas to the south. In the Antarctic there have been no such possibilities. The closest land is the continent of South America separated by the wide Drake Passage which is up to 4000 m deep. From the special flora on the subantarctic islands, adapted to an oceanic climate with mild winter temperatures, only two species may have been adapted, which could – because of the nature of their areas – be called 'hypantarctic'. No high-antarctic species analogous to the high-arctic ones have developed in the Antarctic due to the lack of contact with an alpine flora from where the material for a species formation might be drawn.

The Antarctic region of the polar deserts (Greene's 1964*a*: 'antarctic zone') can be divided into two subregions (Fig. 23): the northern one as expressed on the majority of the islands and the Antarctic Peninsula, where the angiosperms take part in some of the plant associations, and

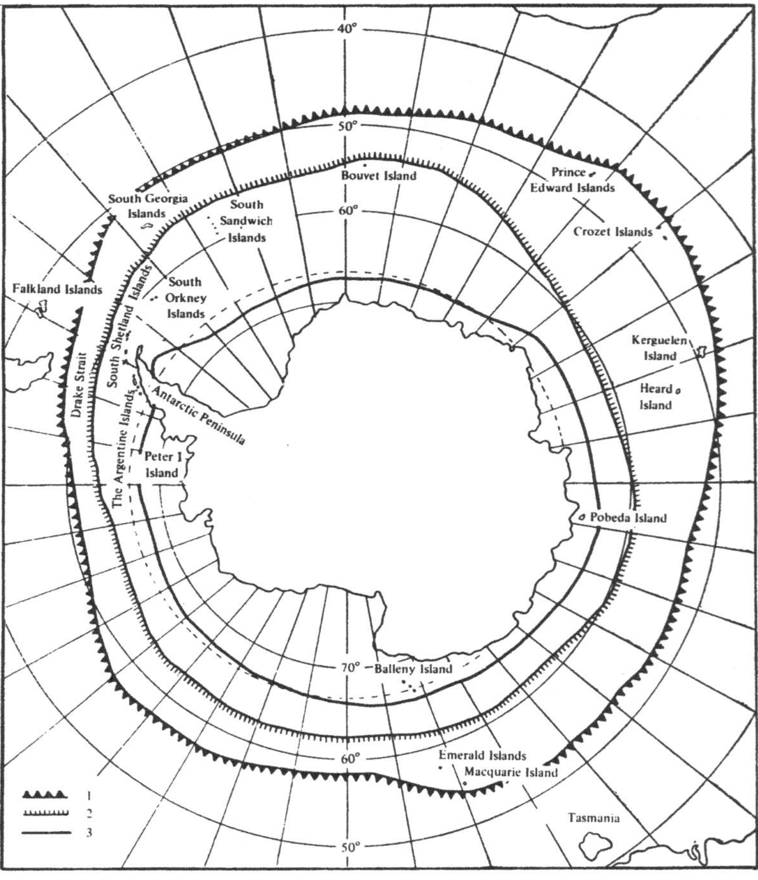

Fig. 23. The geobotanical regions of the Antarctic. The northern boundaries of 1, the region of the subantarctic cushion plants, and 2, the region of the antarctic polar deserts. 3, The southern boundary of the subregion of the northern antarctic polar deserts as well as the northern boundary of the subregion of the southern antarctic polar deserts.

the southern one on the remainder of the Antarctic continent and a number of islands, where there are no angiosperms and the plant associations are represented exclusively by cryptogamic components. The northern subregion agrees mainly with the 'maritime Antarctic' in the sense of Holdgate (1964) and 'the northern subzone of the antarctic polar deserts' sensu Korotkevich (1967: 23). The southern one compares with the 'continental Antarctic' of Holdgate (1964) and 'the middle and southern subzones of the antarctic polar deserts' of Korotkevich (1967: 23–4).

The northern subregion of the Antarctic polar deserts

In contrast to the subantarctic islands, where the average monthly temperatures during the winter are usually positive, the average winter temperature, as well as the mean annual temperature, is below 0 °C on the islands situated in the northern subregion of the polar deserts, the South Shetlands, South Orkneys, South Sandwich, Palmer Archipelago, etc. and also on the northwestern part of the Antarctic Peninsula. The average temperatures of the warmest months there are positive but do not exceed 2 °C. The boundary of this subregion to the north agrees with that of the constant presence of ice-floes in the sea, and to the south with the 0 °C isotherm of the warmest months of the year, which is almost the same as the isoline of the zero radiation balance. The major part of the land surface is covered by ice. The annual amount of precipitation exceeds 500 mm. These territories belong to the cryohumid areas of the polar lands and are reminiscent most of all of the arctic archipelago of Franz Joseph's Land in the northern hemisphere.

Information on the vegetation of the 'maritime antarctic' is available in a number of works (Skottsberg, 1954; Greene & Greene, 1963; Greene, 1964*a*, 1969, 1970; Holdgate, 1964; Korotkevich, 1966, 1967, 1972; Longton, 1966, 1967, 1970; Gimingham & Smith, 1970, 1971; D. M. Greene & Holtom, 1971; Lindsay, 1971; Edwards, 1972, 1973, 1974; Smith, 1972; Allison & Smith, 1973; Collins, 1973; Smith & Corner, 1973; Golubkova, 1974; etc.).

Here, just as in the Arctic, cryogenic formations are widely distributed, causing the appearance of frostcracking, frost-sorting of the ground and solifluction. Big blocks sorted out from the fine soil by frost action form stone nets; smaller stones form polygons of lesser dimensions. On slopes there are stone stripes and various solifluction phenomena, etc. (Gimingham & Smith, 1970: 769, etc.). Because of this the vegetation has a heterogenous structure. However, the heterogeneous complexes have not been described as distinct entities, but the homogeneous parts of the vegetation appear to be studied and classified as units, which are described as sociations and in turn grouped according to the traits of similar specific composition and life-forms into units of higher rank, associations, subformations and formations. Holdgate (1964), describing the vegetation of Signy Island in the South Orkney Archipelago, distinguishes two formations, one uniting all sociations of cryptogams with no individuals of angiosperms, or only single specimens, the other with more significant amounts of angiosperms. This kind of classification has been extended and refined also for other islands and for the Antarctic Peninsula (Longton, 1967; Gimingham & Smith, 1970; Lindsay, 1971; Smith, 1972; Allison & Smith, 1973; Smith & Corner, 1973). There are annotated lists of the coverage and occurrence of every species in a number of papers. Data have also been published on the phytomass (Longton, 1970, 1974; Collins, 1973; Edwards, 1973, etc.).

The herb–moss and herb–lichen–moss formation. Deschampsia antarctica and *Colobanthus quitensis* are met with in the composition of various carpets of mosses and lichens (Lindsay, 1971: 70–4; Allison & Smith, 1973: 202 and table 7; Smith & Corner, 1973: 104, 114, etc.). They appear often as components in a relatively dry carpet of *Drepanocladus uncinatus* and of *Polytrichum alpestre*, *P. alpinum*, *Brachythecium austro-salebrosum* and *Chorisodontium aciphyllum* sward, but also together with other mosses and liverworts such as *Cephaloziella varians*, *Barbilophozia hatcheri* as well as lichens such as *Cladonia furcata*, *C. metacorallifera*, *C. vicaria*, *Stereocaulon alpinum*, *Usnea antarctica*, *Ochrolechia frigida* and others. Both angiosperm species grow in the form of small tufts and cushions, so that an extremely small amount of the surface is covered by these plants. *Colobanthus quitensis* always grows as scattered individuals among the mosses. So usually does also *Deschampsia antarctica*, but it sometimes shows a tendency to form merging mats and fragments of a sward in various protected habitats.

Edwards (1973) has described the amount of phytomass and the seasonal dynamics of such closed mats of *Deschampsia antarctica* from the Signy Island. The largest above-ground phytomass of *D. antarctica* with a 100% coverage was found during March, 327 g/m², while the phytomass of the mosses amounted to 136–233 g/m² (Edwards, 1973: 173).

Non-vascular cryptogam formation of the northern subregion of the antarctic polar deserts. The formation, consisting exclusively of cryptogamic plants, has been subdivided by the authors mentioned above into some subformations.

The fruticose lichen and moss cushion subformation is the most widespread one, met with in the majority of open habitats. Fruticose and foliose lichens are typical as well as apocarpic mosses forming small cushions. Crustose lichens are almost always present, but do not dominate. Sociations of this type occupy stony habitats, the fine soil – if at all present – being found only in cracks and hollows between the stones.

An *Usnea–Andreaea* association is the most frequent one, represented by a number of sociations in which species of *Andreaea* (*A. gainii*, *A. regularis*, and *A. depressinervis*) play the main role. They are usually developed in the form of small, compact but separate cushions; sometimes in especially favorable conditions they have fused forming small mats not more than 50 cm in diameter. In lesser amounts occur *Schistidium antarctici*, *Dicranoweisia grimmiacea*, *Polytrichum alpinum*, *Pohlia nutans*, etc. Among lichens genera such as *Usnea* predominate, mainly *U. antarctica*, less of *U. fasciata*, the latter often associated with the black, fruticose lichen *Himantormia lugubris.* The following lichens are also encountered, *Alectoria pubescens*, *Cladonia vicaria*, *Sphaerophorus globosus*, *Stereocaulon alpinum* as well as the funnel-shaped *Cladonia furcata.* In habitats kept moist by water from melting snow many foliose *Umbilicariae*, such as *U. antarctica* (*Omphalodiscus antarcticus*), *U. decussata*, *U. propagulifera*, and in addition *Parmelia saxatilis*, *Peltigera spuria*, etc., are associated with *Usnea antarctica.* Where it is more humid, the thalli of *Umbilicaria antarctica* can sometimes reach 30 cm in diameter in north-facing localities; in south-facing habitats, they do not exceed 10 cm across. There are many crustose lichens, *Ochrolechia frigida* (often colonizing the decaying cushions of *Andreaea*), *Psoroma follmannii*, *Rhizocarpon geographicum*, *Biatorella antarctica*, *Buellia* spp. and others.

According to data furnished by Allison & Smith (1973, table 7) regarding one of the sociations belonging to this association on Elephant Island in the South Shetland Archipelago and studied on the basis of random selection of 20 plots, 20×20 cm, the coverage of *Usnea antarctica* amounted to 37%, that of *Alectoria pubescens* to 6%, of *Sphaerophorus globosus* to 5%, of *Ochrolechia frigida* to 3%, and of all other lichens to 10%. The coverage of the moss *Andreaea regularis* amounted to 17%, that of *Polytrichum alpinum* to 4%, and of all other mosses to 2%.

The crustose lichen subformation has developed on stony and rocky surfaces. Two groups of sociations are included, nitrophilous sociations found where there is a nitrogen-rich manure produced by bird colonies, and sociations situated in localities poor in nitrogen compounds. The most widely distributed ornithophilous sociation of crustose lichens in the bird colony area appears to be the *Xanthoria elegans–Haematomma erythromma–Mastodia tesselata* sociation in which are also found *Buellia coniops*, *B. russa*, *Rinodina petermannii*, *Catillaria corymbosa*, *Microglaena antarctica*, *Acarospora macrocyclos*, *Xanthoria candelaria*, etc. On slightly moist slopes there is also *Ramalina terebrata.* On the coastal cliffs above the upper tide-water line the ornithophilous lichens form two zones, a lower one with species of *Verrucaria* (*V. psychrophila*, *V. ceuthocarpa*, and *V. tesselatula*) and an upper, orange-colored one with species of *Caloplaca* (*C. cirrochrooides* and *C. regalis*). The latter take over abruptly from the former at a height of about 2 m above the upper tidal limit. On the shores where there is no nitrophilous effect from birds, formation of *Buellia*, *Lecanora*, and *Lecidea* are found. In the interior parts of the land area lichens like *Placopsis contortuplicata* dominate in the sociations of crustose lichens colonizing the boulders in the stone stripes and the polygons, the basaltic cliffs and blocks, and sometimes associating with species of *Andreaea* and *Usnea*, rarely also with *Drepanocladus uncinatus* (Lindsay, 1971; Allison & Smith, 1973; Smith & Corner, 1973; etc.).

The moss turf subformation is distributed over stony slopes of various steepness ranging from gentle, 1–2°, to steep, 30°, most often with a north-facing exposure. *Polytrichum alpestre* appears as the main component in the majority of the sociations (coverage 38–61%). It forms a closed vegetation cover with *Chorisodontium aciphyllum* (coverage 11–49%) together with a number of other mosses, *Pohlia nutans*, rarely *Polytrichum alpinum* and *Drepanocladus uncinatus*, and liverworts such as *Barbilophozia hatcheri* and *Cephaloziella varians* as well as the lichens *Ochrolechia frigida*, *Sphaerophorus globosus*, *Alectoria pubescens*, *Cladonia furcata*, *C. metacorallifera* and others. These sociations were described by Smith (1972), Allison & Smith (1973), and Smith & Corner (1973) under the name *Polytrichum alpestre–Chorisodontium aciphyllum* association. Sometimes a *Chorisodontium aciphyllum* sociation forms a thin carpet on a habitat of stones and fine soil on slopes of various grade; *Barbilophozia hatcheri* and *Cephaloziella varians* are constant associates of *C. aciphyllum*.

Sociations with dominating *Polytrichum alpinum* are also distributed on the South Shetland Islands. Lindsay (1971: 69) describes the sociation *Polytrichum alpinum–P. piliferum* met with on dry stony ground with an admixture of fine soil. Both mosses occur as small colonies but sometimes fuse to form flat, low, irregular mounds up to 1.5 m in diameter and about 15 cm tall. They are usually mixed with *Pohlia nutans*. *Polytrichum alpinum* is also associated with lichens. Allison & Smith (1973, table 7) have described the coverage of *P. alpinum* in a lichen–*Polytrichum* sociation on the basis of randomly selected samples of 60 squares, 20×20 cm, as amounting to 29% with a frequency of 88%. *Chorisodontium* is also mentioned (coverage, 4%, frequency, 25%). The lichens predominate in this sociation, *Usnea antarctica* has a 22% coverage, *Sphaerophorus globosus* 7%, other fruticose lichens 10%, *Ochrolechia frigida* occupies 11%, and other crustose lichens 15% of the surface.

The phytomass of individual plots of *Polytrichum alpestre* on the Signy and Argentine Islands has been described by Longton (1970). He obtained the following figures: on Signy Island, 342 g/m^2, on Argentine Island, in one case 385, in another 421 g/m^2. It should be mentioned that the phytomass of mosses, according to my data from Franz Joseph's Land per one square meter, amounted to about 400 g/m^2. Thus in both cases the size order is about the same.

The moss carpet subformation (mossy, 'mineral' mires) have developed in very wet localities, most often on level, coastal habitats and more infrequently in the inland parts of the islands. They are represented mainly by two associations, *Brachythecium* cf. *antarcticum–Calliergon sarmentosum–Drepanocladus uncinatus*, and an association consisting of almost pure mats of *Drepanocladus uncinatus*. The moss carpets, up to 5 cm thick, sometimes cover relatively large areas. When there is moderate moisture, the moss carpet becomes much thinner and splits up and lichens begin to appear in it, e.g. species of *Psoroma*, *Cladonia* and others, which do not occur in the moss carpet on the well-watered localities, where instead are found the white, round colonies of the ascomycete *Thyrenectria antarctica* var. *hyperantarctica* (Smith, 1972; Smith & Corner, 1973: 103). Around the edges of such patches of mire are found scattered in the moss carpet *Cephaloziella varians*, *Pohlia nutans*, *Bryum algens*, *Polytrichum alpinum* as well as *Deschampsia antarctica*. According to data published by Smith & Corner (1973: 102) in a small, mossy mire the coverage amounted to: for *Brachythecium* cf. *antarcticum* 94%, *Drepanocladus uncinatus* 7%, *Brachythecium austro-salebrosum* 1%, *Pohlia nutans* less than 1%, and for *Prasiola crispa* to less than 1%. In another locality it was for *Drepanocladus uncinatus* 91%, *Brachythecium austro-salebrosum* 4%, *Pohlia nutans* 3%, *Bryum algens* 2%, *Ceratodon* cf. *grossiretis* 1%, *Cephaloziella varians* 1%, *Prasiola crispa* 1% and *Psoroma follmannii* and *Polytrichum alpinum* less than 1%.

The moss hummock subformation is usually met with on steep slopes, where the mosses *Drepanocladus uncinatus* and *Brachythecium austro-salebrosum* associated with *Bryum algens*, *Pohlia nutans*, *P. cruda* var. *imbricata*, *Tortula conferta* and other species form large, loose cushions. Sometimes in favorable habitats (Smith & Corner, 1973: 112–13) there are small amounts of *Deschampsia antarctica* and *Colobanthus quitensis* together with the mosses, as well as some lichens such as *Psoroma follmannii* and others. Locally there are liverworts, *Barbilophozia hatcheri*, and *Cephaloziella varians* among the mosses.

Algal subformation is often represented by the *Prasiola crispa* association. This alga forms thin sheets on level habitats close to bird colonies, because of which the major part of the surface is otherwise naked ground. Along the edges of meltwater rills the alga sometimes forms a thicker mat. Occasionally there are some lichens growing in the algal mat, such as *Mastodia tesselata* and species of *Caloplaca* and

Lecidea. Isolated thalli of *Prasiola crispa* are also found in other sociations, especially on the surface of mosses saturated with water. In the interior part of the land areas on habitats with fine soil saturated by meltwater a *Nostoc* association is found. *Nostoc* forms discrete rosettes about 5 cm in diameter and growing 10–20 cm apart from each other (Lindsay, 1971, etc.).

The southern subregion of the antarctic polar deserts

This subregion (the 'continental Antarctic' according to Holdgate, 1964; the 'middle and southern antarctic polar deserts' according to Korotkevich, 1967, 1972) covers the continent of Antarctica, except for the northwestern part of the Antarctic Peninsula. Included in this subregion are also some islands close to the continent. The northern boundary of the subregion is formed by the 0 °C isotherm for the warmest month, which almost coincides with the isoline for zero radiation balance. Only a fraction of 1% of this territory is free from glaciers. The ice-free areas are the so-called 'oases' and nunataks.

The size of some of these major 'oases' amounts to hundreds of square kilometers. Thus the area of the Bangor 'oasis' equals 263 km^2, that of Getty 748 km^2, and of Vestfold 413 km^2, etc. and in the mountains that of Victoria occupies 487 km^2, Taylor 442 km^2 and Wright 417 km^2, etc. (Korotkevich, 1972: 332). Because of the lack of an ice cover, the soil surface absorbing solar radiation is warmed so much that the local annual radiation balance here will reach 20–40 kcal/cm^2, while over the glacial cover it is always negative (Korotkevich, 1972: 335). As a consequence, there is water in liquid form in the 'oases', a weakly developed hydrographic network and lakes, which, if sufficiently deep, do not freeze all the way to the bottom. During the summer months the average temperature at the surface of the ground and in the near-ground air layer is positive and may even be relatively high, but because of the cooling effect from the surrounding glaciers during January the air at 2 m is in the majority of the 'oases' below or close to 0 °C, in a few cases only slightly higher. Thus, at the research camp 'oasis' the average January temperature during the years it has been measured has equalled 1.8 °C (Korotkevich, 1972: 335).

This amount of heat is sufficient for the development of a vegetation no less than that in the northern subregion of the antarctic polar deserts. The main limiting factor is, however, the lack of moisture.

In contrast to the humid, maritime climate of the islands and the northwestern part of the Antarctic Peninsula, the climate of the 'oases' is characterized by extreme drought. Because of the low relative humidity of the air, on an average 50–60%, but falling to below 10% when föhn winds are blowing, the evaporation, in some 'oases' reaching a magnitude of 500 mm, is considerably higher than the precipitation of 200–300 mm. Not only does the rain evaporate, but also the snow blowing in from the glaciers does so (Korotkevich, 1972: 337). The constant extreme aridity has led to the formation of salt lakes and saline ground in the 'oases', as well as to the appearance of special types of bedrock erosion typical only of areas with extremely arid climates.

The vegetation in the oases is very poor: in the coastal 'oases' only a few percent, in the mountain 'oases' just about 1% of the surface is covered (Korotkevich, 1972: 341). It has been most completely described by Korotkevich (1972: 239–47) on the basis of his personal investigations in the eastern Antarctic between longitudes 0° and 110°E. There are also data on the vegetation of the southern Antarctic from Victoria Land (Rudolph, 1963; Longton, 1972, 1973, 1974). Furthermore, information on the vegetation of the Antarctic is found in the works of Gollerbakh and Syroyechkovsky (1958, 1960), Golubkova (1969, 1974), Savich-Lyubitskaya and Smirnova (1969), etc.

The moss-lichen and moss formations are found comparatively rarely in the most favorable habitats. The most common one is the formation with predominating *Buellia frigida* and *Bryum algens,* with which are associated in various relationships such lichens as *Protoblastenia citrina, Xanthoria candelaria* f. *antarctica, Alectoria miniuscula, Umbilicaria decussata (Omphalodiscus decussatus), Usnea antarctica, U. acromelana, U. sulphurea (Neuropogon sulphureus), Gasparrinia murorum, G. elegans, Lecanora* sp., *Lecidea* sp., and on decaying mosses, also *Rinodina turfacea, Pyrenodesmia mawsonii, Lecanora expectans* and others. Mosses such as *Bryum argenteum, B. antarcticum, B. filicaule, Sacroneurum glaciale* and *Schistidium antarctici* should be mentioned. The associations described by Korotkevich from Enderby Land are rare and dominated by *Alectoria miniuscula* and *Ceratodon purpureus.* There are also such lichens as *Buellia frigida, Xanthoria candelaria* f. *antarctica, Lecanora expectans,* and growing on the mosses *Lepraria neglecta.* The mosses are *Pohlia* and *Bryum* species. More common is the combination of *Alectoria miniuscula* and *Schistidium antarctici*. Some lichens are associated with these two species, such as *Buellia frigida, Umbilicaria decussata, Usnea antarctica, U. acromelana, Protoblastenia citrina, Xanthoria candelaria* f. *antarctica, Physcia* sp., and *Rinodina turfacea*. *Grimmia doniana* is also among the mosses. In this association the vegetation covers up to 30% of the surface, and in gullies up to 50%. In areas where birds are nesting on the cliffs, a closed cover of a similar composition is met with reaching a width of a few meters and up to 10 m in length. From the Vanger 'oasis' associations have been described with predominant *Umbilicaria decussata* and *Sacroneurum glaciale*, with addition of *U. spongiosa, Pyrenodesmia mawsonii, Bryum algens, Bryoerythrophyllum recurvirostre, Grimmia plagiopoda* and *Schistidium antarctici*. In the Grierson 'oasis' Korotkevich discovered one of the richest vegetations in the Antarctic continent on the top of some hills. It had an average coverage of 50%, locally up to 80–90%. There *Usnea antarctica* and *Schistidium antarctici* predominate. The lichens are especially well developed. In addition to the dominating species there are *Usnea acromelana, Alectoria miniuscula, Buellia frigida, Umbilicaria decussata*, and the mosses, *Xanthoria candelaria* f. *antarctica* and *Protoblastenia citrina*. Mixed with the *Schistidium antarctici* cushions are *Ceratodon purpureus* and some other mosses (Korotkevich, 1972: 245–7).

Longton (1973) has described some moss sociations from southern Antarctica in the McMurdo area (Victoria Land). They are dominated by small apocarpic mosses, very common in this area. Often these mosses form small cushions up to 5 cm in diameter growing isolated from each other, but sometimes they form an almost closed mat. Thus, Longton gives a description of strips, 1–2 m broad, on sandy ground, where the coverage of *Bryum antarcticum* reaches 85%; single individuals of *B. argenteum* were mixed in with it. Locally the latter species also builds sociations with up to 75% coverage. Such have been discovered on a north-exposed slope on sandy soil, where the moss covers a surface up to 15 m in length and 2 m in width. In this sociation there is also *Nostoc* in the form of hemispheric lumps up to 2 mm in diameter. But most frequently the mosses form open sociations with less than 2% coverage. Most typical of these sociations in the McMurdo area are those with *Sacroneurum glaciale* in the form of separate cushions hardly exceeding 5 cm in diameter and 2 cm in height. Crustose lichens are common in the decaying parts of the moss cushions, mainly the yellow *Caloplaca darbishiri.* However, moss–lichen sociations are most widespread in this area, where in individual cases the plants may cover up to 50% of the surface. *Usnea antarctica* predominates, or this species associated with *Sacroneurum glaciale, Caloplaca darbishiri, Xanthoria elegans, Candellaria concolor* var. *antarctica* (Longton, 1973: 2342–3).

The lichen formation is the most widely distributed formation of all in the Antarctic continent. It occupies the largest part of the territory free from ice and having any noticeable vegetation cover. As demonstrated by Korotkevich (1972: 243–5), a grey crust – from very pale grey to almost black – is found most frequently. It is formed by the epilithic lichen *Buellia frigida* and grows on rocks and stones. Sometimes the coating of *B. frigida* covers large surfaces, frequently without admixture of any other species, especially in the mountains, where in the most humid localities it forms an almost pure cover and in protected localities is mixed with *Usnea antarctica.* In the oases *B. frigida* is often associated with *Umbilicaria decussata, Alectoria miniuscula, Usnea antarctica, Protoblastenia citrina* and other lichens.

Less widely distributed are lichen covers of *Umbilicaria decussata,* sometimes forming an almost closed cover of small rosettes on the rocks, to which are added *Alectoria miniuscula, Usnea acromelana, U. antarctica,* and *Buellia frigida.* The coverage amounts in most cases to 3–5%. Korotkevich has also discovered lichen formations with a predominance of *Alectoria miniuscula* and an admixture of *Biatorella antarctica* with a total coverage not exceeding 1%. Under more favorable conditions he found *Umbilicaria decussata, Usnea sulphurea, Alectoria miniuscula,* and *Biatorella antarctica* associated with some *Schistidium antarctici* with a total coverage of around 10%. He has also described sociations, where *Umbilicaria decussata* and *Alectoria miniuscula* grow with an admixture of *Buellia frigida* and odd specimens of *Sacroneurum glaciale.* Longton (1973: 2341–2) has also described from the McMurdo area on Victoria Land, in addition to the sociations already mentioned with *Usnea antarctica* and *Umbilicaria decussata,* widespread crusts of crustose lichens such as *Xanthoria elegans, Caloplaca darbishiri, Candelaria concolor* var. *antarctica, Lecanora rubina* var. *melanophthalma, Blastenia sparsa,* and *Lecidea cancriformis.* The coverage usually does not exceed 1–5%, but may in individual cases grow up to 50%.

Lichen–algae and lichen–moss–algae formations are found in areas with bird colonies on grounds moderately manured by organic animal products. Most common are *Prasiola crispa* and *Buellia frigida,* which have been observed even in the areas most severely affected by stormy petrels. Associations of *Prasiola crispa* with the foliose lichen *Umbilicaria decussata* and the fruticose *Usnea antarctica* relate to the most favorable conditions. The coverage of the soil by plants usually amounts to only a few percent, and rarely exceeds 30%, but may in a few cases reach 100%. *Nostoc commune* is associated with these lichens in moderately wet localities. This alga is also found together with *Alectoria miniuscula.* Furthermore, associations of *Nostoc commune–Buellia frigida* are found high in the mountains in the southern nunataks.

Under more favorable conditions the lichens and the algae are joined by mosses: *Ceratodon purpureus, Bryum algens, B. argenteum, Grimmia doniana* and others. They enter into the composition of sociations with *Prasiola crispa* and *Buellia frigida.* Here it is possible to find, combined in some way, species such as *Biatora flava, Umbilicaria decussata, Usnea antarctica, U. sulphurea, U. acromelana, Pyrenodesmia mawsonii, Protoblastenia citrina, Xanthoria mawsonii, Parmelia coreyi* and others. Rudolph (1963) has described a sociation from the Hallett area in Victoria Land with predominating *Prasiola crispa* and some *Bryum argenteum, Buellia frigida, Xanthoria mawsonii* and *Parmelia coreyi,* covering 15% of the surface, of which the algae occupy 13%, the mosses 2%, and the lichens less than 1%. Longton (1973) has also mentioned the sociation of *Prasiola crispa* with *Bryum* spp. from the McMurdo area on Victoria Land.

Communities with *Nostoc commune,* mosses and lichens are found far from bird colonies. The richest ones of this type have been described by Korotkevich from Enderby Land. Here patches of *Nostoc,* but also of some other blue-green and green algae, have developed in well watered depressions, where there are also found small cushions of the mosses *Bryum algens* and *Ceratodon purpureus.* Colonies of *Lepraria neglecta* and *Physcia caesioides* are found on decaying mosses. They occur on patches of fine soil. On the stones have developed such crustose lichens as *Buellia frigida, Biatora flava, Lecidea physciella* and

Acarospora petalina, foliose ones like *Umbilicaria decussata* and *U. spongiosa,* the small fructicose *Alectoria miniuscula* and in cracks between the stones colonies of *Usnea antarctica* and *U. acromelana.* The total coverage by the vegetation does not exceed 20%. Under more severe conditions *Nostoc commune, Buellia frigida and Sacroneurum glaciale* cover less than 10%. *Nostoc* and cushions of *Sacroneurum glaciale* occur in the wettest localities (Korotkevich, 1972: 241–3).

Algal and bacterial–algal formations. Korotkevich found patches of vegetation with *Nostoc commune* in both the Grierson and the Bangor oases. Longton (1973) found some rare ones on Victoria Land, usually close to brooks or lakes, where they grow in patches of uneven outlines, usually mixed with other algae and sometimes with small mosses and lichens. In other cases small lumps of *Nostoc* in the form of spheres or rosettes a few centimeters in diameter have colonized slopes watered by melting snow. From afar they look like a black, closed cover, but the total coverage rarely exceeds 50%. In places manured by bird droppings, mainly in penguin colonies, a green mat of *Prasiola crispa* has formed. Where there are organic remains, some mosses such as *Bryum algens, B. argenteum* and lichens associate with the algae. In addition thin films of algae and bacteria are common in Antarctica, in some cases mixed with microscopic fungi. They are also found in the most southerly nunataks and high up in the mountains (Korotkevich, 1972: 239–40).

5

Conclusions

In compiling this review I have attempted to accomplish two things: on one hand to present an outline of a division into geobotanical areas of the polar lands, while considering the vegetation of both the Arctic and the Antarctic from a single point of view, and on the other hand to give a brief summary of the data available on the vegetation cover in the areas so divided.

The division suggested should be considered only as a step towards solving the problem, as an hypothesis launched, which may serve as a departure point for debate and discussion and for further refinement and elucidation. The vegetation of the polar lands is still too unevenly and incompletely known; on the one hand there is an acute shortage of material about the full composition and other characteristics of the phytocoenoses in many cases, and on the other hand, as a consequence of this, the classification of the vegetation at all taxonomic levels is still insufficiently worked out.

The system suggested in the present work for the vegetation types of the Arctic is based mainly on ecological–physiognomic principles and especially on a combination into associations of this or that kind of *ecobiomorphs,* i.e. life-forms adopted to certain environmental factors such as temperature, moisture, etc. To the number of diagnostic characteristics have also been referred such matters as the degree of participation in the composition of communities by this or that geographical element of species, hyparctic, moderately arctic, or high-arctic, and the character of the structure, i.e. the structural mosaic of vegetation, or the lack thereof, the degree of closedness of the root systems, etc. Great emphasis is placed on the structural characteristics for the delimitation of the vegetation types. Thus the presence or absence of a cryogenic mosaic is one of the main bases for the differentiation of associations belonging to the tundra type of vegetation or to the type of hyparctic shrub thickets. The completely closed root systems of the flowering plants in the zonal, mesic tundra phytocoenoses and the lack thereof in the plant communities of the polar deserts, indicating the dissolution of the phytocoenotic connection between the angiosperms, are of fundamental importance for the delimitation of tundra type and polar desert type of vegetation. In the polar deserts a closed root system may be found under a few groups of plants, but here it is not a question of the completely closed system, which is typical of the tundra sward (cf. Figs. 19 and 21).

The ideas expressed by Sochava (1948*a*, 1952, etc.) on the important rank of the boundary between the subarctic and the arctic tundras have been used as a basis for the differentiation of the areas suggested above. Gorodkov had considered the arctic tundras as a subzone, equally important in rank to the other subzones and, in accordance with his opinions, the geobotanical provinces had cut all the subzones in a meridional direction (this point of view is held at present by Andreyev, cf. Andreyev & Nakhabtseva, 1974; Andreyev, 1975). Sochava (1948*a*) has carried out the division into 'geobotanical fields', i.e. the subdivisions at the rank of 'province', different in the 'arctic belt', within which he placed the arctic tundras, from the 'temperate belt' (Sochava, 1952), in which he included the subarctic tundras together with the adjacent more southerly territories. Sochava divided the arctic tundras of the USSR into three 'geobotanical fields' (the Atlantic, the Asiatic, and the Chukotka fields); the subarctic tundras together with the adjacent southern parts of the temperate belt (including some areas of the forest zone) were put into seven geobotanical 'fields' (the North European, The Uralian, The Ob-Irtysh, the Middle-Siberian, the Yana-Kolyma, and the Beringian fields). The necessity of separating the arctic tundras from the subarctic ones by a boundary of higher rank than, e.g. that separating the subzone of the southern subarctic tundras from the subzone of the middle subarctic tundras, has been emphasized in recent works by Yurtsev (1966, etc.). He has not only elucidated the differences between the 'arctic' and 'hyparctic' botanical–geographical belts, but has also convincingly demonstrated the principal differences in the historical origin of their floras (Tolmachev & Yurtsev, 1970).

I have used other arguments and evaluations, different from those of Sochava and Yurtsev, for weighing the characteristics separating areas of higher rank. Thus, in distinguishing regions and subregions I have placed great emphasis on the characters of a definite combination of life-forms, while Sochava in defining geobotanical belts and Yurtsev in defining botanical–geographical belts do not consider this characteristic as diagnostic. Sochava unites the subarctic tundras with the northern taiga forests into one region of high rank, 'the temperate geobotanical belt' based on the character of distribution of certain 'fratriae of formations', phytocoenogenetically substantiated; Yurtsev unites the subarctic tundras with the northern taiga into a single botanical–geographical 'hyparctic belt' based on the character of the high activity of the hyparctic elements in their floras. This disparity in the evaluation of the characteristics used for the division into areas also explains why the boundary between the arctic and the subarctic tundra subregions, as suggested here, has a lower rank than that used for the division made by Sochava and Yurtsev. This means that I separate the arctic and the subarctic tundras at the rank of subregions of the tundra region. The distinction into geobotanical provinces has, then, been made separately.

The comparative characterization of the arctic and the antarctic vegetation has led to the conclusion that tundras as a vegetation type are lacking in the Antarctic, and that the vegetation of the subantarctic islands must be distinguished in the form of a special geobotanical region, the subantarctic cushion plant region, which has no analog in the northern hemisphere. At the same time, in the 'antarctic zone' according to Greene (1964*a*), including the islands within the area where ice-floes are constantly present in the ocean as well as the Antarctic continent itself, there are plant communities which, according to their basic characters, may be comparable to the plant communities in the most high-latitude areas of the Arctic. Together these two form a single type of vegetation: that of the polar deserts.

The second purpose of the present work has been the compilation of a review of the vegetation based on the distinctions used for the division into geobotanical areas. For this I have drawn on the basic available literature and in a number of cases on my own original material. The expediency of such a survey seems evident. However, a number of difficulties were encountered while compiling it. One of them concerned the extreme poverty of data for particular areas, e.g. the area between the Yana and Indigirka Rivers and in the non-Soviet Arctic on the western part of the Canadian Arctic Archipelago. Assembling the review was also hampered by the lack of a classification embracing the taxonomic units at all levels and based on all the varieties of the plant associations which has already been mentioned. This is the cause of the empiricism in the description of the vegetation due to the impossibility of arranging the available data into a definite system. The small size of this booklet, suitable for the publication in the Komarov Lecture Series, did not allow of more detailed descriptions where they might have been desirable, or a more complete bibliography. I do hope, however, that this review as published – a first attempt of its kind in the polar literature – will be of interest to the investigators who are concerned with studying and mastering the vegetation cover in the polar lands.

REFERENCES

Abbe, E. C. (1938). Phytogeographical observations in northernmost Labrador. *Amer. Geogr. Soc. Spec. Publ.* 22.

Acock, A. M. (1940). Vegetation of a calcareous inner fjord in Spitsbergen. *J. Ecol.*, 28 (1).

Ahti, T., Hämet-Ahti, L., & Jalas, J. (1968). Vegetation zones in northwestern Europe. *Ann. Bot. Fenn.* 5 (3).

Aleksandrova, V. D. (1937). Tundry pravoberezh'ya reki Popigay. [The tundras on the right bank of Popigay River.] *Tr. Arkt. Inst.* 63.

Aleksandrova, V. D. (1945). Kratky ocherk rastitel'nosti Novoy Zemli po linii peresechecheniya zaliv Rogashev – Bukhta Savina. [Short sketch on the vegetation of Novaya Zemlya along a line intersecting Rogachev Inlet and Sabin Bay.] *Probl. Arktiki,* Vol. 5–6.

Aleksandrova, V. D. (1946). Rastitel'ny pokrov ostrova Mezhdusharskova (Novaya Zemlya). [The vegetation cover of Mezhdusharsky Island (Novaya Zemlya).] *Probl. Arktiki,* Vol. 5.

Aleksandrova, V. D. (151). Rastitel'nost' Grenlandii. [The vegetation of Greenland] *Priroda,* 10.

Aleksandrova, V. D. (1953). O granitsakh rastitel'nykh podzon v nizov'yakh r. Yany. [On the boundaries between vegetation subzones in the lowlands of Yana River.] *Izv. VGO,* 85 (1).

Aleksandrova, V. D. (1956). Rastitel'nost' Yuzhnova Ostrova Novoy Zemli mezhdu 70°56′ i 72° 12′ s. sh. [The vegetation of the South Island of Novaya Zemlya between latitudes 70°56′ and 72°12′ N.] In: *Rastitel'nost'Kraynevo Severa SSSR i eyo os voyenie.* [The vegetation of the Far North and its utilization.] Vol. 2, Moscow–Leningrad.

Aleksandrova, V. D. (1957). Voprosy razgranichenia arkticheskikh pustyn' i tundr kak tipov rastitel'nosti. [Problems of the delimitation of the arctic deserts and tundras as vegetation types.] [*Tez. dokl. Delegatsk. s'ezda VBO, May 1957.*] Lecture. Conference of the VBO, May 1957, Vol. 4. Leningrad.

Aleksandrova, V. D. (1958). Opyt opredeleniya nadzemnoy i podzemnoy massy rasteniy v arkticheskoy tundre. [An attempt at determining the above-ground and below-ground mass of plants in the arctic tundra.] *Bot. Zh.* 43 (12).

Aleksandrova, V. D. (1960*a*). Fenologiya rasteniy i sezonnye aspekty v podzone arkticheskikh tundr. [Phenology of the plants and seasonal aspects in the subzone of the arctic tundras.] *Tr. Fenologichesk. soveshch.*, Leningrad.

Aleksandrova, V. D. (1960*b*). Flora sosudistykh rasteniy ostrova Bol'shovo Lyakhovskovo, (Novosibirskie Ostrova). [The vascular plants on the Big

Lyakhovsky Island (Novosiberian Islands).] *Bot. Zh.* 45 (11).
Aleksandrova, V. D. (1962). O podzemnoy strukture nekotorykh rastitel'nykh soobshchestv arkticheskoy tundry na o. B. Lyakhovskom. [On the below-ground structure of some plant associations in the arctic tundra on the Big Lyakhovsky Island.] *Probl. Bot.* 6.
Aleksandrova, V. D. (1963). Ocherk flory i rastitel'nosti o. B. Lyakhovskovo, [Sketch on the flora and vegetation of Big Lyakhovsky Island.] *Tr. Arkt. i Antarkt. n-issl. inst.* 224.
Aleksandrova, V. D. (1964). *Arkticheskye Tundry SSSR.* [Arctic Tundras of the USSR.] Autoref. Dokt. Diss., Leningrad.
Aleksandrova, V. D. (1969*a*). *Klassifikatiya rastitel'nosti.* [The classification of vegetation.] Leningrad.
Aleksandrova, V. D. (1969*b*). Nadzemnaya i podzemnaya massa rasteniy polarnoy pustyni ostrova Zemlya Aleksandry (Zemlya Frantsa Iosifa). [The above-ground and below-ground mass of plants in the polar deserts on Aleksandra Land (Frantz Joseph's Land).] *Probl. Boy.*, 11:
Aleksandrova, V. D. (1970*a*). Dinamika mosiachnosti rastitel'nykh soobshchestv pyatnistikh tundr v arkticheskoy Yakutii. [Dynamics of the mosaic of the plant associations of the spotty tundras in arctic Yakutia.] In: *Mosaichnost' rastitel'nykh soobshchestv i eyo dinamika.* [The mosaic of plant associations and its dynamics.] Vladimir.
Aleksandrova, V. D. (1970*b*). The vegetation of the tundra zones in the USSR and data about its productivity. In: *Productivity and conservation in northern circumpolar lands,* ed. W. A. Fuller & P. G. Kevan. IUCN N. Ser. 16, Morges, Switzerland.
Aleksandrova, V. D. (1971*a*). Opyt analiza structury rastitel'novo pokrova na granitse fytosenosov pyatnostoy i bugorkovoi tundry v Zapadnom Taimyre. [An attempt at analysis of the structure of the plant cover for the delimitation of spotty and hummocky tundras in Western Taimyr. In: *Biogeotsenozy Taimyrskoy tundry i ikh produktovnost'*. [The Taimyr tundras and their productivity.] Leningrad.
Aleksandrova, V. D. (1971*b*). Printsipy zonal'novo deleniya rastitel'nosti Arktiki. [Principles for the zonal division of the vegetation of the Arctic.] *Bot. Zh.*, 56 (1).
Aleksandrova, V. D. (1973). Russian approaches to classification of vegetation. In: *Handbook of Vegetation Science. Part V. Ordination and classification of vegetation*, ed. R. Whittaker.
Aleksandrova, V. D. (1974). Svyaz' mezhdu klassifikatsiey rastitel'nosti i geobotanicheskim rayonirovaniem na primere rayonirovaniya Arktiki. [Comparison between the classification of the vegetation and the geobotanical division on examples of division of the Arctic.] In: *IV. Vsesoyuznoye soveshchanie po klassifikatsii rastitel'nosti.* [IV All-Soviet Conference on the classification of vegetation.] Lvov.
Aleksandrova, V. D. (1977). Struktura rastitel'nykh gruppirovok polarnoy pustyni ostrova Zemlya Aleksandry (Zemlya Frantsa-Iosifa). [The structure of plant associations of the polar deserts on the Aleksandra Land (Franz Joseph's Land).] In: *Problemy ekologii, geobotaniki, botanicheskoy geografii i floristiki.* [Problems of ecology, geobotany, botanical geography and floristics.] Leningrad.
Aleksandrova, V. D. & Zhadrinskaya, N. G. (1963). Smena aspektov v tundrakh o. Bol'shovo Lyakhovskovo. [Changing aspects of the tundras on the Big Lyakhovsky Island.] *Tr. Arkt. i Antarkt. n-issl. inst.* 224.

Aleksandrova, V.D. & Zubkov, A. I. (1937). Fiziko-geografichesky ocherk Novoy Zemli. [Physical-geographical sketch of Novaya Zemlya.] In: *Novozeml'skaya ekskursiya. 1.* [Excursions in Novaya Zemlya. 1.] Leningrad.

Allison, J. S. & Smith, R. I. L. (1973). The vegetation of Elephant Island, South Shetland Islands. *Brit. Antarct. Surv. Bull.* 33–4.

Anderson, J. H. (1975). Vegetation mapping in Alaska using earth resources technology satellite imagery. Abstracts of the papers presented at the XII Internat. Bot. Congress, 3–10 July, 1975. Leningrad.

Andreev, V. N., see Andreyev, V. N.

Andreyev, V. N. (1931). Rastitel'nost' tundry severnovo Kanina. [The vegetation of the tundras of North Kanin Peninsula.] In: *Olen'i pastbishcha Severnovo Kraya.* [Reindeer pastures in the Far North.] Archangelsk.

Andreyev, V. N. (1932*a*). Podzony tundry Severnovo Kraya. [The tundra subzones of the Far North.] *Priroda,* 10.

Andreyev, V. N. (1932*b*). Tipy tundr zapada Bol'shoy Zemli. [Tundra types in the western Bol'shaya Zemlya.] *Tr. Bot. Muz. AN SSSR,* 25.

Andreyev, V. N. (1934). Kormovaya baza Yamal'skovo olenevodstva. [Fodder basis for reindeer herding in the Yamal.] *Sov. olenev.* 1.

Andreyev, V. N. (1935). Rastitel'nost' i prirodnye rayony vostochnoy chasti Bol'shezemel'skoy tundry. [The vegetation and natural geobotanical areas in the eastern part of the Bol'shezemelya tundras.] *Tr. polyarn. komissii AN SSSR,* 22.

Andreyev, V. N. (1938). Obsledovanie tundrovykh olen'ikh pastbishch c pomoshch'yu samoleta. [Investigation of reindeer pastures in the tundra by the aid of airplanes.] *Tr. n-issl. in-ta polyarn. zemled., zhivotn. i promysl. Khoz., sev. Olenevodstvo, 1.*

Andreyev, V. N. (1954). *Rastitel'ny pokrov vostochnoevropeyskoy tundry i meropriyatiya po yevo ispol'zovaniyu i preobrazovaniyu.* [The vegetation cover of the East-European tundras and arrangements for its utilization and re-establishment.] Autoref. Dokt. Diss., Leningrad.

Andreyev, V. N. (1966). Osobennosti zonal'novo raspredeleniya nadzemnoy fitomassy na vostochnoevropeyskom Severe. [The peculiarities of zonal distribution of above-ground phytomass in the East-European North.] *Bot. Zh.* 51 (10).

Andreyev, V. N. (1971). [Andreev, V. N] Methods of defining above-ground phytomass on vast territories of the Subarctic. *Rep. Kevo Subarctic Res. Stat.* 8.

Andreyev, V. N. (1975. *Geobotanicheskoye rayonirovanie tundrovoy zony SSSR.* [The geobotanical division of the tundra zone in the USSR.] Manuscript, Inst. Bot. Yakutskovo filiala SO, AN SSSR. Yakutsk.

Andreyev, V. N. (1976). *Poyezdka po Alyaske. Doklad na seminare Otdela Geobotanika Bot. Inst. AN SSSR.* [Journey around Alaska. Lecture at the Geobotany Department of the Botanical Institute of the Academy of the Sciences.] Manuscript, Bot. Inst. AN SSSR, Leningrad.

Andreyev, V. N., Igoshina, K. N. & Leskov, A. I. (1935). Olen'i pastbishcha i rastitel'ny pokrov Polyarnovo Priural'ya. [Reindeer pasture and vegetation cover in the Polar Urals.] *Sov. Olenev.* 5.

Andreyev, V. N. & Nakhabtseva, S. F. (1974). Podzony Yakutskoy tundry. [Subzones in the Yakutian tundra.] In: *Biologicheskie problemy Severa.* [Biological problems of the North.] *IV. Symposium,* Yakutsk.

Andreyev, V. N. & Perfil'yeva, V. I. (1975). Fragmenty stepnoy rastitel'nosti na severo-vostoke Yakutii. [Fragments of steppe vegetation in northeastern Yakutia.]

In: *Biologicheskie problemy Severa. Byulleten' nauchnoteknicheskoy informatsii.* [Biological problems of the North. Bulletin of scientific-technological information.] Yakutsk.

Arkticheskaya Flora SSSR. [The Arctic Flora of the USSR.] Ed. A. I. Tolmachev. Vol. 1: 1960, Vol. II: 1964, Vol. III: 1966, Vol. IV: 1963, Vol. V: 1966, Vol. VI: 1971, Vol. VII: 1975. AN SSSR. Leningrad.

Atlas Antarktiki. [*Atlas of the Antarctic.*] Vol. I. 1966. Moscow–Leningrad.

Avramchik, M. N. (1937). Geobotanicheskaya i pastbishchnaya kharakteristika rayona reki Dudypty. [Geobotanical and pasture characteristics of the area along Dudypta River.] *Tr. Arkt. Inst.* 63.

Avramchik, M. N. (1969). K podzonal'noy kharakteristike rastitel'novo pokrova tundry, lesotundry i taygi Zapadno Sibirskoy nizmennosti. [On the subzonal characteristics of the vegetation cover of tundra, forest tundra, and taiga of the West Siberian Lowlands.] *Bot. Zh.* 54 (3).

Babb, T. A. & Bliss, L. C. (1974*a*). Effects of physical disturbance of arctic vegetation in the Queen Elizabeth Islands. *J. appl. Ecol.* 11 (2).

Babb, T. A. & Bliss, L. C. (1974*b*). Susceptibility to environmental impact in the Queen Elizabeth Islands. *Arctic,* 27 (3).

Baer, K. (1838). Végétation et climat de Novaia Semlia. *Bull. Acad. Imp. Sci. St.-Pétersb.* 3 (11–12).

Banfield, A. W. F. (1954). Preliminary investigations of the Barren Ground caribou. *Wildlife Bull. Ser. 1,* 10*a, b.*

Berg, L. C. (1928). Zona tundr. [The tundra zone.] *Izv. Leningrad. Univ.* 1.

Beschel, R. E. (1961). Botany and some remarks on the history of vegetation and glacialization. In: *Expedition to Axel-Heiberg Island. Preliminary Report, 1959–1960.* Montreal.

Beschel, R. E. (1963*a*). Geobotanical studies on Axel-Heiberg Island in 1962. In: *Axel-Heiberg Island.* McGill University Press, Montreal.

Beschel, R. E. (1963*b*). Vegetation map of the western part of the Austro Ridge. *The Canadian Surveyor,* 17.

Beschel, R. E. (1969). Floristicheskie sootnosheniya na ostrovakh Neoarktiki. [Floristic correlation on the Neoarctic Islands.] *Bot. Zh.,* 54 (6).

Beschel, R. E. (1970). *Head of Expedition Fjord, Axel-Heiberg Island, N. W. T.* Report on CCSBP–CT. Edmonton.

Bird, C. D., Scotter, G. W., Steere, W. C. & Marsh, A. H. (1977). Bryophytes from the area drained by the Peel and Mackenzie rivers, Yukon and Northwest Territories, Canada. *Can. J. Bot.,* 55 (23).

Birulya, A. (1902). Otchet o botanicheskikh rabotakh za letny sezon 1901. [Account of botanical explorations during the summer season of 1901.] *Izv. Imp. AN* 16 (5).

Birulya, A. (1907). Ocherki iz zhizni ptits polyarnovo poberezh'ya Sibiri. [Sketches on the birds of the polar coast of Siberia.] *Zap. Imp. AN, ser. 8,* 18 (2).

Bliss, L. C. (1956). A comparison of plant development in microenvironments of arctic and alpine tundra. *Ecol. Monogr.* 26.

Bliss, L. C. (1971). Devon Island, Canada, high-arctic ecosystem. *Biol. Conserv.* 2.

Bliss, L. C. (1972). IBP high arctic ecosystem study, Devon Island. *Arctic,* 25 (2).

Bliss, L. C. & Cantlon, J. E. (1957). Succession on river alluvium in Northern Alaska. *Amer. Midl. Nat.* 58.

Bliss, L. C. & Kerik, J. (1973). Primary production of plant communities of the Truelove Lowland, Devon Island, Canada – Rock outcrops. In: *Tundra biome.*

Primary production and production processes. Ed. L. C. Bliss & F. E. Wielgolaski. Stockholm.

Bliss, L. C., Kerik, J. & Peterson, W. (1977). Primary production of dwarf shrub heath communities. In: *Truelove Lowland, Devon Island, Canada: a high arctic ecosystem.* Ed. L. C. Bliss. Edmonton.

Bobrov, E. G. (1972). *Istoriya i sistematika listvennits.* [The history and systematics of the larches.] Leningrad.

Boch, M. S. (1974). Bolota tundrovoy zony Sibiri (printsipy tipologii). [Mires in the tundra zone of Siberia (principles of a typology).] In: *Tipy bolot SSSR i printsipy ikh klassifikatii.* [Bog-types in the USSR and principles of their classification.] Leningrad.

Boch, M. S. (1975). Bolota nizov'yev r. Indigirki (v predelakh tundrovoy zony). [Mires in the lower reaches of Indigirka River (within the limits of the tundra zone).] In: *Flora, sistematika i filogeniya rasteniy.* [Flora, systematics, and phylogeny of plants.] Kiev.

Boch, M. S., Gerasimenko, T.V. & Tolchel'nikov, Yu.S. (1971*a*). Bolota Yamala. [The mires of Yamal.] *Bot. Zh.* 56 (10).

Boch, M. S., Gerasimenko, T. V. & Tolchel'nikov, Yu.S. (1971*b*). O nekotorykh osobennostyakh rastitel'nosti i pochv tundrovoy zony Yamala. [On some peculiarities in the vegetation and the soils of the tundra zone in Yamal.] *Izv. VGO* 103 (6).

Boch, M. S. & Solonevich, N. G. (1972). Bolota i zabolochennye redkoles'ya i tundry. [Mires and swampy forests and tundras.] In: *Pochvy i rastitel'nost' vostochnoevropeyskoy lesotundry.* [Soils and vegetation in the East European forest tundras.] Leningrad.

Boch, M. S. & Tolchel'nikov, Yu. S. (1973). Deshifrovochnye priznaki bolot tundrovoy zony na aerofotosnimkakh [Interpretable characteristics of bogs in the tundra zone on aero-photographs.] In: *Primenenie aerofotosnimkov pri izuchenii lesnikh i bolotnikh meliorativnykh fondov.* [Application of aero-photography for the study of forest and bog amelioration reserves.] Leningrad.

Boch, M. S. & Tsareva, V. T. (1974). K flore nizov'yev r. Indigirki (v predelakh tundrovoy zony). [On the flora of the lower reaches of the Indigirka River (within the limits of the tundra zone).] *Bot. Zh.* 59 (6).

Boch, M. S. & Vasilyevich, V. I. (1975). Opyt krupnomasshtabnovo kartigovaniya rastitel'nosti tundry. [Attempt at large-scale mapping of the tundra vegetation.] In: *Aerometody izucheniya lesnykh landshaftov.* [*Aerial methods for the study of forest landscapes.*] Krasnoyarsk.

Böcher, T. W. (1933*a*). Phytogeographical studies of the Greenland flora. *Medd. om Grønl.* 104 (3).

Böcher, T. W. (1933*b*). Studies on the vegetation of the east coast of Greenland. *Medd. om Grønl.* 104 (4).

Böcher, T. W. (1938). Biological distributional types in the flora of Greenland. *Medd. om Grønl.* 106 (2).

Böcher, T. W. (1948). Contribution to the flora and plant geography of West Greenland, I. *Medd. om Grønl.* 147 (3).

Böcher, T. W. (1950). Contribution to the flora and plant geography of West Greenland, II. *Medd. om Grønl.* 147 (7).

Böcher, T. W. (1952). Contribution to the flora and plant geography of West Greenland, III. *Medd. om Grønl.* 147 (9).

Böcher, T. W. (1954). Oceanic and continental vegetational complexes in Southwest Greenland. *Medd. om Grønl.* 148 (1).
Böcher, T. W. (1959). Floristic and ecological studies in middle West Greenland. *Medd. om Grønl.* 156 (5).
Böcher, T. W. (1963*a*). Phytogeography of Greenland in the light of recent investigations. In: *North Atlantic biota and their History.* Ed. Á. Löve & D. Löve. Oxford.
Böcher, T. W. (1963*b*). Phytogeography of middle West Greenland. *Medd. om Grønl.* 1948 (3).
Böcher, T. W., Holmen, K., & Jakobsen, K. (1968). *The flora of Greenland.* Copenhagen.
Bogdanovskaya-Gienef, I. D. (1938). Prirodnie usloviya i olen'i pastbishcha o. Kolguyeva. [Natural conditions and reindeer pastures on Kolguyev Island.] *Tr. N-issl. Inst. Polyarn. zemled. zhivotn., i promysl. khoz., ser. Olenevodstvo,* 2.
Borisov, A. A. (1965). *Paleoklimaty territorii SSSR.* [The paleoclimates of the USSR territories.] Moscow–Leningrad.
Brassard, G. R. & Beschel, R. E. (1968). The vascular flora of Tanquary Fjord, Northern Ellesmere Island, N.W.T. *Canad. Field Naturalist,* 82 (2).
Brassard, G. R. & Longton, R. E. (1970). The flora and vegetation of Van Hauen Pass, Northwestern Ellesmere Island. *Canad. Field Naturalist,* 84.
Braun-Blanquet, J. (1964). *Pflanzensoziologie.* 3rd ed. Vienna–New York.
Breslina, I. P. (1969). Flora i rastitel'nost' Semi Ostrovov i prilegayushchevo poberezh'ya Vostochnovo Murmana. [Flora and vegetation of the Sem' (Seven) Islands and the coasts situated in eastern Murman.] *Tr. Kandalakshsk. gos zap.* 7.
Breslina, I. P. (1970). Flora i rastitel'nost' Semi Ostrovov i prilegayushchevo poberezh'ya Vostochnovo Murmana. [Flora and vegetation of the Semi (Seven) Islands and the coasts situated in eastern Murman.] Autoref. Dokt. Diss., Leningrad.
Britton, M. E. (1957). Vegetation of the Arctic tundra. In: *Arctic biology.* Ed. H. P. Hansen. Oregon State Univ. Press.
Britton, M. E. (1966). Vegetation of the arctic tundra. Oregon State Univ. Press.
Brown, J. (1968). Barrow, Alaska. In: *Working meeting on analysis of ecosystems. Tundra zone.* Ustaoset, Norway.
Brown, R. J. E. & Kupsch, W. O. (1974). Permafrost terminology. National Research Council of Canada 14274, Tech. Mem. No. 111.
Bruggemann, P. E. & Calder, J. A. (1953). Botanical observations in northeast Ellesmere Island, 1951. *Canad. Field Naturalist,* 67.
Buks, I. I. (1973). Printsipy sostavleniya legendy melkomasshtabnoy obzornoy karty rastitel'nosti Aziatskoy Rosii (na primere tundrovoy rastitel'nosti). [Principles for the composition of legends to the small-scale survey map of the vegetation of Asiatic Russia (e.g. the tundra vegetation).] In: *Pochvy i rastitel'nost' merzlotnykh rayonov SSSR.* [Soils and vegetation of the permafrost areas in the USSR.] Magadan.
Buks, I. I. (1974). Rabochy variant melkomasshtabnoy geobotanicheskoy karty severa Dal'nevo Vostoka. [Draft copy of legends to the small-scale geobotanical map of the northern Far East.] In: *Biologicheskie problemy Severa.* [Biological problems of the North.] VI. Symposium 3, Yakutsk.
Cajander, A. K. (1903). Beiträge zur Kentniss der Vegetation der Alluvionen des nördlichen Eurasiens. 1. Die Alluvionen des unteren Lena-Thales. *Acta Soc. Sci. Fennica,* 32 (1).

Chekanovsky, A. L. (1896). Dnevnik ekspeditsii po rekam Nizhney Tunguske, Oleneky i Lene v 1873–1875 godakh. [Diary of the expedition along the lower reaches of Tunguska, Olenek, and Lena Rivers during the years of 1873–1875.] SPb.

Cherepanov, S. K. (1966). *Betulaceae* S. F. Gray – Beryozovie. [*Betulaceae* S. F. Gray – The Birches.] In: *Arkticheskaya Flora SSSR.* [Arctic Flora of the USSR.] Vol. V. Moscow–Leningrad.

Chernyavsky, F. B. (1975). Ekologicheskie i zoogeograficheskie osobennosti nazemnoy teriofauny Chukotki i ostrova Vrangelya. [Ecological and zoogeographical peculiarities of the ground terio-fauna in Chukotka and Wrangel Island.] *Tr. 2-vo Vsesoyuzn. melkopitayushchim.* Moscow.

Churchill, E. D. (1955). Phytosociological and environmental characteristics of some plant communities in the Umiat region of Alaska. *Ecology,* 36 (4).

Collins, N. J. (1973). Productivity of selected bryophyte communities in the maritime Antarctic. In: *Tundra biome. Primary production and production processes.* Ed. L. C. Bliss & F. E. Wielgolaski. Stockholm.

Corns, I. G. W. (1974). Arctic plant communities east of the Mackenzie delta. *Canad. J. Bot.* 52 (7).

De Candolle, Alph. (1874). Constitution dans le régne végétal de groupes physiologiques. *Arch. Sci. Phys. Nat.* 50. Geneva.

Dedov, A. A. (1933). Letnie olen'i pastbishcha vostochnoy chasti Malozemel'skoy tundry. [Summer pastures of reindeer in the eastern parts o the Malozemel'skaya tundra.] In: *Olen'i pastbishcha Severnovo Kraya,* 2. [Reindeer pastures in the North District.] Leningrad.

Dennis, J. G. & Johnson, P. L. (1970). Shoot and rhizome-root standing crops of tundra vegetation at Barrow, Alaska. *Arctic and Alpine Research,* 2 (4).

Derviz-Sokolova, T. G. (1964). Rastitel'nost' kraynevo vostoka Chukotskovo polyostrova. [The vegetation of the far eastern Chukotka peninsula.] *Problemy Severa,* 8.

Dibner, B. D. (1961). Zarosli ivy mokhnatoy za 75-y parallel'yu. [Willow thickets at the 75th parallel.] *Izv. VGO* 93 (4).

Dobbs, C. G. (1939). The vegetation of Cape Napier, Spitsbergen. *J.Ecol.* 27 (1).

Dostovalov, B. N. (1960). Zakonomerností razvitiya tetragonal'nykh sistem ledyanykh i gruntovykh zhil v dispersnykh porodakh. [Natural development of the tetragonal systems of ice-veins in dispersed sediments.] In: *Periglyatsial'nye yavleniya na territorii SSSR.* [Periglacial appearances in the USSR territories.] Moscow.

Drew, J. V., Tedrow, J. C. F., Shanks, R. E. & Koranda, J. E. (1958). Rate and depth of thaw in arctic soils. *Amer. geophys. Union Transact.* 39 (4).

Drury, W. H. (1962). Patterned ground and vegetation on southern Bylot Island, N.W.T., Canada, *Contrib. Gray Herbarium,* 190.

Dushechkin, V. I. (1937). Olen'i pastbishcha v Kharaulakhskikh gorakh (Yakutiya). [Reindeer pastures in the Kharaulakh mountains (Yakutia).] *Tr. Arkt. Inst.* 63.

Edwards, J. A. (1972). Studies in *Colobanthus quitensis* (Kunth) Bartl. and *Deschampsia antarctica* Desv. V. Distribution, ecology and vegetative performance on Signy Island. *Brit. Antarct. Surv. Bull.* 28.

Edwards, J. A. (1973). Vascular plant production on the maritime Antarctic. In: *Tundra Biome. Primary production and production processes.* Ed. L. C. Bliss & F. E. Wielgolaski. Stockholm.

Edwards, J. A. (1974). Studies in *Colobanthus quitensis* (Kunth) Bartl. and *Deschampsia antarctica* Desv.: VI. Reproductive performance on Signy Island. *Brit. Antarct. Surv. Bull.* 39.

Ekstam, O. (1897). Einige blütenbiologische Beobachtungen auf Nowaja Semlja. *Tromsø Museum Aarsheft,* 18.

Engelskjön, T. & Schweitzer, H. J. (1970). Studies on the flora of Bear Island (Björnöja). I. Vascular plants. *Astarte,* 3 (1).

Eurola, S. (1968). Über die Fjeldheidevegetation in den Gebieten von Isfjorden und Hornsund in Westspitzbergen. *Aquilo, Ser. Bot.* 7.

Eurola, S. (1971*a*). Die Vegetation einer Sturzhalde (Sveagruva, Spitzbergen, 77° 53′ n. Br.). *Aquilo, Ser. Bot.* 10.

Eurola, S. (1971*b*). The middle arctic mire vegetation in Spitsbergen. *Acta Agralia Fennica,* 123.

Fränkl, E. (1955). Rapport über die Durchquerung von Nord PearyLand (Nordgrönland) im Sommer 1953. *Medd. om Grønl.* 103 (8).

Fredskild, B. (1961). Floristic and ecological studies near Jakobshavn, West Greenland. *Medd. om Grønl.* 163 (4).

Fredskild, B. (1966). Contribution to the flora of Peary Land, North Greenland. *Medd. om Grønl.* 178 (2).

Fristrup, B. (1952). Physical geography of Peary Land. *Medd. om Grønl.* 127 (4).

Gelting, P. (1934). Studies on the vascular plants of East Greenland between Franz Joseph Fjord and Dove Bay (Lat. 73° 15′ – 76° 20′ N.). *Medd. om Grønl.* 101 (2).

Gelting, P. (1955). A West Greenland *Dryas integrifolia* community rich in lichens. *Svensk Bot. Tidskr.* 49 (1).

Gimingham, C. H. & Smith, R. I. L. (1970). Bryophyte and lichen communities in the maritime Antarctic. In: *Antarctic ecology.* Ed. M. W. Holdgate. London.

Gimingham, C. H. & Smith, R. I. L. (1971). Growth form and water relations of mosses in the maritime Antarctic. *Brit. Antarct. Surv. Bull.* 25.

Giterman, R. E., Golubeva, L. V., Zaklinskaya, E. D., Koreneva, E. V., Matveyeva, O. V. & Skiba, L. A. (1968). *Osnovnye etapi razvitiya rastitel'nosti Severnoy Azii v antropogene.* [Basic stages in the development of the vegetation of North Asia in the Quaternary period.] Moscow.

Gollerbakh, M. M. & Syroyechkovsky, Ye. Ye. (1958). Biogeograficheskie issledovaniya v Antarktide v 1957 g. [Biogeographical investigations in the Antarctic during 1957.] *Izv. AN SSSR, ser. geogr.* 6.

Gollerbakh, M. M. & Syroyechkovsky, Ye.Ye. (1960). Biogeograficheskie issledovaniya v Vostochnoy Antarktide v letny sezon 1957 g. [Biogeographical investigations in the Eastern Antarctic during the summer season of 1957.] *Tr. Sovietsk. Antarkt. Eksped.* 9, Leningrad.

Golubkova, N. S. (1969). Lishayniki. [The lichens.] In: *Atlas Antarktiki* 2. [Atlas of the Antarctic, 2.] Leningrad.

Golubkova, N. S. (1974). Zhiznennye formy lishaynikov Antarktidy. [Lichen life-forms in the Antarctic.] In: *Novosty sistematiki nizshikh rasteniy.* [News in the systematics of lower plants.] 11. Leningrad.

Gorbatsky, G. V. (1967). *Fiziko–geograficheskoye rayonirovanie Arktiki.* [Physical–geographical division of the Arctic.] Part I. Leningrad.

Gorbatsky, G. V. (1970). *Fiziko–geograficheskoye rayonirovanie Arktiki.* [Physical–geographical division of the Arctic.] Part II. Leningrad.

Gorbatsky, G. V. (1973). *Fiziko–geograficheskoye rayonirovanie Arktiki.* [Physical–geographical division of the Arctic.] Part III. Leningrad.

Gorchakovsky, P. L., Gribova, S. A., Isachenko, T. I., Karpenko, A. S., Nikonova, N. N., Famelis, T. V., Fedorova, I. T. & Sharafutdinov, M. I. (1975). Rastitel'nost' Urala na novoy geobotanischeskoy karte. [The vegetation of the Urals on the new geobotanical map.] *Bot. Zh.* 60 (10).

Gorchakovsky, P. L. & Trotsenko, G. V. (1974). Rastitel'nost' stationara 'Kharp'. [The vegetation around the research station 'Kharp'.] *Tr. Inst. ekologii rast. i zhivotn. Ural'sk. nauchn. tsentra SO AN SSSR* 88. Sverdlovsk.

Gorodkov, B. N. (1916). Opyt deleniya Zapadno-Sibirskoy nizmennosti na botaniko-geograficheskie oblasti. [Attempt at division of the western Siberian lowland into botanical–geographical regions.] *Yezhegodn. Tobol'sk. gub. muzeya*, 27.

Gorodkov, B. N. (1924). Zapadno-Sibirskaya ekspeditsiya Akademiya Nauk i Russkovo geograficheskovo obshchestva. [The West-Siberian expedition of the Academy of Sciences and the Russian Geographical Society.] *Priroda*, 7–12.

Gorodkov, B. N. (1926). Polyarny Ural v verkhnem techenii r. Sobi. [Polar Urals around the upper course of Sob' River.] *Tr. bot. muz. AN SSSR* 19.

Gorodkov, B. N. (1927). Botanicheskie issledovaniya Gydanskoy ekspeditsii. [Botanical investigations of the Gydan Expedition.] *Osvedomitel'ny byulleten' AN SSSR* 11–12.

Gorodkov, B. N. (1929). Gydanskaya komplekskaya ekspeditsiya pod rukovodstom B. N. Gorodkova. [The Gydan combined expeditions under the leadership of B. N. Gorodkov.] *Otchet o deyatel'nosti AN SSSR za 1928 g.* [Account of the activities of the Academy of Sciences of the USSR during 1928.] Leningrad.

Gorodkov, B. N. (1932). Pochvi Gydanskoy Tundry. [The soils of the Gydan tundras.] *Tr. Polyarn. Komissii AN SSSR*, 7.

Gorodkov, B. N. (1933). Yestyestvennye pastbishchye ugod'ya tundrovoy zony DVK. [Natural pasture lands in the tundra zone of the Far East of the USSR.] *Sov. olenev.* 2.

Gorodkov, B. N. (1935*a*). Geobotanichesky i pochvenny ocherk Penzhinskovo rayona Dal'nevostochnovo Kraya. [Geobotanical and soils essay on the Penzhina area of the Far East District.] *Tr. Dal'nevost. fil. AN SSSR, ser. bot.* 1. Moscow–Leningrad.

Gorodkov, B. N. (1935*b*). Materialy dlya poznaniya gornykh tundr Urala. [Materials for the knowledge of the mountain tundras of the Urals.] In: *Priroda Urala.* [The Nature of the Urals.] Sverdlovsk.

Gorodkov, B. N. (1935*c*). *Rastitel'nost' tundrovoy zony SSSR.* [The vegetation of the tundra zone of the USSR.] Moscow–Leningrad.

Gorodkov, B. N. (1938). Rastitel'nost' Arktiki i gornykh tundr SSSR. [The vegetation of the arctic and mountain tundras of the USSR.] In: *Rastitel'nost' SSSR*, vol. 1 [The vegetation of the USSR, 1.] Moscow–Leningrad.

Gorodkov, B. N. (1939). Botaniko–geografichesky ocherk Chukotskovo poberezh'ya. [Botanical–geographical essay on the Chukotka coast.] *Uch. zap. Len. gos. Ped. inst. im. A. I. Gertsen*, 21. *Kaf. fiz. geogr.*

Gorodkov, B. N. (1943). Polyarnye pustyni ostrova Vrangela. [The polar deserts on Wrangel Island.] *Bot. Zh.* 28 (4).

Gorodkov, B. N. (1944). Tundry Ob'-Yeniseyskovo vodorazdela. [The tundras of the Ob-Yenisey watershed.] *Sov. bot.* 3, 4–5.

Gorodkov, B. N. (1946*a*). Botanico–geografichesky ocherk Kraynevo Severa i Arktiki SSSR. [Botanical–geographical essay on the Far North and the Arctic of the USSR.] *Uch. zap. Len. gos. ped. inst. im. A. I. Gertsena*, 49. *Kaf. fiz. geogr.*

Gorodkov, B. N. (1946*b*). Opyt klassifikatsii rastitel'nosti Arktiki. [An attempt at classification of the vegetation in the Arctic.] Parts 1, 2. *Sov. Bot.* 1, 2.

Gorodkov, B. N. (1952). Proiskhozhdenie arkticheskikh pustyn' i tundr. [The origin of the arctic deserts and the tundras.] *Tr. Bot. Inst. AN SSSR, ser. III. Geobot.* 8.

Gorodkov, B. N. (1956). Rastitel'nost' i pochvy o. Kotel'novo. [The vegetation and the soils of Kotel'ny Island.] In: *Rastitel'nost' Kraynevo Severa SSSR i yeyo osvoyenie*, 2. [The vegetation of the Far North of the USSR and its utilization.] Moscow–Leningrad.

Gorodkov, B. N. (1958*a*). Analiz rastitel'nosti zony arkticheskhikh pustyn' na primere ostrova Vrangelya. [An analysis of the vegetation of the arctic desert zone using Wrangel Island as an example.] In: *Rastitel'nost' Kraynevo Severa SSSR i yeyo osvoyenie.* [The vegetation of the Far North of the USSR and its utilization.] Vol. 3. Moscow–Leningrad.

Gorodkov, B. N. (1958*b*). Pochvenno-rastitel'ny pokrov ostrova Vrangelya. [The soil and vegetation cover of Wrangel Island.] In: *Rastitel'nost' Kraynevo Severa SSSR i yeyo osvoyenie.* [The vegetation of the Far North of the USSR and its utilization.] Vol. 3. Moscow–Leningrad.

Govorukha, L. S. (1968). Landshaftno–geograficheskaya kharakteristika Zemli Frantsa-Iosifa. [Landscape–geographical characteristics of Franz Joseph's Land.] *Tr. Arkt. i Antarkt. Inst.* 285.

Govorukha, L. C. (1970). Zemlya Frantsa–Iosifa. [Franz Joseph's Land.] In: *Sovietskaya Arktika.* [The Soviet Arctic.] Moscow.

Govorukha, L. S. & Bogdashevsky, B. I. (1970). Klimat. [Climate.] In: *Taymyro-Severo-zemel'skaya oblast'*. [The Taimyr Northland District.] Leningrad.

Govorukhin, V. S. (1933). Ocherk rastitel'nosti letnikh pastbishch severnovo olenya v tundrakh Obsko-Tasovskovo poluostrova. [Essay on the vegetation of summer pastures for the northern reindeer in the Obsk-Tazov tundras.] *Zemlevedenie* 35 (1).

Govorukhin, V. S. (1960). Pyatnistye tundry i plikativnye pochvy Severa. [The spotty tundras and the cryoturbated soils of the North.] *Zemlevedenie*, 5.

Greene, D. M. & Holtom, A. (1971). Studies in *Colobanthus quitensis* (Kunth) Bartl. and *Deschampsia antarctica* Desv.: III. Distribution, habitats and performance in the Antarctic botanical zone. *Brit. Antarct. Surv. Bull.* 23.

Greene, S. W. (1964*a*). Plants of the land. In: *Antarctic Research.* London.

Greene, S. W. (1964*b*). The vascular flora of South Georgia. *Brit. Antarct. Surv. Sci. Rep.* 45.

Greene, S. W. (1969). New records for South Georgia vascular plants. *Brit. Surv. Bull.* 22.

Greene, S. W. (1970). Studies in *Colobanthus quitensis* (Kunth.) Bartl. and *Deschampsia antarctica* Desv.: I. Introduction. *Brit. Antarct. Surv. Bull.* 23.

Greene, S. W. & Greene, D. M. (1963). Check list of the sub-Antarctic and Antarctic vascular flora. *Polar Record*, 11 (73).

Greene, S. W. & Walton, D. W. H. (1975). An annotated check list of the sub-Antarctic and Antarctic vascular flora. *Polar Record*, 17 (110).

Gribova, S. A. (1972). Opyt sostavleniya melkomasshtabnoy karty rastitel'nosti tundry i lesotundry (na primere severo-vostoka Russkoy ravniny). [An attempt at construction of a small-scale map of the vegetation of the tundra and the forest tundra (using the northeastern Russian lowland as an example).] In: *Geobotanicheskoye kartografirovaniye.* [Geobotanical cartography, 1972.] Leningrad.

Gribova, S. A. (1974). *O sootnoshenii regional'nykh i tipologicheskikh yedinits rastitel'novo pokrova v vostochnoevropeyskoy tundre.* [On the correlation of regional and typological units of the vegetation cover in the east-European tundra.] *Vses. sovyets. po klass. rast. IV.* [All-Soviet conf. on Vegetation classification.] Lvov. Lecture.

Gribova, S. A. (1975). *Rastitel'nost' Arktiki.* [The vegetation of the Arctic.] Manuscript. Bot. Inst. Acad. Sci. USSR.

Gribova, S. A. & Ignatenko, I. V. (1970). Pochvenno-rastitel'ny pokrov basseyna r. Korotaikhi. [Soils and vegetation cover in the Korotaikha River basin.] In: *Biologicheskie osnovy ispol'zovaniya priroda Severa.* [Biological basis for the utilization of the nature of the North.] Syktyvkar.

Gribova, S. A., Isachenko, T. I. & Karpenko, A. S. (1972). O zonal'nykh granitsakh na 'Karte rastitel'nosti yevropeyskoy chasti SSSR, m. 1:2 500 000'. [On zonal boundaries on the 'Map of the vegetation in the European parts of the USSR, scale 1:2 500 000'.] In: *Geobotanicheskoye kartografirovanie* 1972. [Geobotanical cartography, 1972.] Leningrad.

Gribova, S. A., Isachenko, T. I., Karpenko, A. S., Lavrenko, E. M., Lipatova, V. V. & Yurkovskaya, T. K. (1970). Legenda k 'Karte rastitel'nosti yeropeyskoy chasti SSSR', m. 1:2 500 000 v predelakh vostochno-Yevropeyskoy ravniny. [Legends for the 'Map of the vegetation in the European parts of the USSR in limits of the East-European Plain'.] *Bot. Zh.* 55, (11).

Grigor'yev, A. A. (1946). *Subarktika.* [The Subarctic.] Moscow–Leningrad.

Hadač, E. (1946). The plant communities of the Sassen Quarter, Vestspitsbergen. *Studia Bot. Československaka*, 7 (2–4).

Hadač, E. (1960). The history of the flora of Spitsbergen and Bear Island and the age of some arctic plant species. *Preslia*, 32.

Hanson, H. C. (1951). Characteristics of some grasslands, marsh and other plant communities in western Alaska. *Ecol. Monogr.* 21.

Hanson, H. C. (1953). Vegetation types in northwestern Alaska and comparison with communities in other arctic regions. *Ecology*, 34 (1).

Hanssen, O. & Holmboe, J. (1925). The vascular plants of Bear Island. *Nytt Magasin f. Naturvidenskapene* 62.

Hanssen, O. & Lid, J. (1932). Flowering plants of Franz Josef Land. *Skrift. om Svalbard og Ishavet*, 39.

Hare, F. K. (1955). Mapping of physiography and vegetation in Labrador–Ungava, a review of reconnaissance methods. *Canad. Geogr.* 5.

Harper, F. (1964). Plant and animal associations in the interior of the Ungava Peninsula. *Univ. Kansas Mus. Nat. Hist. Miscell. Publ.* 38.

Hart, H. C. (1880). On the botany of the British Polar Expedition of 1875–6. *J. Bot., N.S.* 9.

Hartz, N. (1895). Østgrønlands Vegetationsforhold. [Conditions of the East Greenland vegetation.] *Medd. om Grønl.* 18.

Hartz, N. & Kruuse, C. (1911). The vegetation of northeast Greenland 69°25′–75° lat. N. *Medd. om Grønl.* 30.

Heezen, B. C. & Tharp, M. (1963). The Atlantic floor. In: *The North Atlantic Biota and their History*. Ed. Á. Löve & D. Löve. Oxford.

Hofmann, W. (1968). Geobotanische Untersuchungen in Südost-Spitzbergen 1960. *Ergebn. Stauferland Expedition* 1959/1960, 8. Wiesbaden.

Holdgate, M. W. (1964). Terrestrial ecology in maritime Antarctic. In: *Biologie Antarctique*. Ed. E. Carrick *et al.* Paris.

Holmen, K. (1955). Notes on the bryophytes vegetation of Peary Land, North Greenland. *Mitt. Thüringisch. Bot. Ges.* 1 (2/3).

Holmen, K. (1957). The vascular plants of Peary Land, North Greenland. *Medd. om Grønl.* 124 (9).

Holmen, K. (1960). The mosses of Peary Land, North Greenland. *Medd. om Grønl.* 162 (2).

Holttum, R. E. (1922). The vegetation of West Greenland. *J. Ecol.* 10 (1).

Hultén, E. (1964). *The circumpolar plants*, I. Stockholm.

Hultén, E. (1968). *Flora of Alaska and neighboring territories.* Stanford, California.

Huntley, B. J. (1967). A preliminary account on the vegetation of Marion and Prince Edward Island. *South Afric. J. Sci.* 63 (6).

Hustich, I. (1949). Phytogeographical regions of Labrador. *Arctic*, 2 (1).

Hustich, I. (1957). On the phytogeography of the Subarctic Hudson Bay Lowland. *Acta Geografica* 16 (1).

Hustich, I. (1966). On the forest–tundra and the northern tree-lines. *Ann. Univ. Turku, A* 2 (36).

Ignatenko, I. V. (1966). O pochvakh ostrova Vaygach. [On the soils of Vaigach Island.] *Pochvovedenie*, 9.

Ignatenko, I. V., Knorre, A. V., Lovelius, N. V. & Norin, B. N. (1973). Zapasy fitomassy v tipichnykh rastitel'nykh soobshchestvakh lesnovo massiva Ary-Mas. [The phytomass reserve in typical plant associations in the forest massive at Ary-Mas.] *Ekologia*, 3.

Ignatenko, I. V. & Norin, B. N. (1969). Dinamika pyatnistykh tundr vostochnoyeropeyskovo Severa. [The dynamics of the spotty tundras in the East-European North.] *Probl. Bot.* 11.

Ignat'yev, G. M. (1963). Severnaya Amerika. [North America.] In: *Fizicheskaya geografiya chastey sveta.* [Physical geography of the parts of the World.] Moscow.

Igoshina, K. N. (1933). Botanicheskaya i khozyaistvennaya kharakteristika olen'ikh pastbishch v rayone Obdorskoy zonal'noy stantsii. [Botanical and agricultural characteristics of the reindeer pastures in the area of the Obdorsky zonal station.] *Sov. olenev.* 1.

Igoshina, K. N. (1944). Olen'i pastbishcha Polyarnovo Urala v verkhov'yakh rr. Longot-Yugan i Shchuch'ey. [Reindeer pastures of the Polar Urals at the upper reaches of the rivers Longot-Yugan and Shchuch'a.] *Sov. olenev.* 5.

Igoshina, K. N. (1961). Opyt botaniko-geograficheskovo rayonirovaniya Urala na osnove zonal'nykh floristicheskikh grupp. [An attempt at botanico-geographical division of the Urals on the basis of zonal, floristic groups.] *Bot. Zh.* 46 (2).

Il'ina, I. S. (1974). Geobotnicheskoye rayonirovanie na osnove strukturno – dinamichaskovo analiza rastitel'nosti. [Geobotanical division on the basis of a structural-dynamic analysis of the vegetation.] In: *IV. Vsesouyznoye soveshchanie po klassifikatsii rastitel'nosti.* [IV All-Soviet conference on the classification of the vegetation.] Lvov.

Il'ina, I. S. (1975). *Novaya obzornaya karta Zapado-Sibirskoy ravniny.* [New survey map of the West Siberian Lowlands.] Lecture. *Inst. Geogr. Sib. i. Dal'n. Vost.* 46.

Isachenko, A. G. (1965). *Osnovy landshaftovedeniya i fiziko-geograficheskovo rayonirovaniya.* [Basis for landscape management and physico-geographical regionalization.] Moscow.

Ivanova, E. I. (Ed.) (1958). *Pochvy Komi ASSR.* [Soils of the Komi ASSR.] Moscow.

Ivanova, E. I. (1962). Nekotorye zakonomernosti stroyeniya pochvennovo pokrova v tundre i lesotundre poberezh'ya Obskoy guby. [Some regularities in the structure of the soil cover in the tundra and forest tundra on the coast of the Ob Inlet.] In: *O pochvakh Urala, Zapadnoy i Tsentral'noy Sibiri.* [On the soils of the Urals, western and central Siberia.] Moscow.

Johnson, P. L. (1969). Arctic plants, ecosystems and strategies. *Arctic*, 22.

Karamysheva, Z. V., Karpenko, A. S., Malinovsky, K. A. & Odinak, Ya. P. (1975). IV Vsesoyuznoye soveshchanie po klassifikatsii rastitel'nosti. [The IV All-Soviet conference on vegetation classification.] *Bot. Zh.* 60 (7).

Karamysheva, Z. V., Lavrenko, E. M. & Rachkovskaya, E. I. (1969). Granitsa mezhdu steppnoy i pustynnoy oblastyami v Tsentral'nom Kazakhstane. [Boundaries between steppe and desert regions in Central Kazakhstan.] *Bot. Zh.* 54 (4).

Karamysheva, Z. V. & Rachkovskaya, E. I. (1973). *Botanicheskaya geografiya steppnoy chasti Tsentral'novo Kazakhstana.* [Botanical geography of the steppe areas of Central Kazakhstan.] Leningrad.

Karavayeva, N. A. (1962). K kharakteristike arkto-tundrovikh pochv o-va B. Lyakhovskovo (Novosibirskie ostrova). [On the characteristics of the arctic–tundra soils of the Big Lyakhovsky Island (Novosiberian Islands).] In: *O pochvakh Vostochnoy Sibiri.* [On the soils of Eastern Siberia.] Moscow.

Karavayeva, N. A. (1969). *Tundrovye pochvy Severnoy Yakutii.* [Tundra soils of northern Yakutia.] Moscow.

Kartushin, V. M. (1963). O rastitel'nosti o. Benneta. [On the vegetation of Bennett Island.] *Tr. Arkt. i Antarkt. n-issl. inst.* 224.

Katenin, A. E. (1970). Zonal'noye polozhenie i obshchie zakonomernosti rastitel'novo pokrova. [Zonal position and general regularity in the vegetation cover.] In: *Ekologiya i biologia rasteniy vostochnoyevropeyskoy lesotundry.* [Ecology and biology of the East European forest tundra.] Leningrad.

Katenin, A. E. (1972*a*). Printsipy klassifikatsii rastitel'nykh soobshchestva lesotundrovo statsionara. [Principles of classification of plant communities in the forest tundra research station.] In: *Pochvy i rastitel'nost' vostochnoyevropeyskoy lesotundry.* [Soils and vegetation of the East European forest tundra.] Leningrad.

Katenin, A. E. (1972*b*). Rastitel'nost' lesotundrovo statsionara. [The vegetation at the forest-tundra research station.] In: *Pochvy i rastitel'nost'vostochnoyevropeyskoy lesotundry.* [Soils and vegetation of the East European forest tundra.] Leningrad.

Katenin, A. E. (1974). Geobotanicheskie issledovaniya na Chukotke. I Rastitel'nost' srednevo techeniya reki Amguyemy. [Geobotanical investigations on Chukotka. I. The vegetation along the middle course of Amguyema River.] *Bot. Zh.* 59 (11).

Kats, N. Ya. (1939). *Bolota nizov'ye Obi.* [Mires along the lower reaches of Ob River.] *Sbornik, posvyashchennyy 70-letniyu Akademika V. L. Komarov.* [Articles dedicated to the 70th birthday on Academician V. L. Komarov.] Moscow–Leningrad.

Kats, N. Ya. (1948). *Tipy bolot SSSR i Zapadnoy Yevropy i ikh geograficheskoye rasprostranenie.* [Mire types of the USSR and West Europe and their geographical distribution.] Moscow.

Kats, N. Ya. (1966). O bolotakh Subantarktiki i kholodno-umerennoy zony Yuzhnovo polyshariya. [On the mires in the Subantarctic and cold-temperate zone of the southern hemisphere.] *Pochvovedenie*, 2.

Kats, N. Ya. (1971). *Bolota zemnovo shara.* [Bogs of the world.] Moscow.

Khodachek, E. A. (1969). Rastitel'naya massa tundrovykh fitosenozov Zapadnovo Taymyra. [The phytomass in the tundra phytocoenoses of Western Taimyr.] *Bot. Zh.* 54 (7).

Kihlman, A. O. (1890). Pflanzenbiologische Studien aus Russischen Lappland. *Acta Soc. Fauna et Flora Fennica*, 6 (3).

Kitsing, L. I., Koroleva, T. M. & Petrovsky, V. V. (1974). Flora sosudistikh rasteny okrestnostey bukhty Rodzhers (ostrov Vrangelya). [The flora of vascular plants in the area of Roger's Bay (Wrangel Island).] *Bot. Zh.* 59 (7).

Kjellman, F. R. (1883). Über den Pflanzenwuchs an der Nordküste Sibiriens. In: *Die wissenschaftliche Ergebnisse der Vega-Expedition.* I. [The scientific results of the Vega Expedition. I.] Leningrad.

Knapp, R. (1965). *Die Vegetation von Nord- und Mittelamerika.* Jena.

Knorre, A. V. (1974). Tipy listvennichnykh redkolesy na polyarnom predele lesov (Taymyr, 'Ary-Mas'). [Types of open larch forest at the polar limit of forests (Taimyr, 'Ary-Mas'.] In: *Biologicheskie Problemy Severa. VI. Symposium* [The biological problems of the North.] 5. Yakutsk.

Komarov, V. L. (1926). Vvedenie v izuchenie rastitel'nosti Yakutii. [Introduction to the study of the vegetation of Yakutia.] *Tr. Komissii po izuch. Yakutskoy ASSR* 1. Leningrad.

Komarov, V. L. (1936). Rod *Alnus* Gaertn. Sektsiya 1. *Alnobetula* W. D. Koch. [The genus *Alnus* Gaertn. Section 1. *Alnobetula* W. D. Koch.] *Flora USSR* V.

Korchagin, A. A. (1933). Ob osnovnykh ponyatiyakh tundrovedeniya. [On the basic concepts of tundra-science.] *Sov. Bot.* 2.

Korchagin, A. A. (1937). Rastitel'nost' morskikh allyuviev Mezenskogo Zaliva Cheshskoy guby. [The vegetation on the marine alluvium at Mezensky Inlet of Cheshska Bay.] *Tr. Bot. Inst. AN SSSR, ser. III, Geobotanika*, 2.

Korotkevich, E. S. (1958*a*). Rastitel'nost' Severnoy Zemli. [The vegetation of Severnaya Zemlya.] *Bot. Zh.* 43 (5).

Korotkevich, E. S. (1958*b*). Fiziko–geograficheskaya kharakteristika rayona rabot sovietskoy Antarkticheskoy ekspeditsii 1955–1957 gg. [Physico–geographical characteristics of the work area of the Soviet Antarctic expedition 1955–1957.] *Izv. VGO* 90 (3).

Korotkevich, E. S. (1966). Geobotanicheskoye rayonirovanie. [*Geobotanical zonation.*] In: *Atlas Antarktiki* 1. [Atlas of the Antarctic.]

Korotkevich, E. S. (1967). Polyarnye pustyni. [The Polar deserts.] *Informats. byull. Sovietsk. antarktich. ekspeditsii*, 65.

Korotkevich, E. S. (1970). Antarktika. [Antarctica.] *BSE*, 2.

Korotkevich, E. S. (1972). *Polyarnye pustyni.* [The Polar deserts.] Leningrad.

Kozhevnikov, Yu. P. (1973). Botaniko–geograficheskie nablyudeniya na zapade Chukotskovo poluostrova v 1971–1972 g. [Botanical–geographical discoveries on the western Chukotka peninsula during 1971–1972.] *Bot. Zh.* 58 (7).

Kruchinin, Yu. A. (1963). Fiziko–geograficheskie nablyudeniya na o. Kotel'nom

(rayon bukhti Temp). [Physical–geographical discoveries on Kotel'ny Island (in the area of Temp Bay).] *Tr. Arkt. i Antarkt. n-issl. inst.* 224.
Kuc, M. (1969). Plants from the nunataks of Torell Land, Vestspitsbergen. *Årbok Norsk Polarinst.* 1967. Oslo.
Kuc, M. (1973). Addition to the arctic moss flora. VII. Altitudinal differentiation of the moss cover at Purchase Bay, Melville Island, N.W.T. *Rev. Bryol. et Lichénol.* 39 (4).
Küchler, A. W. (1966). Potential natural vegetation of Alaska. In: *National Atlas of the U.S.* Washington.
Ladyzhenskaya, K. I. & Zhukova, A. L. (1971). Ekologo–morfologicheskie osobennosti pechenochnykh mkhov v usloviyakh vysokoshirotnoy Arktiki. [Ecological–morphological peculiarities of the hepatic mosses under the conditions in the high-latitude Arctic.] *Ekologia*, 3.
Ladyzhenskaya, K. I. & Zhukova, A. L. (1972). Pechenochnye mkhi (*Hepatici*) ostrova Zemlya Aleksandry. [Hepatic mosses of Alexandra Land.] *Bot. Zh.* 57 (3).
Larsen, J. A. (1965). The vegetation of the Ennedai Lake Area, N.W.T. *Ecol. Monographs*, 35 (1).
Lavrenko, E. M. (Ed.) (1939). *Karta rastitel'nosti SSSR, m. 1:5000 000.* [Vegetation map of the USSR, scale 1:5 000 000.] Moscow–Leningrad.
Lavrenko, E. M. (1946). Znachenie rabot V. V. Dokuchayev dlya razvitiya russkoy geobotaniki. [The importance of the works by V. V. Dokuchayev for the development of Russian geobotany.] In: *Dokuchayev i geografiya.* [Dokuchayev and the geography.] Moscow.
Lavrenko, E. M. (1947). Printsipy i yedinitsy geobotanicheskovo rayonirovaniya. [Principles and units of geobotanical division.] In: *Geobotanicheskoye rayonirovanie SSSR.* [Geobotanical division of the USSR.] Moscow–Leningrad.
Lavrenko, E. M. (1950). Osnovnye cherty botaniko-geograficheskovo razdeleniya SSSR i sopredel'nykh stran. [Basic traits of the botanical-geographical division of the USSR and neighboring countries.] *Probl. Bot.* 1.
Lavrenko, E. M. (1968). Ob ocherednykh zadachakh izucheniya geografii rastitel'novo pokrova v svyazi s botaniko-geograficheskim rayonirovaniem SSSR. [On the next problems in the study of the geography of the vegetation cover in the connection with the botanical-geographical division of the USSR.] In: *Osnovnye problemy sovremennoy geobotaniki.* [Basic problems of modern geobotany.] Leningrad.
Lavrenko, E. M. & Isachenko, T. I. (1976). Zonal'noye i provintsial'noye botaniko–geograficheskoye razdelenie yevropeyskoy chasti SSSR. [Zonal and provincial botanical–geographical division of the European parts of the USSR.] *Izv. VGO,* 108 (6).
Lavrenko, E. M. & Sochava, V. B. (Ed.) (1949). *Karta rastitel'nosti Yevropeyskoy chasti SSSR, m. 1:2 500 000.* [Map of the vegetation of the European parts of the USSR, scale 1:2 500 000.] Moscow–Leningrad.
Lavrenko, E. M. & Sochava, V. B. (Ed.) (1954). *Geobotanicheskaya karta SSSR, m. 1:4 000 000.* [Geobotanical map of the USSR, scale 1:4 000 000.] Moscow–Leningrad.
Lee, H. A. (1960). Late glacial and postglacial Hudson Bay sea episode. *Science*, 131, p. 3413.
Leskov, A. I. (1947). A. Arkticheskaya tundrovaya oblast'. B. Yevropeysko-Sibirskaya kustarnikovaya (lesotundrovaya) oblast'. C. Beringiyskaya kustarnikovaya

(lesotundrovaya) oblast'. [A. The arctic tundra region. B. The European-Siberian shrub (forest-tundra) region. C. The Beringian shrub (forest-tundra) region.] In: *Geobotanicheskoye rayonirovanie SSSR.* [Geobotanical division of the USSR.] Moscow–Leningrad.

Lid, J. (1967). Synedria of twenty vascular plants from Svalbard. *Bot. Jahrb.* 86.

Lindsay, D. C. (1971). Vegetation of the South Shetland Islands. *Brit. Antarct. Surv. Bull.* 25.

Lindsay, D. C. (1973). Effects of reindeer on plant communities in the Royal Bay area of South Georgia. *Brit. Antarct. Surv. Bull.* 35.

Lindsay, D. C. (1975). The macrolichens of South Georgia. *Brit. Antarct. Surv. Sci. Reports,* 89.

Lindsey, A. A. (1952). Vegetation of the ancient beaches above Great Bear and Great Slave Lakes. *Ecology*, 33 (4).

Longton, R. E. (1966). Alien vascular plants on Deception Island, South Shetland Islands. *Brit. Antarct. Surv. Bull.* 9.

Longton, R. E. (1967). Vegetation in the maritime Antarctic. *Phil. Trans. Roy. Soc. London, Ser. B.* 252.

Longton, R. E. (1970). Growth and productivity of the moss *Polytrichum alpestre* Hoppe in Antarctic regions. *Antarct. Ecol.* 2. London.

Longton, R. E. (1972). Studies of the classification, biomass and microclimate of vegetation near McMurdo Sound. *Antarct. J., U.S.* 7.

Longton, R. E. (1973). A classification of terrestrial vegetation near McMurdo Sound, continental Antarctica. *Canad. J. Bot.* 51.

Longton, R. E. (1974). Microclimate and biomass in communities of the *Bryum* association on Ross Island, Continental Antarctica. *Bryologist*, 77 (2).

Löve, Á. & Löve, D. (1974). Origin and Evolution of the arctic and alpine floras. In: *The Arctic and Alpine Environment*, ed. R. C. Barry & J. D. Ives, pp. 571–603. Methuen, London.

Löve, Á. & Löve, D. (1975). Cryophytes, polyploidy and continental drift. *Phytocoenologia*, **2**, 54–65.

Löve, Á. & Löve, D. (1979). The history and geobotanical position of the Icelandic flora. *Phytocoenologia*, **6**, 94–5.

Lundager, A. (1912). Some notes concerning the vegetation of Germania Land, N. E. Greenland. *Medd. om Grønl.* 43.

Lynge, B. (1923). Vascular plants from Novaya Zemlya. *Rep. Sci. Res. Norweg. Expedition to Novaya Zemlya 1921*, 13. Kristiania.

Lynge, B. (1928). Lichens from Novaya Zemlya. *Rep. Sci. Res. Norweg. Expedition to Novaya Zemlya 1921*, 43. Kristiania.

Lynge, B. (1931). Lichens collected on the Norwegian scientific expedition to Franz Josef Land 1930. *Skrifter om Svalbard og Ishavet*, 38.

Maini, J. S. (1966). Phytoecological study of sylvotundra at Small Tree Lake, N.W.T. *Arctic*, 19 (3).

Malyshev, L. I. (Ed.) (1972). *Vysokogornaya flora Stannovovo Nagor'ya.* [*The alpine flora of the Stannovoye Nagorye Mountains.*] Novosibirsk.

Manakov, K. N. (1972). *Produktivnost' i biologichesky krugovorot v tundrovykh biogeotsenozakh Kol'skovo poloustrova.* [Productivity and biological rotation in the tundra biogeocoenoses of the Kola peninsula.] Leningrad.

Marr, J. W. (1948). Ecology of the forest–tundra ecotone of the east coast of Hudson Bay. *Ecol. Monogr.* 18.

Matveeva, N. V., *see* Matveyeva, N. V.
Matveyeva, N. V. (1968). Osobennosti structury rastitel'nosti osnovnykh tipov tundr v srednem techenii reki Pyasiny (zapadnyi Taymyr). (The structure of vegetation of the basic tundra types along the middle course of the Pyasina River (western Taimyr).) *Bot. Zh.* 53 (11).
Matveyeva, N. V. (Matveeva, N. V.) (1972). The Tareya world model. In: *Tundra biome. Proc. IV Intern. meet. on the biol. product. of tundra.* Ed. F. E. Wielgolaski & Th. Rosswall. Stockholm.
Matveyeva, N. V. & Chernov, Yu. I. (1976. Polarnye pustyni poloustrova Taymyr. [The polar deserts of the Taimyr peninsula.] *Bot. Zh.* 61 (3).
Matveyeva, N. V., Polozova, T. G., Blagodatskikh, L. S. & Dorogostaiskaya, E. V. (1973). Kratky ocherk rastitel'nosti okrestnostey Taymyrskovo biogeotsenologicheskovo statsionara. [Short essay on the vegetation in the surroundings of the Taimyr biogeocoenological research station.] In: *Biogeotsenozy taymyrskoy tundry i ikh productivnost'.* [Biogeocoenoses of the Taimyr tundras and their productivity.] 2.
Mel'tser, L. I. (1973). Osnovye zakonomernosti raspredeleniya rastitel'novo pokrova Zapadno-Sibirskikh tundr. [Basic regularity in the distribution of the vegetation cover of the west Siberian tundra.] In: *Priroda i prirodniye resursy Tyumenskoy oblast.* [Nature and natural resources in the Tyumen' region.] Tyumen'.
Meusel, H. Jäger, E. & Weinert, E. (1965). *Vergleichende Chorologie der zentraleuropäischen Flora.* Jena.
Michelmore, A. P. G. (1934). Botany of the Cambridge expedition to Edge Island, S. E. Spitsbergen, in 1927. Part II. Vegetation. *Ecol.* 22(1).
Mikhailichenko, V. S. (1936). Eskiz roslinnosti pivnichnoskhidnoy okrayin p-va Yamalu. [Sketch of the vegetation of the north-eastern part of the Yamal peninsula.] *Zh. Inst. Bot. Ukrain. AN* 7(15).
Mikhailov, I. S. (1960). Nekotorye osobennosti dernovikh arkticheskikh pochv o. Bolshevik. [Some peculiarities of the arctic tundra soils on Bolshevik Island.] *Pochvovedenie*, 6.
Mikhailov, I. S. (1963*a*). Pochvy severo-vostochnoy chasti o. Faddeeyvskovo. [Soils in the northeastern parts of Faddeyevsky Island.] *Probl. Arkt. i Antarkt.* 14.
Mikhailov, I. S. (1963*b*). Evolyutsia pochv i rastitel'novo pokrova severo – vostochnoy chasti o Faddeyevskovo. [The evolution of soils and vegetation cover in the northeastern parts of Faddeyevsky Island.] *Tr. Arkt. i Antarkt. n-issl. inst.* 224.
Mikhailov, I. S. & Govorukha, L. S. (1962). Pochvy Zemli Frantsa-Iosifa. [Soils of Franz Joseph's Land.] *Vestn. MGU*, 6.
Mineyev, A. I. (1946). *Ostrov Vrangelya.* [Wrangel Island.] Moscow – Leningrad.
Minyayev, N. A. (1963). *Structura rastitel'nikh assotsiatsiy.* [The structure of plant associations.] Moscow – Leningrad.
Molenaar, J. C. de (1974). Vegetation of the Angmagssalik district, Southeast Greenland. I. Littoral vegetation. *J. Ecol.* 22 (1).
Muc, M. (1972). Vascular plant production in the sedge meadows of the Truelove Lowland. In: *Devon Island IBP project, higharctic ecosystem.* Ed. L C. Bliss. Edmonton.
Muc, M. (1973). Primary production of plant communities of the Truelove Lowland, Devon Island, Canada, Sedge meadows. In: *Tundra Biome. Primary production and production processes.* Ed. L. C. Bliss & F. E. Wielgolaski. Stockholm.
Muc, M. (1977). Ecology and primary production of sedge–moss meadow

communities, Truelove Lowland. In: *Truelove Lowland, Devon Island, Canada: a high arctic ecosystem.* Ed. L. C. Bliss. Edmonton.

Muc, M. & Bliss, L. C. (1977). Plant communities of Truelove Lowland, In: *Truelove Lowland, Devon Island, Canada: a high arctic ecosystem.* Ed. L. C. Bliss. Edmonton.

Neilson, A. M. (1968). Vascular plants from the northern part of Nordaustlandet, Svalbard. *Norsk Polarinst. Skrifter*, 143.

Nikolayeva, M. G. (1941). Kustarnikovy tip rastitel'nosti yuzhnoy chasti Bol'shovo i Malovo Yamala. [The shrub vegetation in the southern parts of Big and Little Yamal.] *Bot. Zh.* 26 (1).

Nitsenko, A. A. (1966). O kriteryakh vydeleniya rastitel'nykh assotsiatsiy. [On the criteria for the determination of plant associations.] *Bot. Zh.* 51 (8).

Norin, B. N. (1957). Mesto lesotundry v sisteme rastitel'nykh zon i problema vydeleniya lesotrundrovo tipa rastitel'nosti. [The position of the forest tundra in the system of vegetation zones and the problem of determination of the forest tundra type of vegetation.] Lecture. *Delegatsk. s"yezda VBO (May 1957)*, 4. Leningrad.

Norin, B. N. (1961). Chto takoye lesotundra? [What is the forest tundra?] *Bot. Zh.* 46 (1).

Norin, B. N. (1966). O zonal'nykh tipakh rastitel'novo pokrova v Arktike i Subarktike. [On the zonal types of the vegetation cover in the Arctic and the Subarctic.] *Bot. Zh.* 51 (11).

Norin, B. N. (1972). The main ecological surveys at the station Ary-Mas In: *Tundra biome. Proc. IV Intern. meet. on the biol. product of tundra.* Ed. F. E. Wielgolaski & Th. Rosswall. Stockholm.

Norin, B. N. (1974). Nekotorye problemy izucheniya vzaimo-otnosheniya lesnykh i tundrovykh ekosistem. [Some problems in the study of inter-relationships between the forest and the tundra ecosystems.] *Bot. Zh.* 59 (9).

Norin, B. N. (1976). *Struktura rastitel'nykh soobshchesty Vostochnoyevropeyskoy lesotundry.* [The structure of the plant communities in the East-European forest tundra.] Manuscript. Bot. Inst. AN SSSR. Leningrad.

Norin, B. N. (1978). Rastitel'ny pokrov urochishcha Ary-Mas. [Vegetation of Ary-Mas.] In: *Ary-Mas, prirodnye usloviya, flora i rastitel'nost'.* [Ary-Mas, the natural conditions, flora and vegetation.] Leningrad.

Norin, B. N., Druzin, A. B., Boch, M. S. & Solonevich, N. G. (1970). Rastitel'ny pokrov i pochvy. [The plant cover and the soils.] In: *Ekologiya i biologiya rasteniy vostochno-yevropeikoy lesotundry.* [Ecology and biology of the plants in the East-European forest tundra.] Leningrad.

Nosova, L. I. (1964). Ocherk tundrovoy i lesotundrovoy rastitel'nosti mezhdu-rech'ya Yany i Omoloya (severnaya Yakutia). [Essay on the tundra and forest tundra vegetation of the interfluvial area between Yana and Omoloy Rivers (northern Yakutia).] *Bot. Zh.* 49 (5).

Novichkova-Ivanova, L. N. (1963). Smeny sinuziy pochvennykh vororosley zemli Frantsa-Iosifa. [The successions in the synusia of soil algae on Franz Joseph's Land.] *Bot. Zh.* 48 (1).

Novichkova-Ivanova, L. N. (1972). Soil and aerial algae in the polar deserts and arctic tundras. In: *Tundra biome. Proc. IV Internat. meet. on the biol. prod. of tundra.* Ed. F. E. Wielgolaski & Th. Rosswall. Stockholm.

Oosting, H. J. (1948). Ecological notes on the flora of East Greenland and Jan Mayen. *Amer. Geogr. Soc. Spec. Publ.* 30.

Ostenfeld, C. H. (1925). Vegetation of North Greenland. *Bot. Gaz.* 80.
Ostenfeld, C. H. (1927). The vegetation of the north-east Greenland, based upon Dr. Th. Wulff's collections and observations. *Medd. om Grønl.* 64.
Pakarinen, P. (1974). Tundrasioden kasvillisuudesta Devon-saarella. [Tundra vegetation of Devon Island.] *Suo*, 25 (3–4).
Pakarinen, P. & Vitt, D. H. (1973). Primary production of plant communities of the Truelove Lowland, Devon Island, Canada. Moss communities. In: *Tundra biome. Primary production and production processes.* Ed. L. C. Bliss & F. E. Wielgolaski. Stockholm.
Palibin, I. V. (1903). Botanicheskie rezultaty plavaniya ledokola 'Yermak' v Severnom Ledovitom Okeane letom 1901 g. [Botanical results from the navigation of the icebreaker 'Yermak' in the North Polar Sea in the summer of 1901.] *Izv. SPb Bot. Sada*, 3 (2, 3, 5).
Palmer, L. J. & Rouse, C. H. (1945). Study of the Alaska tundra with reference to its reaction to reindeer and other grazing. *U.S. Fish and Wildlife Serv., Res. Rep.* 10.
Passarge, S. (1921). *Vergleichende Landschaftskunde.* Vol. 1, 2. Munich.
Pavlova, E. B. (1969). O rastitel'noy masse tundr Zapadnovo Taymyra. [On the phytomass of the tundra in Western Taimyr] *Vestn. MGU, biol. pochvoved.* 5.
Perfil'yev, I. A. (1928). *Materialy k flore o. o. Novoy Zemli i Kolguyeva.* [Material for a flora of Novaya Zemlya and Kolguyev Island.] Archangelsk.
Perfil'yev, I. A. (1934). *Flora Severnovo Kraya.* [Flora of the North District.] Part 1. Archangelsk.
Perfil'yeva, V. I. & Rykova, Yu. V. (1975). Arkticheskaya tundra v ust'ye r. Chukochyey. [The arctic tundra at the mouth of Chukochya River.] In: *Botanicheskie issledovaniya v Yakutii.* [Botanical investigations of Yakutia.] Yakutsk.
Petrovsky, V. V. (1959). O strukture rastitel'nykh assotsiatsiy valikovykh polygonal'nykh bolot v nizov'yakh r. Leny. [On the structure of the plant associations on the raised edges of the low centre polygons in polygonal mires at lower Lena River.] *Bot.Zh.* 44 (10).
Petrovsky, V. V. (1967). Ocherk rastitel'nykh soobshchestv tsentral'noy chasti ostrova Vrangelya. [Essay on the plant communities in the central parts of Wrangel Island.] *Bot. Zh.* 52 (3).
Petrovsky, V. V. (1973). Spisok sosudistykh rastenny ostrova Vrangelya. [List of the vascular plants on Wrangel Island.] *Bot. Zh.* 58 (1).
Petrovsky, V. V. & Yurtsev, B. A. (1970). Znachenie flory o. Vrangelya dlya rekonstruktsii landshaftov shel'fovykh territoriy. [Importance of knowledge of the flora of Wrangel Island for the reconstruction of the landscape on the shelf territory.] In: *Severny Ledovitiy Okean i yevo poberezh'ye v kaynozoe.* [The North Polar Sea and its shore during the Coenozoic.] Leningrad.
Philippi, G. (1973). *Moosflora und Moosvegetation des Freeman-Sund-Gebietes (Südost Spitsbergen).* [The moss flora and moss vegetation in the Freeman Sund area (southeast Spitzbergen).] Wiesbaden.
Pohle, R. (1903). Pflanzengeographische Studien über die Halbinsel Kanin und das angrenzende Waldgebiet. *Tr. SPb. Bot. Sada*, 21 (1).
Pohle, R. (Pole) (1910). Programma dlya botanico-geograficheskikh issledovaniy tundry. [Program for the botanical-geographical investigation of the tundra.] In: *Programma dlya botanico–geograficheskikh issledovaniy* [Program for botanical–geographical investigations.] 2 SPb.

Pohle, R. (1918). Beiträge zur Kenntniss der Westsibirischen Tiefebene. *Zschr. Ges. für Erdkunde*, 1–2, 1918.
Pohle, R. (1919). Beiträge zur Kenntniss der Westsibirischen Tiefebene. *Zschr., Ges. für Erdkunde*, 9–10, 1919.
Polozova, T. G. (1961). O samykh severnykh mestonakhozhdeniyakh listvennitsy (*Larix dahurica* Turcz.) i kustarnoy ol'khi (*Alnaster fruticosa* Ldb.) v nizov'yakh r. Leny. [On the northernmost localities of larch (*Larix dahurica* Turcz.) and shrub alder (*Alnaster fruticosa* Ldb.) along the lower Lena River.] In: *Materialy po rastitel'nosti Yakutii.* [Material on the vegetation of Yakutia.] Leningrad.
Polozova, T. G. (1966). K biologii i ekologii karlikovoy beryozki (*Betula nana* L.) v vostochnoyevropeyskoy lesotundre. [To the biology and ecology of the dwarf birch (*Betula nana* L.) in the East European tundra.] In: *Prisposoblenie rasteniy Arktiki k usloviyam sredy.* [Adaptation of plants to the Arctic and the conditions of the environment.] Moscow–Leningrad.
Polozova, T. G. (1970). Biologicheskie osobennosti *Eriophorum vaginatum* L. kak kochko-obrazovatelya (po nablyudeniyam v tundrakh Zapadnovo Taymyra). [Biological peculiarities of *Eriophorum vaginatum* L. as a tussock former (in the tundras of Western Taimyr).] *Bot. Zh.* 55 (3).
Polozova, T. G. (1976). Zhiznennye formy komponentov rastitel'nykh soobshchestv v tipichnoy tundre Zapadnovo Taymyra. [Life-forms of the components of the plant communities in typical tundra of Western Taimyr.] In: *Biologicheskie problemy Severa.* [Biological Problems of the North.] Petropavlovsk.
Polunin, N. (1934). The vegetation of Akpatok Island, 1. *J. Ecol.* 22 (2).
Polunin, N. (1935). The vegetation of Akpatok Island, 2. *J. Ecol.* 23 (1).
Polunin, N. (1945). Plant life in Kongsfjord, West Spitsbergen. *J. Ecol.* 33 (1).
Polunin, N. (1948). Botany of the Canadian Eastern Arctic. 3. Vegetation and ecology. *Bull. Nat. Mus. Canada*, 104.
Polunin, N. (1951). The real Arctic. *J. Ecol.* 39 (2).
Polunin, N. (1960). *Introduction to plant geography and some related sciences.* London.
Popov, A. I. (1967). *Merzlotnye yavleniya y zemnoy kore (kriolitologia).* [Permafrost development in the earth's crust (cryolithology).] Moscow.
Porsild, A. E. (1951). Plant life in the Arctic. *Canad. Geogr. J.* 42 (3).
Porsild, A. E. (1955). The vascular plants of the Western Canadian Arctic Archipelago. *Nat. Mus. Canada Bull.* 135.
Porsild, A. E. (1957*a*). *Illustrated flora of the Canadian Arctic Archipelago.* Ottawa.
Porsild, A. E. (1957*b*). Natural vegetation. In: *Atlas of Canada.* Ottawa.
Porsild, A. E. & Cody, W. J. (1968). *Checklist of the vascular plants of Continental Northwest Territories, Canada.* Ottawa.
Porsild, M. P. & Porsild, A. E. (1920). The flora of Disko Island and adjacent coast of West Greenland from 66°–71°N. lat. with remarks on phytogeography, ecology, flowering, fructification and hibernation. *Medd. om Grønl.* 58.
Pospelova, E. B. (1972). Vegetation of the Agapa station and productivity of the main plant communities. In: *Tundra biome. Proc. IV Internat. meet. on the biol. product. of tundra.* Ed. F. E. Wielgolaski & Th. Rosswall. Stockholm.
Pospelova, E. B. (1974). Structura i prostranstvennoye raspredelenie rastitel'noy massy v osnovnykh rastitel'nykh soobshchestvakh stationara 'Agapa'. [The structure and spatial distribution of the phytomass in the main plant communities

at the Agapa station.] In: *Pochvy i produktivnost' rastitel'nykh soobshchestv*. [The soils and the productivity of plant associations.] 2. Moscow.

Pospelova, E. B. & Zharkova, Yu.G. (1972). Rastitel'ny pokrov i fitomassa osnovnykh rastitel'nykh soobshchestv stationara 'Agape'. [The vegetation cover and phytomass of the main plant communities at the 'Agapa' station.] In: *Pochvy i produktivnost' rastitel'nykh soobshchestv*. [The soils and the productivity of plant associations.] 1. Moscow.

Price, L. W., Bliss, L. C. & Svoboda, J. (1974). Origin and significance of wet spots on scraped surfaces in the high Arctic. *Arctic*, 27 (4).

P'yavchenko, N. I. (1955). *Bugristye torfyaniki*. [Palsa bogs.] Moscow.

Rakhmanina, A. T. (1971). Nadzemnaya i podzemnaya fitomassa nekotorykh soobshchestv Vostochnoyevropeyskoy lesotundry. [Above and below-ground phytomass of some plant communities in the East European forest-tundra.] In: *Biologicheskaya produktivnost' i krugovorot khimicheskikh elementov v rastitel'nykh soobshchestvakh*. [Biological productivity and circulation of chemical elements in plant communities.] Leningrad.

Rakhmanina, A. T. (1974). Zapasy nadzemnoy i podzemnoy fitomassa v tipichnykh soobshchestvakh Vostochnoyevropeyskoy lesotundry. [Reserve of above and below-ground phytomass in typical plant communities in East-European forest-tundra.] *Bot. Zh.* 59 (6).

Ramenskaya, M. L. (1972). Rastitel'nost' Pechengskikh tundr. [The vegetation of the Pechengsky tundras.] In: *Flora i rastitel'nost' Murmanskoy oblasti*. [Flora and vegetation in the Murmansk area.] Leningrad.

Raup, H. (1965). The flowering plants and ferns of the Mesters Vig district, Northeast Greenland. *Medd. om Grønl.* 166 (2).

Raup, H. (1969). The relation of the vascular flora to some factors of site in the Mesters Vig district, Northeast Greenland. *Medd. om Grønl.* 176 (5).

Rebristaya, O. V. (1966). Spisok sosudistykh rasteniy ostrova Muostakh (guba Buorkhaya, Arkticheskaya Yakutiya). [List of the vascular plants of Muostakh Island (the Bay of Buorkhaya, Arctic Yakutia).] In: *Rasteniya severa Sibiri i Dal'nevo Vostoka*. [The plants of northern Siberia and the Far East.] Moscow–Leningrad.

Rebristaya, O. V. (1970*a*). K kharakteristike flory vostochnoy chasti Bol'shezemel'skoy tundry. [To the description of the flora in the eastern part of the Bol'shezemelya tundra.] In: *Biologicheskie osnovy ispol'zovaniya prirody Severa*. [Biological basis for the utilization of the nature in the North.] Syktykvar.

Rebristaya, O. V. (1970*b*). Sibirskie elementy vo flore severovostokaYevropy i ikh proiskhozhdenie. [Siberian elements in the flora of northeast Europe and their origin.] In: *Severny Ledovitiy Okean i yevo poberezh'ye v kaynozoe*. [The North Polar Sea and its shore during the Coenozoic.] Leningrad.

Rebristaya, O. V. (1977). *Flora vostoka Bol'shezemel'skoy tundry*. [Flora of eastern Bol'shezemelya Tundra.] Leningrad.

Rebristaya, O. V. & Yurtsev, B. A. (1973). *Floristicheskoye rayonirovanie Arktiki*. [Floristic division of the Arctic.] Bot. Inst. AN SSSR. Leningrad.

Regel, K. W. (1923). Die Pflanzendecke des Halbinsel Kola. *Mêm. Fac. Sci. Univ. Lithuania*, 1.

Regel, K. W. (1927). Die Pflanzendecke des Halbinsel Kola. *Mêm. Fac. Sci. Univ. Lithuania*, 2.

Regel, K. W. (1928). Die Pflanzendecke des Halbinsel Kola. *Mêm. Fac. Sci. Univ. Lithuania*, 3.

Regel, K. W. (1932). Die Fleckentundra von Nowaja Semlja. *Beitr. Biol. Pflanzen, Breslau*, 20.

Regel, K. W. (1935). Die Moore von Nowaja Semlja. *Beitr. Biol. Pflanzen, Breslau*, 23.

Reutt, A. T. (1970). Rastitel'nost'. [Vegetation.] In: *Sever Dal'nevo Vostoka.* [Northern Far East.] Leningrad.

Richardson, D. H. S. & Finegan, E. J. (1973). Primary production of plant communities of the Truelove Lowland, Devon Island, Canada. Lichen communities. In: *Tundra biome. Primary production and production processes.* Ed. L. C. Bliss & F. E. Wielgolaski. Stockholm.

Richardson, D. H. S. & Finegan, E. (1977). Studies on the lichens of Truelove Lowland. In: *Truelove Lowland, Devon Island, Canada: a high arctic ecosystem.* Ed. L. C. Bliss. Edmonton.

Ritchie, J. C. (1960). The vegetation of northern Manitoba. V. *Arctic*, 13.

Romanovsky, N. N. (1960). Veneering frost structures. *Buletyn Peryglacjalny*, 7. Lódź.

Rønning, O. (1963). Phytogeographical problems in Svalbard. In: *North Atlantic biota and their History.* Ed. Á. Löve & D. Löve. Oxford.

Rønning, O. (1964). *Svalbards flora.* Oslo.

Rønning, O. (1965). Studies in Dryadion of Svalbard. *Norsk Polarinst. Skrifter*, 134.

Rønning, O. (1969). Features of the ecology of some arctic Svalbard (Spitsbergen) plant communities. *Arctic and Alp. Res.* 1 (1).

Rønning, O. (1970). Synopsis of the flora of Svalbard. *Årbok Norsk Polarinst.* 1969. Oslo.

Rousseau, J. (1968). The vegetation of the Quebec–Labrador peninsula between 55° and 60° N. *Naturaliste Canad.* 95 (2).

Rubinshtein, E. S. (1953). O vliyanii raspredeleniya osadkov sushi na zemnom share na temperatury vosdukha. [On the effect on the air temperature of the distribution of the precipitation over the globe.] *Izv. VGO*, 85 (4).

Rudolph, E. D. (1963). Vegetation of Hallett Station area, Victoria Land, Antarctica. *Ecology*, 44 (3).

Safronova, I. N. (1975). Novye dannye o rastitel'nosti ostrova Oktyabr'skoy Revolutsii (Severnaya Zemlya). [New data on the vegetation of the October Revolution Island (Severnaya Zemlya).] Manuscript. Bot. Inst. AN SSSR. Leningrad.

Salazkin, A. S. (1934). Yestyestvennye kormovye ugod'ya Murmanskovo okruga. [The natural pastureland in the Murmansk district.] *Sov. olenev.* 1.

Salazkin, A. S., Sambuk, F. V., Pol'yanskaya, O. S. & Pryakhin, M. I. (1936). Olen'i pastbishcha i rastitel'ny pokrov Murmanskovo okruga. [The reindeer pastures and the vegetation cover in the Murmansk District.] *Tr. Arkt. in-ta*, 72.

Sambuk, F. V. (1931). Geobotanicheskaya kharakteristika olen'ikh pastbishch u ust'ya r. Pechory. [Geobotanical description of the reindeer pastures at the mouth of Pechora River.] In: *Olen'i pastbishcha Severnovo Kraya.* [Reindeer pastures in the North District.] Archangelsk.

Sambuk, F. V. (1933). Pastbishchnye ugod'ya pervovo nenetskovo olenevodcheskovo kolkhoza. [Pasture lands of the first Nenets reindeer kolkhoz.] In: *Olen'i pastbishcha Severnovo Kraya.* [Reindeer pastures in the North District.] 2. Leningrad.

Sambuk, F. V. (1934). Kormovye ugod'ya tundr nenetskovo okruga Severnovo Kraya. [Pasture land in the tundras of the Nenets area of the North District.] *Sov. Olenev.* 1.

Sambuk, F. V. (1937). Kratky ocherk rastitel'nosti Taymyra. [Short essay on the vegetation of Taimyr.] *Probl. Arktiki*, 1.

Sambuk, F. V. & Dedov, A. A. (1934). Podzony Pripercherskikh tundr. [The subzones of the Pechora tundras.] *Tr. Bot. Inst. AN SSSR, ser. III. Geobotanika*, 1.

Savich-Lyubitskaya, L. I. & Smirnova, Z. N. (1969). Mkhi vostochnoy Antarktidy. [Mosses of the eastern Antarctic.] In: *Atlas Antarktiki.* [Atlas of the Antarctic.] 2. Leningrad.

Savile, D. B. O. (1959). The botany of Somerset Island, District of Franklin. *Canad. J. Bot.* 37 (5).

Savile, D. B. O. (1961). The botany of the northernmost Queen Elizabeth Islands. *Canad. J. Bot.*, 39 (4).

Savile, D. B. O. (1964). General ecology and vascular plants of the Hazen Camp area. *Arctic*, 17 (4).

Schmithüsen, J. (1968). *Allgemein Vegetationsgeographie.* 3rd ed. Berlin.

Schofield, W. B. (1959). The salt marsh vegetation of Churchill, Manitoba, and its phytogeographical implication. *Nat. Mus. Canad. Bull.* 160.

Scholander, P. E. (1934). Vascular plants from Northern Svalbard with remarks on the vegetation in North-East Land. *Skrifter om Svalbard og Ishavet*, 62.

Schrenk, A. G. (1848). *Reise nach dem Nordosten des europäischen Russlands.* I. Dorpat.

Schrenk, A. G. (1854). *Reise nach dem Nordosten des europäischen Russlands.* II. Dorpat.

Schwarzenbach, F. H. (1960). Die arktische Steppe in den Trockengebiet Ost- und Nordgrönlands. *Ber. Geobot. Inst. Rübel für 1959.*

Schwarzenbach, F. H. (1961). Botanische Beobachtungen in der Nunatakerzone Ostgrönlands zwischen 74° und 75° n. Br. *Medd. om Grønl.* 163 (5).

Scotter, G. W. & Telfer, E. S. (1975). Potential for red meat production from wildlife in Boreal and Arctic regions. In: *Proc. Circumpolar Conference on Northern Ecology.* Ottawa.

Sdobnikov, V. M. (1937). Raspredelenie mlekopitayushchikh i ptits po tipam mestoobitaniy v Bol'shezemel'skoy tundre i na Yamale. [Distribution of mammals and birds according to habitat in the tundras of Bol'shezemlya and Yamal.] *Tr. Arkt. Inst.* 92.

Sdobnikov, V. M. (1959). Biotopy Severnovo Taymyra i plotnost' populatsiy naselyayushchikh ikh zhivotnikh. [Biotopes of northern Taimyr and the population density of the wildlife inhabitants.] *Zool. Zh.* 38 (2).

Seidenfaden, G. (1931). Moving soil and vegetation in East Greenland. *Medd. om Grønl.* 87 (2).

Seidenfaden, G. & Sørensen, Th. (1937). The vascular plants of Northeast Greenland from 74° 30′ to 79° 00′ N. lat. and a summary of all species found in East Greenland. *Medd. om Grønl.* 163 (4).

Semenov, I. V. (1966*a*). Vnutrilandshaftnoye rayonirovanie Severnoy Zemli. [Landscapes of the inland area of Severnaya Zemlya.] *Izv. VGO*, 98 (6).

Semenov, I. V. (1966*b*). Fiziko-geograficheskoye rayonirovanie Severnoy Zemli. [Physical-geographical division of Severnaya Zemlya.] *Izv. VGO*, 98 (1).

Semenov, I. V. (1968). O zakonomernostyakh differentsiatsii prirodnykh usloviy ostrovov Sovietskoy Arktiki. [On the regular differentiation of the natural conditions on the islands of the Soviet Arctic.] *Tr. Arkt. i Antarkt. n-issl. in-ta*, 285.

Semenov, I. V. (1970). Severnaya Zemlya. In: *Sovietskaya Arktika.* [The Soviet Arctic.] Moscow.

Serebryakov, I. G. (1960). Materialy k flore doliny r. Pyasiny. [Materials for a flora of the Pyasina River valley.] *Uch. zap. Mosk. gos. ped. inst. im. Potemkina*, 57.

Serebryakov, I. G. (1963). Nekotorye dannye o ritme sezonnovo razvitiya rasteniy doliny r. Pyasiny v yevo nizhnem Techenii. [Some data on the seasonal rhythmic development of the plants in the lower course of the Pyasina River valley.] *Byull. MOIP, otd. biologich.* 68 (2).

Shamurin, V. F. (1960). Sezonny ritm i ekologiya tsveteniya rasteniy v rayony bukhty Tiksi. [Seasonal rhythm and ecology of the angiosperms in the area of Tiksi Bay.] In: *Tundry fenologicheskovo soveshchaniya.* [The tundra phenology conference.] Leningrad.

Shamurin, V. F. (1966). Sezonny ritm i ekologiya tsveteniya rasteniy tundrovykh soobshchestv na severe Yakutii. [The seasonal rhythm and ecology of the angiosperms in the plant tundra communities in northern Yakutia.] In: *Prisposovlenie rasteniy Arktiki k usloviyma sredy.* [Adaptation of the plants to the Arctic and the environmental conditions.] Moscow–Leningrad.

Shamurin, V. F. (1970). Zapas fitomassy v nekotorykh tundrovykh soobshchestvakh rayona Vorkuty. [The phytomass reserve in some tundra communities in the Vorkuta area.] In: *Biologicheskie osnovy ispol'zovaniya priroda Severa.* [Biological basis for the utilization of the nature in the North.] Syktyvkar.

Shamurin, V. F., Aleksandrova, V. D. & Tikhomirov, B. A. (1975). Produktivnost' tundrovykh soobshchestv. [The productivity of tundra communities.] In: *Resursy biosfery.* [Resources of the Biosphere.] 1.

Shchelkunova, R. P. (1968). Kartirovanie rastitel'nosti severnykh rayonov Yakutskoy ASSR pri provedenii zemle'usroustva olen'ikh pastbishch. [Mapping of the vegetation in the northern part of the Yakutia ASSR for organization of land exploitation in the reindeer pasture areas.] *Bot. Zh.* 53 (6).

Shchelkunova, R. P. (1969). Rastitel'nost' tundrovoy zony Yakutskoy ASSR. [The vegetation of the tundra zone in the Yakutia ASSR.] Autoref. Dokt. Diss. Irkutsk.

Shchelkunova, R. P. (1975). Legenda k srednemasshtabnoy geobotanicheskoy karte Taymyra. [Legends to the middle-scale geobotanical map of Taimyr.] *Tr. N-issl. Inst. Sel'sk. Khoz. Krayn. Severa* 21. Norilsk.

Sheludyakova, V. A. (1938). Rastitel'nost' basseyna r. Indigirki. [The vegetation of the Indigirka River basin.] *Sov. Bot.* 4–5.

Shennikov, A. P. (1938). Lugovaya rastitel'nost' SSSR. [The meadow vegetation of the USSR.] In: *Rastitel'nost' SSSR.* [The vegetation of the USSR. 1.] Moscow–Leningrad.

Shennikov, A. P. (1940). Printsipy geobotnicheskovo rayonirovaniya. [Principles of geobotanical regionalization.] *Tr. Bot. Inst. AN SSSR, ser. III, Geobotnika*, 4.

Shennikov, A. P. (1964). *Vvedenie v geobotaniku.* [Introduction to geobotany.] Leningrad.

Sisko, A. K. (1970). Novosibirsky Archipelag. [The Novosiberian Archipelago.] In: *Sovietskaya Arktika.* [The Soviet Arctic.] Moscow.

Skottsberg, C. (1912*a*). Einige Bemerkungen über die Vegetationsverhältnisse des Graham-Landes. *Wiss. Erg. Schwed. Südpolar. Exped.* 4 (13).

Skottsberg, C. (1912*b*). The vegetation in South Georgia. *Wiss. Erg. Schwed. Südpolar Exped.* 4 (12).

Skottsberg, C. (1954). Antarctic flowering plants. *Bot. Tidsskr.* 54.

Skottsberg, C. (1960). Remarks on the plant geography of the southern cold temperate zone. *Proc. Roy. Soc. B.*, 152.

Skvortsov, A. K. (1966). *Salicaceae* Lindl. – Ivovye. [The willows.] *Arkticheskaya Flora SSSR.* [The Arctic flora of the USSR] V. Moscow–Leningrad.

Skvortsov, A. K. (1968). *Ivy SSSR.* [The willows of the USSR.] Moscow.

Skvortsov, E. F. (1930). V pribrezhnykh tundrakh Yakutii. [On the coastal tundras of Yakutia.] *Tr. Komissii po izuch. Yakutsk ASSR.* 15.

Smirnova, Z. N. (1938). Rastitel'nye assotsiatsii ostrova Kolguyeva. [Plant associations of Kolguyev Island.] *Bot. Zh.* 23 (5–6).

Smith, R. I. L. (1972). Vegetation of the South Orkney Islands with particular reference to Signy Island. *Brit. Antarct. Surv. Sci. Rep.* 68.

Smith, R. I. L. (1973). The occurrence of *Empetrum rubrum* Vahl ex Willd. on South Georgia. *Brit. Antarct. Surv. Bull.* 33–34.

Smith, R. I. L. & Corner, R. W. M. (1973). Vegetation of the Arthur Harbour–Argentine Islands region of the Antarctic Peninsula. *Brit. Antarct. Surv. Bull.* 33–34.

Sochava, V. B. (1930*a*). Predely lesov v gorakh Lyapinskovo Urala. [The forest limit in the mountains of Lyapinsky Urals.] *Tr. bot. muz. AN SSSR.* 22.

Sochava, V. B. (Soczawa, V.) (1930*b*). Das Anadyrgebiet. Botanisch–geographische Beobachtungen im äussersten Nordosten Asiens. *Zschr. Ges. für Erdkunde*, 7–8.

Sochava, V. B. (1931). Nekotorye osnovnye ponyatiya i terminy tundrovedeniya. [Some basic concepts and terms of tundra-science.] *Zh. Bot. obshch.* 16 (1).

Sochava, V. B. (1932). Po tundram basseyna Penzhinskoy guby. [Along the tundras of the Penzhina Bay basin.] *Izv. GGO*, 64 (4–5).

Sochava, V. B. (1933*a*). Yestyestvennye kormovye ugod'ya tundrovoy zony Yakutii. [The natural pasture lands in the Yakutia tundra zone.] *Sov. olenev.* 2.

Sochava, V. B. (1933*b*). Na istokakh rr. Shugora i Severnoy Sosvy. [At the sources of the Shugor and North Sosva Rivers.] *Izv. GGO*, 65 (6).

Sochava, V. B. (1933*c*). Tundry basseyna reki Anabary. [On the tundra of basin of Anabar River.] *Izv. GGO*, 65 (4).

Sochava, V. B. (1934*a*). Botaniko–geograficheskie podzony v zapadnykh tundrakh Yakutii. [Botanical–geographical subzones in the western tundras of Yakutia.] *Tr. I. Vsesoyuz. s"ezda geografov.* Leningrad.

Sochava, V. B. (1934*b*). Rastitel'nye assotsiatsii Anabarskoy tundry. [The plant associations of the Anabar tundras.] *Bot. Zh.* 19 (3).

Sochava, V. B. (1944*a*). O proiskhozhdeniya flory severnykh polyarnykh stran. [On the origin of the flora of the North Polar lands.] *Priroda*, 4.

Sochava, V. B. (1944*b*). Opyt filotsenogeneticheskoy sistematiki rastitel'nykh assotsiatsiy. [An attempt on the phylocoenogenetical systematics of plant associations.] *Sov. Bot.* 1.

Sochava, V. B. (1945). Fratrii rastitel'nykh formatsiy SSSR i ikh filotsenogeniya. [Fratriae of plant formations in the USSR and their phylocoenogenesis.] *DAN*, 45 (1).

Sochava, V. B. (1948*a*). Geograficheskie svyazi rastitel'novo pokrova na territorii SSSR. [Geographical relations of the vegetation cover in the USSR.] *Uch. zap. Len. gos. ped. inst. im. A. I. Gertsen*, 73. Kaf. fiz. geogr.

Sochava, V. B. (1948*a*). Razlichnye puti geobotanicheskovo rasdeleniya zemnoy poverkhnosti i ikh samostoyatel'noy znachenie. [Various ways of geobotanical division of the land surface and their independent importance.] *Bot. Zh.* 33 (1).

Sochava, V. B. (1952). Osnovnye polosheniya geobotaniceskovo rayonirovaniya. [The basic position of geobotanical regionalization.] *Bot. Zh.* 37 (3).

Sochava, V. B. (1956*a*). Zakonomernosti geografii rastitel'novo pokrova gornykh tundr SSSR. [Regularity in the geography of the vegetation cover in mountain tundras of the USSR.] In: *Akademiky V. N. Sukachevy k 75-letiyu so dnya rozhdeniya.* [To the Academician V. N. Sukachev on his 75th birthday.] Moscow–Leningrad.

Sochava, V. B. (1956*b*). Ural'skie (gornye tundry). [Mountain tundras of the Urals.] In: *Rastitel'ny pokrov SSSR. Poyasnitel'ny tekst k 'Geobotanicheskoy karte SSSR', m. 1:4 000 000.* [Vegetation cover of the USSR. Explanatory text to the 'Geobotanical map of the USSR, scale 1:4 000 000.] Part 1. Moscow–Leningrad.

Sochava, V. B. (Ed.) (1964*a*). Karta rastitel'nosti SSSR., m 1:15 000 000. [Vegetation map of the USSR, scale 1:15 000 000.] In: *Fiziko–geograficheskiy atlas mira.* [Physical–geographical atlas of the world.] Moscow–Leningrad.

Sochava, V. B. (1964*b*). Maket novoy karty rastitel'nosti mira. [Model for a new vegetation map of the world.] In: *Geobotanicheskoye kartografirovanie 1964.* [Geobotanical cartography 1964.] Moscow–Leningrad.

Sochava, V. B. (1965). Sovremennye zadachi kartografii rastitel'nosti v krupnom masshtabe. [Contemporary problems of the cartography of vegetation on a big scale.] In: *Geobotanicheskoye kartografirovanie* 1965. [Geobotanical cartography 1965.] Moscow–Leningrad.

Sochava, V. B. (1966). Rayonirovanie i kartografiya rastitel'nosti. [Regionalization and cartography of vegetation.] In: *Geobotanicheskoye kartografirovanie 1966.* [Geobotanical cartography 1966.] Moscow–Leningrad.

Sochava, V. B. (1967). Razvitiye teoreticheskikh polozheniy geobotanicheskovo rayonirovaniya na sovremennom etape. [The development of the theoretical position on geobotanical division into areas at a contemporary stage.] In: *Materialy mezhvuzovskoy konferentsii po geobotanicheskomu rayonirovaniyu SSSR.* [Material for an interuniversity conference on the geobotanical division into areas of the USSR.] Moscow.

Sochava, V. B. (1968). Rastitel'nye soobshchestva i dinamika prirodnykh sistem. [Plant communities and the dynamics of natural systems.] In: *Doklady Inst. Geogr. Sib. i Dal'n. Vost. 20.* [Reports of Institute of Siberia and the Far East, vol. 20.]

Sochava, V. B. (1972). Klassifikatsiya rastitel'nosti kak ierarkhiya dinamicheskikh sistem. [Classification of the vegetation as a hierarchy of dynamic systems.] In: *Geobotanicheskoye kartografirovanie 1972.* [Geobotanical cartography 1972.] Leningrad.

Sochava, V. B., Bachurin, G. V., Borob'yev, V. V., Mikhailov, Yu. P., Prokhorov, B. B. & Shotsky, V. P. (1972). Geograficheskie problemy sovietskoy Subarktiki. [Geographical problems of the Soviet Subarctic.] *Doklady Inst. Geogr. Sib. i Dal'n. Vost.* 35. [Reports of the Institute of Siberia and the Far East, vol. 35.]

Sochava, V. B. & Gorodkov, B. N. (1956). Arkticheskie pustyni i tundry. [Arctic deserts and tundras.] In: *Rastitel'ny pokrov SSSR. Poyasnitel'ny tekst k 'Geobotnicheskoy karte SSSR', m. 1:4 000 000.* [Vegetation cover of the USSR. Explanatory text to 'Geobotanical map of the USSR', scale 1:4 000 000.] Part 1. Moscow–Leningrad.

Sochava, V. B. & Timofeyev, D. A. (1968). Fiziko–geograficheskie oblasti severnoy Azii. [Physical–geographical regions of Northern Asia.] *Doklady Inst. Geogr. Sib. i Dal'n. Vost.* 19.

Soczawa, V. *see* Sochava, V. B.

Sokolovsky, V. (1905). O flore Novoy Zemli. [On the flora of Novaya Zemlya.] *Izv. Imp. lesn. inst.* 13.

Soper, J. D. (1930*a*). Exploration in Baffin Island. *Geogr. J.* 75.

Soper, J. D. (1930*b*). Exploration in Foxe Peninsula and along the west coast of Baffin Island. *Geogr. Rev.* 20.

Soper, J. D. (1936). The Lake Harbour region, Baffin Island. *Geogr. Rev.* 26 (3).

Sørensen, Th. (1933). The vascular plants of East Greenland from 71° 00′ to 73° 30′ N. lat. *Medd. om Grønl.* 101 (3).

Sørensen, Th. (1935). Bodenformen und Pflanzendecke in Nordgrönland. *Medd. om Grønl.* 93 (4).

Sørensen, Th. (1937). Remarks on the flora and vegetation of Northern Greenland. 74° 30′–79° N. lat. *Medd. om Grønl.* 101 (4).

Sørensen, Th. (1941). Temperature relation and phenology of the northeast Greenland flowering plants. *Medd. om Grønl.* 125 (9).

Sørensen, Th. (1945). Summary of the botanical investigations in N.E. Greenland. *Medd. om Grønl.* 144 (3).

Spetzman, L. A. (1959). Vegetation of the Arctic slope of Alaska. *U.S. Geol. Surv., Prof. Paper*, 302–B.

Steffen, H. (1928). Beiträge zur Flora und Pflanzengeographie von Nowaja Semlja, Waigatch und Kolguew. *Beih. Bot. Zentralbl.* 44 (2).

Størmer, P. (1940). Bryophytes from Franz Josef Land and eastern Svalbard. *Norges Ishav. Undersøgelse. Medd.* 47. Oslo.

Sukachev, V. N. (1911). K voprosu o vliyanii merzloty na pochvu. [On the problem of the effect of permafrost on the soils.] *Izv. Imp. AN*, ser. 6, 5.

Sumina, O. I. (1975). Rastitel'nost' baydzharakhov o. Kotel'novo (Novosibirskie ostrova). [The vegetation on the bayadzerakhs ('cemetery mounds') in the thermokarst areas on the Kotel'ny Island (Novosiberian Islands).] *Bot. Zh.* 60 (9).

Summerhayes, V. S. & Elton, C. S. (1923). Contribution to the ecology of Spitsbergen and Bear Island. *J. Ecol.* 11 (2).

Summerhayes, V. S. & Elton, C. S. (1928). Further contribution to the ecology of Spitsbergen. *J. Ecol.* 16 (2).

Sunding, P. (1966). Plantefunn fra Vestspitsbergen sommeren 1964. [Plant finds from West Spitsbergen the summer of 1964.] *Årbok Norsk Polarinst. 1964*. Oslo. 1966.

Svatkov, N. M. (1958). Pochvy ostrova Vrangelya. [Soils of Wrangel Island.] *Pochvovedenie* 1.

Svatkov, N. M. (1961). Priroda ostrova Vrangelya. [The nature of Wrangel Island.] *Probl. Severa*, 4.

Svatkov, N. M. (1970). Ostrov Vrangelya. [Wrangel Island.] In: *Sovietskaya Arktika.* [The Soviet Arctic.] Moscow.

Svoboda, J. (1972). Vascular plant productivity studies of raised beach ridges (semi-polar desert) in Truelove Lowland. In: *Devon Island IBP project, high-arctic ecosystem.* Ed. L. C. Bliss. Edmonton.

Svoboda, J. (1973). Primary production of plant communities of the Truelove Lowland, Devon Island, Canada. Beach ridges. In: *Tundra biome. Primary*

production and production processes. Ed. L. C. Bliss & F. E. Wielgolaski. Stockholm.
Svoboda, J. (1977). Ecology and primary production of raised beach communities. In: *Truelove Lowland, Devon Island, Canada: a high arctic ecosystem*. Ed. L. C. Bliss. Edmonton.
Takhtadzhyan, A. L. (1966). *Sistema i filogeniya tsvetkovykh rasteniy*. [System and phylogeny of the flowering plants.] Moscow–Leningrad.
Takhtadzhyan, A. L. (1970). *Proiskhozhdenie i rasselenie tsvetkovykh rasteniy*. [Origin and dispersal of the flowering plants.] Leningrad.
Tanfil'yev, G. I. (1894). Po tundram timanskhikh samoyedov letom 1892 g. [Over the tundras of the Timan Samoyeds during the summer of 1892.] *Izv. RGO*, 30 (1).
Targul'yan, V. O. (1971). *Pochvoobrazovanie i vyvetrivanie v kholodnykh gumidnykh oblastyakh*. [Soil formation and erosion in cold, humid regions.] Moscow.
Targul'yan, V. O. & Karavayeva, N. A. (1964). Opyt pochvenno-geokhimicheskovo razdeleniya polyarnykh oblastey. [An attempt of soil-chemical division of the polar regions.] *Probl. Severa* 8.
Taylor, B. W. (1955). The flora, vegetation and soils of Macquarie Island. *Austral. Nat. Antarct. Res. Exp., Rep. B* 2.
Tedrow, J. C. F. (1966). Polar desert soils. *Proc. Soil Sci. Soc. Amer.* 30 (3).
Tedrow, J. C. F. (1968). Pedogenic gradient of the Polar regions. *J. Soil. Sci.* 19 (1).
Tedrow, J. C. F. (1970). Soil investigation in Inglefield Land, Greenland. *Medd. om Grønl.* 188 (3).
Tedrow, J. C. F. (1972). Soil morphology as an indicator of climatic changes in the arctic areas. *Acta Univ. Ouluensis, Ser. A*. 3.
Tedrow, J. C. F., Bruggemann, P. F. & Walton, G. F. (1968). Soils of Prince Patrick Island. *Arct. Inst. North Amer., Techn. Rep.* ONR–352.
Tedrow, J. C. F. & Cantlon, J. E. (1958). Concepts of soil formation and classification in arctic regions. *Arctic*, 11 (3).
Tedrow, J. C. F. & Douglas, L. A. (1964). Soil investigations on Banks Island. *Soil Sci.* 98 (1).
Tedrow, J. C. F. & Harries, H. (1960). Tundra soil in relation to vegetation, permafrost and glaciation. *Oikos*, 11 (2).
Tedrow, J. C. F. & Ugolini, F. C. (1965). Antarctic soils. *Antarct Res., Ser.* 8.
Terentjew, P. V. *see* Terent'yev, P. V.
Terent'yev, P. V. (Terentjew, P. V.) (1931). Biometrische Untersuchungen über die morphologischen Merkmale von *Rana ridibunda* Pall. *Biometrica*, 23.
Terent'yev, P. V. (1959). Metod korrelyatsionnykh pleyad. [Methods of correlation swarms.] *Vestn. LGU*, 9.
Thannheiser, D. (1975). Beobachtungen zur Küstenvegetation auf dem westlichen kanadischen Arctic–Archipel. *Polarforsch.* 45 (1).
Tikhomirov, B. A. (1948*a*). K poznaniyu flory kraynikh polyarnykh predelov Yevrazii. [To the knowledge of the flora of the extreme polar parts of Eurasia.] *Byull. MOIP, otd. biol.* 53 (4).
Tikhomirov, B. A. (1948*b*). *K kharakteristike flory zapadnovo poberezh'ya Taymyra*. [To the description of the flora of coastal West Taimyr.] Petrozavodsk.
Tikhomirov, B. A. (1956). Nekotorye voprosy strukture rastitel'nykh soobshchestv Arktiki. [Some questions on the structure of the plant communities of the Arctic.]

In: *Akademiku V. N. Sukachevu k 75-letiyu so dnya rozhdeniya.* [To Academician V. N. Sukachev on his 75th birthday.] Moscow–Leningrad.

Tikhomirov, B. A. (1957). Dinamicheskie yavleniya v rastitel'nosti pyatnistykh tundr Arktiki. [Dynamical phenomena in the vegetation of the spotty tundras of the Arctic.] *Bot. Zh.* 42 (11).

Tikhomirov, B. A., Petrovsky, V. V. & Yurtsev, B. A. (1966). Flora okrestnostye bukhty Tiksi (arkticheskaya Yakutiya). [The flora around Tiksi Bay (Arctic Yakutia).] In: *Rasteniya severa Sibiri i Dal'neva Vostoka.* [Plants of northern Siberia and the Far East.] Moscow–Leningrad.

Tolmachev, A. I. (1923). Botaniko–geograficheskie raboty v rayone Yugorskovo Shara v 1921 i 1922 gg. [Botanical–geographical works in the area of Yugorsky Shar during 1921 and 1922.] *Izv. GBS*, 22(2).

Tolmachev, A. I. (Tolmachew, A. I.) (1929). Beiträge zur Kenntniss des Gebietes von Matotschkin Shar und der Ostküste Nowaja Semlja's. *Izv. AN SSSR, ser.* 7, 4.

Tolmachev, A. I. (1930). O proiskhozhdenii flory Vaygacha i Novoy Zemli. [On the origin of the floras of Vaigach and Novaya Zemlya.] *Tr. Bot. Muz. AN SSSR*, 22.

Tolmachev, A. I. (1932*a*). Flora tsentralnoy chasti vostochnovo Taymyra, 1. [The flora of the central part of eastern Taimyr.] *Tr. Polyarn. Komissii*, 8.

Tolmachev, A. I. (1932*b*). Flora tsentralnoy chasti vostochnovo Taymyra, 2. [The flora of the central part of eastern Taimyr.] *Tr. Polyarn. Komissii* 13.

Tolmachev, A. I. (1935). Flora tsentralnoy chasti vostochnovo Taymyra, 3. [The flora of the central part of eastern Taimyr.] *Tr. Polyarn. Komissii*, 25.

Tolmachev, A. I. (1936). Obzor flory Novoy Zemli. [Survey of the flora of Novaya Zemlya.] *Arctica*, 4.

Tolmachev, A. I. (1939). O nekotorykh zakonomernostyakh raspredeleniya rastitel'nykh soobshchestv v Arktike. [On some regularities in the distribution of plant communities in the Arctic.] *Bot. Zh.* 24 (5–6).

Tolmachev, A. I. (1959). K flore ostrova Benneta. [On the flora of Bennett Island.] *Bot. Zh.* 44 (4).

Tolmachev, A. I. (1962). Avtokhtonnoye yadro arkticheskoy flory i yeyo svyazi s vysokogornymu florami Severnoy i Tsentral'noy Azii. [The autochtonous nucleus of the Arctic flora and its connections with the alpine floras of northern and Central Asia.] *Probl. bot.* 6.

Tolmachev, A. I. (1966). *Eriophorum* L. – Pushitsa. [The cottongrasses.] In: *Arkticheskaya Flora SSSR*, [The Arctic Flora of the USSR.] III. Moscow–Leningrad.

Tolmachev, A. I. & Shukhtina, G. G. (1974). Novye dannye o flore Zemli Frantsa–Iosifa. [New data on the flora of Franz Joseph's Land.] *Bot. Zh.* 59 (2).

Tolmachev, A. I. & Tokarevskikh, S. A. (1968). Issledovanie rayona 'lesnovo ostrova' u reki Morye-Yu v Bol'shezemel'skoy tundre. [Investigations in the area of the 'forest island' at Morye-Yu River in the Bol'she–zemelya tundra.] *Bot. Zh.* 53 (4).

Tolmachev, A. I. & Yurtsev, B. A. (1970). Istoriya arkticheskoy flory v yeyo svazi c istoriey Severnovo Ledovitovo Okeana. [The history of the arctic flora and its relation to the history of the North Polar Sea.] In: *Severny Ledovity Okean i yevo poberezh'ye v kaynozoe.* [The North Polar Sea and its coast during the Coenozoic.] Leningrad.

Tolmatschew, A. I. *see* Tolmachev, A. I.

Trapnell, G. O. (1933). Vegetation types in Godthaab Fjord. *J. Ecol.* 21.

Trautvetter, E. R. (1850). *Die Pflanzengeographische Verhältnisse der europäischen Russlands*, 2. Riga.
Trautvetter, E. R. (Trautvetter, R. E.) (1851). O rastitel'no-geograficheskikh okrugakh Yevropeyskoy Rossii. [On the plant–geographical districts of European Russia.] *Tr. Komissii pri un–te sv. Vladimira dlya opis. gub. Kievsk. uchebn. okruga.* Kiev.
Triloff, E. G. (1943). Verbreitung und Ökologie der Gefässpflanzen im Gebiete des Hornsundes: ein Beitrag zur Vegetationskunde Spitzbergens. *Bot. Jahrb.* 73 (3).
Tsinzerling, Yu. D. (1932). Geografiya rastitel'novo pokrova severo–zapada yevropeyskoy chasti SSSR. [The geography of the plant cover in the northwestern European part of the USSR.] *Tr. Geomorfolog. inst., ser. fiz.-geogr.* 4.
Tsinzerling, Yu. D. (1935). Materialy po rastitel'nosti sever-vostoka Kol'skovo poluostrova. [Material on the vegetation of the north-eastern Kola Peninsula.] *Tr. SOPS AN SSSR, Kol'sk. ser.*, 10.
Tuomikoski, R. (1942). Untersuchungen über die Untervegetation der Bruchmoore in Ostfinnland. 1. Zur Metodik der pflanzensociologischen Systematik. *Ann. Bot. Soc. 'Vanamo'*, 17 (1).
Tyrtikov, A. P. (1958). Nekotorye svedeniya o rastitel'nosti nizov'yev r. Indigirki. [Some information on the vegetation along the lower Indigirka River.] *Byull. MOIP, otd. biol.* 63 (1).
Tyrtikov, A. P. (1969). *Vlianie rastitel'novo pokrova na promerzanie i protaivanie gruntov.* [The effect of the vegetation cover on the freezing and thawing of the substrate.] Moscow.
Tyulina, L. N. (1937) Lesnaya rastitel'nost' Khatangskovo rayona u yeyo severnovo predela. [The forest vegetation in the Khatanga area at its northern limit.] *Tr. Arkt. Inst.* 63.
Vasil'yev, V. N. (1936). Olen'i pastbishcha Anadyrskovo kraya. [Reindeer pastures in the Anadyr District.] *Tr. Arkt. Inst.* 62.
Vasil'yev, V. N. (1956). *Rastitel'nost' Anadyrskovo kraya.* [The vegetation of the Anadyr District.] Moscow–Leningrad.
Vasilyevich, V. I. (1964). Yestyestvennaya klassifikatsiya v fitotsenologii. [Natural classification in phytocoenology.] *Programma i tez. dokl. nauchn. konf., posv. 100-letiyu so dnya rozhdeniya prof. N. I. Kuznetsova.* [Program and lectures of a scientific conference on the occasion of the 100th anniversary of the birth of Prof. N. I. Kutsnetsov.] Tartu.
Vasilyevich, V. I. (1966). Chto schtitat' yestyestvennoy klassifikatsiey. [How to interpret natural classification.] In: *Filosofskie problemy sovremennoy biologii.* [Philosophical problems in modern biology.] Moscow–Leningrad.
Vasilyevich, V. I. (1975). Vyyavlenie granits v rastitel'nom pokrove. [Investigation of boundaries in a vegetation cover.] *Byull. MOIP, otd. biol.* 80 (3).
Vas'kovsky, A. P. (1958). Novye dannye o granitsakh rasprostraneniya derev'yev i kustarniko-tsenozobrazovateley na kraynem severo-vostoke SSSR. [New data on the limits of distribution of tree- and shrub-coenosis-formants in the northeastern USSR.] *Mater. po geol. i polezn. iskop. Sev.-Vost. SSSR.* 13. Magadan.
Viereck, L. A. & Little, E. L. (1972). Alaska trees and shrubs. *Agric. Handbook*, 410. Washington.
Vikhireva-Vasil'kova, V. V., Gavrilyuk, V. A. & Shamurin, V. F. (1964). Nadzemnaya i podzemnaya massa nekotorykh kustarnichkovykh soobshchstv

Koryakskoy Zemli. [Above and below ground mass of some dwarf-shrub communities in the Koryak Land.] *Probl. Severa* 8.

Vinogradov, B. V. (1970). Melkomasshtabnoye geobotanicheskoye rayonirovanie i kartirovanie po kosmicheskim izobrazheniyam Zemli. [Small-scale geobotanical regionalization and mapping by the aid of satellite photography of the Earth.] In: *Geobotanical kartigrafirovanie.* [Geobotanical mapping.] 1970.

Vinogradova, A. N. (1937). Geobotanichesky ocherk olen'ikh pastbishch rayona reki pyasiny. [Geobotanical essay on the reindeer pastures in the area of Pyasina River.] *Tr. Arkt. Inst.* 63.

Vitt, D. H. & Pakarinen P. (1977). The bryophyte vegetation, production, and organic compounds of Truelove Lowland. In: *Truelove Lowland, Devon Island, Canada: a high arctic ecosystem.* Ed. L. C. Bliss. Edmonton.

Vize, N. Yu. (1924). *Novozeml'skaya bora.* [The Novaya Zemlya strong gales.] Leningrad.

Vize, N. Yu. (1940). *Klimat morey Sovietskoy Arktiki.* [The marine climate in the Soviet Arctic.] Moscow–Leningrad.

Vul'f E. V. (1944). *Istoricheskaya geografiya rasteniy.* [Historical plant geography.] Moscow.

Walter, H. (1968). *Vegetation der Erde. 2. Die gemässigten und arktischen Zonen.* Jena.

Walton, D. H. W. (1973). Changes in standing crop and dry matter production in an *Acaena* community on South Georgia. In: *Tundra biome. Primary production and production processes.* Ed. L. C. Bliss & F. E. Wielgolaski. Stockholm.

Walton, D. W. H. & Smith, R. I. L. (1973). Status of the alien vascular flora of South Georgia. *Brit. Antarct. Surv. Bull.* 36.

Walton, J. (1922). A Spitsbergen salt marsh; with observations on the ecological phenomena attendant on the emergence of land from sea. *J. Ecol.* 10.

Warming, E. (1888). Om Grønlands Vegetation. *Medd. om Grønl.* 12.

Warming, E. (1889). Über Grönlands Vegetation. *Engler's Jahrb.* 10.

Warming, E. (1912). The structure and biology of arctic flowering plants, 1. *Medd. om Grønl.* 36.

Warming, E. (1928). The Vegetation of Greenland, In: *Greenland.* Ed. M. Vahl *et al.*, vol. 1 Copenhagen–London.

Webber, P. J. (1964). Geobotanical studies around the northwestern margin of the Barnes Ice Cap, Baffin Island, N. W. T. *Canada Dept. Mines and Techn. Surv., Geogr. Branch.* 905–18–8.

Wielgolaski, F. E. (1971). Vegetation types and primary production in tundra. In: *IBP Tundra Biome. Proc. IV Intern. Meet., Leningrad, 1971.* Ed. F. E. Wielgolaski & Th. Rosswall. Stockholm.

Wielgolaski, F. E. (Vielgolaski, F. E.) (1973). Tipy rastitel'nosti i biomassa tundry. [Vegetation types and biomass of the tundras.] *Ekologiya*, 2.

Wiggins, I. L. (1951). The distribution of vascular plants on polygonal ground near Point Barrow, Alaska. *Contr. Dudley Herbarium*, 4.

Wiggins, I. L. & Thomas, J. H. (1962). *A flora of the Alaskan Arctic Slope.* Toronto.

Yegorova, T. V. (1966). *Carex.* L. – Osoka. [The sedges.] In: *Arkticheskaya flora SSSR.* [The Arctic Flora of the USSR.] III. Moscow–Leningrad.

Yesipov, V. K. (1933). Ostrov Vaigach. Fiziko–geograficheskii ocherk. [Vaigach Island. Physical–geographical essay.] In: *Ostrova Sovietskoy Arktiki.* [Islands of the Soviet Arctic.] Ed. V. K. Yesipov & N. V. Pinegin. Archangelsk.

Young, S. B. (1971). The vascular flora of St. Lawrence Island with special reference to floristic zonation in the Arctic Region. *Contr. Gray Herb. of Harvard Univ.* 201.

Yurtsev, B. A. (1959). Vysokogornaya flora gory Sokuydakh i yeyo mesto v ryadu gornykh flor arkticheskoy Yakutii. [The alpine flora of the Sokuydy Mountains and its place among the number of mountain floras of arctic Yakutia.] *Bot. Zh.* 44 (8).

Yurtsev, B. A. (1962). Botaniko–geograficheskie nablyudeniya u severnovo predela rastprostraneniya listvennitsy na r. Olenek. [Botanical–geographical discoveries at the northern limit of the larches along Olenek River.] *Probl. bot.* 6.

Yurtsev, B. A. (1964). O sootnoshenii okeanicheskikh i kontinental'nykh elementov v gol'tsovykh florakh Vostochnoy Sibiri. [On the correlation between oceanic and continental elements in the 'goltsy' [alpine] floras of Eastern Siberia.] *Probl. Severa* 8.

Yurtsev, B. A. (1966). Gipoarktichesky botaniko–geografichesky poyas i proiskhzhdenie yevo flory. [The hyparctic botanical–geographical belt and the origin of its flora.] *Komarovskie chteniya XIX.* Moscow–Leningrad.

Yurtsev, B. A. (1967). Botaniko–geograficheskie issledovaniya na zapadnoy i tsentral'noy Chukotke v 1964–1966 gg. [Botanical–geographical investigations in western and central Chukotka during the years 1964–1966.] *Bot. Zh.* 52 (7).

Yurtsev, B. A. (1968*a*). Stepnye soobshchestva Chukotskoy tundry i vopros ó pleystotsenovoy 'tundrostepi'. [Steppe communities in the Chukotka tundra and the problem of Pleistocene 'tundra steppes'.] In: *Problemy izucheniya chetvertichnovo perioda.* [Problems of the Quaternary period.] Khabarovsk.

Yurtsev, B. A. (1968*b*). *Flora Suntar–Khayata.* Leningrad.

Yurtsev, B. A. (1970*a*). Botaniko–geograficheskie issledovaniya na Chukotke. [Botanical–geographical investigations in Chukotka.] In: *Biologicheskie osnovy ispol'zovaniya prirody Severa.* [Biological basis for the study of nature in the North.] Syktyvar.

Yurtsev, B. A. (1970*b*). O polozhenii polyarnovo poberezh'ya severo–vostochnoy Sibiri v pleystotsene. [On the position of the shore of northeastern Siberia during Pleistocene.] In: *Severny Ledovity Okean i yevo poverezh'ye v kaynozoe.* [The North Polar Sea and its coast during the Cenozoic.] Leningrad.

Yurtsev, B. A. (1972). Phytogeography of northeastern Asia and the problem of trans-Beringian floristic interrelations. In: *Floristics and paleofloristics of Asia and eastern North America.* Amsterdam.

Yurtsev, B. A. (1973*a*). Botaniko–geograficheskaya zonal'nost' i floristicheskoye rayonirovanie Chukotskoy tundry. [Botanical–geographical zonation and floristic division into areas of the Chukotka tundras.] *Bot. Zh.* 58 (7).

Yurtsev, B. A. (1973*b*). Problemy pozdne kaynozoyskoy paleogeografii Beringii (v botaniko–geograficheskom osveshchenii). [Problems of the Late Coenozoic paleogeography of Beringia (in botanical-geographical interpretation).] In: *Beringiskaya susha i yeyo znachenie dlya golarkticheskikh flor i faun v kaynozoe.* [The Bering landarea and its importance for the development of flora and fauna during the Coenozoic.] Khabarovsk.

Yurtsev, B. A. (1974*a*). *Problemy botanicheskoy geografii severo-vostochnoy Azii.* [Problems of botanical geography in northeastern Asia.] Leningrad.

Yurtsev, B. A. (1974*b*) Stepnye soobshchestva Chukotskoy tundry i pleystotsenovaya 'tundrostep''. [The steppe associations of the Chukotka tundra and the Pleistocene 'tundra steppe'.] *Bot. Zh.* 59 (4).

Yurtsev, B. A. (1976). Problemy pozdnekaynozoyskoy paleogeografii Beringii v svete botaniko–geograficheskikh dannykh. [Problems of the Late Coenozoic paleogeography of Beringia in the light of botanical–geographical data.] In: *Beringiya v kaynozoye.* [Beringia during the Coenozoic.] Vladivostok.

Yurtsev, B. A. (1977). O sootnoshenii arkticheskoy i vysokogornykh subarkticheskikh flor. [On the interrelation between the arctic and the alpine subarctic floras.] In: *Problemy ekologii, geobotaniki, botanicheskoy geografii i floristiki.* [Problems of ecology, geobotany, botanical geography and floristics.] Leningrad.

Yurtsev, B. A., Kozhevnikov, Yu. P. & Nechayev, A. L. (1972). Intersnye floristicheskikiye nakhodki na vostoke Chukotskovo poluostrova. [Interesting floristic finds on the eastern Chukotka Peninsula.] *Bot. Zh.* 57 (7).

Yurtsev, B. A., Tolmachev, A. I. & Rebristaya, O. V. (1975). *Problemy floristicheskovo ogranicheniya i razdeleniya Arktiki.* [Problems of floristic delimitation and division of the Arctic into areas.] *Lecture to XII Mezhdunarodn. bot. Kongresse.* [XII Internat. Bot. Congr.] Manuscript, Bot. Inst. AN SSSR. Leningrad.

Zhitkov, B. M. (1913). Poluostrov Yamal. [The Yamal Peninsula.] *Zap. imp. Russk. geogr. obshch.* 49.

Zhukova, A. L. (1973*a*). Pechenochnye mkhi polyarnykh pustyn' Zemli Frantsa-Iosifa. [The hepatic mosses of the polar deserts on Franz Joseph's Land.] Autoref. Dokt. Diss. Leningrad.

Zhukova, A. L. (1973*b*). Floristichesky analiz pechenochnykh mkhov (*Hepatici*) arkhipelaga Zemlya Frantsa-Iosifa. [Floristic analysis of the liverworts (*Hepatici*) on the Franz-Joseph's Land Archipelago.] *Bot. Zh.* 58 (4).

Zubakov, V. A. (1965). Pleystotsenovoye oledenenie severnovo polushariya (statika kriosfery). [Pleistocene glaciation of the northern hemisphere (cryospheric statics).] *Tr. N-issl. inst. geolog. Arkt.* 143. Moscow.

Zubkov, A. I. (1932). Tundry Gusinoy Zemli. [The tundras of Gusina Land, Novaya Zemlya.] *Tr. Bot. Muz. AN SSSR*, 25.

Zubkov, A. I. (1934). Kratky predvaritel'ny otchet o geobotanicheskikh rabotakh na severnom ostrove Novoy Zemli. [Short preliminary account of the geobotanical works on the North Island of Novaya Zemlya.] Manuscript. Fondy Inst. Sel'sk. Khoz. Krayn. Severa. Norilsk.

Zubkov, A. I. (1935*a*). Olenevodstvo i olen'i pastbishcha na Novoy Zemle. [Reindeer husbandry and pastures on Novaya Zemlya.] *Tr. Arkt. Inst.* 22.

Zubkov, A. I. (1935*b*). Prodozhitel'nost' vegetatsionnovo perioda na Severnom ostrove Novoy Zemli. [Duration of the vegetative period on the North Island of Novaya Zemlya.] *Arktika*, 3.

LIST OF LATIN PLANT NAMES

VASCULAR PLANTS

Acaena Mutis ex L.
  adscendens Vahl 171
  magellanica (Lam.) Vahl 171
Aconitum L.
  septentrionale Koelle 30
Alchemilla L.
  alpina L. 86, 88
  filicaulis Bus. 88
  glomerulans (Rupr.) Ldb. 86, 88
Alnaster Spach
  fruticosa (Rupr.) Ldb. 11, 38
Alnobetula Pouzar 11
Alnus Mill. 11, 64
  crispa (Ait.) Pursh 11, 24, 67, 69, 75–6, 83, 86–7, 93
  fruticosa Rupr. 11, 24, 26–7, 33, 36, 40, 50, 56–7, 62, 65, 70, 76–8, 93
Alopecurus L.
  alpinus Sm. 14, 22, 53, 93, 106–8, 110–12, 114, 116, 118, 120, 123–4, 130, 132, 135, 140, 164–5, 167–8
  pratensis L. 30, 51
Andromeda L.
  polifolia L. 76–7
Androsace L.
  gorodkovii Ovcz. & Karav. 54
  ochotensis Willd. 57, 116
  septentrionalis L. 123
  triflora Adams 161
Antennaria Gaertn.
  glabrata (J. Vahl) Greene 87, 122
Anthoxanthum L.
  alpinum A. & D. Löve 30
Arabideae Hayek 15
Archangelica Hoffm.
  norvegica Rupr. 32, 86
Arctagrostis Griseb.
  latifolia (R. Br.) Griseb. 23, 69, 75, 77, 80, 108, 115, 119, 123, 125–6, 130, 132
× Arctodupontia Tzvelev 15
Arctophila Anderss. 15
  fulva (Trin.) Anderss. 23, 57, 97, 108, 112, 118–19, 124
Arctostaphylos Adans.
  uva-ursi (L.) Spreng. 85, 87
Arctous Niedenzu
  alpina (L.) Niedenzu 32, 53, 65, 68, 75–6, 79, 80, 119
  erythrocarpa Small 77
Arenaria L.
  pseudofrigida (Ostenf. & Dahl) Juz. 97, 128–9
  tschuktschorum Rgl. 64
Armeria Willd.
  maritima (Mill.) Willd. 127, 131
Artemisia L. 16
  arctica Less. 64
  comata Rydb. 70
  furcata M.B. 116
  glomerata Ldb. 116
  lagopus Fisch. ssp. abbreviata Krasch. 54
  tilesii Ldb. 72, 97
Astragalus L.
  alpinus L. 11, 119
  arcticus Bge. 72
  gorodkovii Jurtz. 38
  pseudadsurgens Jurtz. 54
  schelichovii Turcz. 64
  umbellatus Bge. 119
Azorella Lam.
  selago Hook. f. 171

Betulaceae 19

Betula L.
cajanderi Sukacz. 59, 66
callosa Notö ex Lindq. 33
ermanii Cham. 59, 66
exilis Sukacz. 10, 11, 13, 21, 23–4, 26, 44, 46–7, 50–3, 55–7, 62–3, 65–9, 70, 92, 95, 106, 108
glandulosa Michx. 11, 21, 23–4, 26, 67–8, 70, 75–7, 79, 80, 83, 85, 87, 92
middendorffii Trautv. & Mey. 23–4, 26, 65, 92
nana L. 4, 10, 11, 13, 21, 23–4, 26, 29, 30, 32, 34–44, 46, 70, 82, 85–6, 90, 92, 95
papyrifera Marsh 60, 70
subarctica Orlova 33
tortuosa Ldb. 14, 26, 30–8, 84, 87
tundrarum Perf. 44, 46, 90
Braya Sternb. & Hoppe
intermedia Th. Sør. 122, 128
purpurascens (R. Br.) Bunge 22, 127, 131, 134
thorild-wulffii Ostenf. 122, 127, 131, 134, 167

Calmagrostis Adans.
canadensis (Michx.) Beauv. 67, 68
groenlandica (Schrank) Kunth 85
langsdorfii (Link) Trin. 23, 86
lapponica (Wahlenb.) Hartm. 34
lapponica (Wahlenb.) Hartm. var. groenlandica Lge 87
Caltha L.
artica R. Br. 72
Campanula L.
uniflora L. 97, 117
Caragana L.
jubata (Pall.) Poir. 54
Cardamine L.
bellidifolia L. 22, 38, 64, 119, 124, 131, 160
digitata Richards. 118
pratensis L. 80
pratensis L. ssp. angustifolia (Hook.) O. E. Schulz 140, 160
sphenophylla Jurtz. 64
Carex L.
aquatilis Wahlenb. 51, 68
atrofusca Schkuhr 119, 130, 132
bicolor Bell. ex All. 86
bigelowii Torr. ex Schwein. (C. bigelowii Torr. ex Schwein. ssp. bigelowii) 32–3, 66–7, 76, 79, 80, 85, 88, 118, 122–3, 125–6, 129, 130–1, 136
capillaris L. 37, 131
chordorrhiza Ehrh. 13, 40, 51, 54, 77
consimilis Holm 66–7, 69, 95
duriuscula C. A. Mey. 12, 63
ensifolia Turcz. 66
ensifolia (Turcz. ex. Gorodk.) V. Krecz. ssp. arctisibirica Jurtz. (C. bigelowii Torr. ex Schwein. ssp. arctisibirica (Jurtz.) Á. & D. Löve) 23, 27, 29, 35, 37–42, 50, 54, 65–7, 95–8, 100–7, 136, 147
glacialis Mackenzie 131
glareosa Wahlenb. 88
globularis L. 34, 38, 40, 147
gynocrates Wormskj. 80, 86
hepburnii Boott 114–15, 125–8, 130–1, 133, 135, 161
lachenalii Schkuhr 142
lugens Holm 23, 26, 57, 61–7, 69, 70, 72, 75, 95, 107–8, 112, 114–15, 118
magellanica Lam. 76
maritima Gunn. 120, 130–3
membranacea Hook. 77, 80, 119, 122, 124, 126
microglochin Wahlenb. 86
miliaris Michx. 78, 80
misandra R. Br. 114, 119, 124, 126–7, 130–4, 161
obtusata Liljebl. 12, 115
parallela (Laest.) Sommerf. 97, 128
pediformis C. A. Mey. 12, 57, 64
podocarpa R. Br. 54, 68
rariflora (Wahlenb.) Smith 37, 40, 54, 77, 86
rotundata Wahlenb. 13, 51, 54, 66, 68, 72, 76, 78
rupestris Bell. ex All. 12, 22, 37, 80, 86, 88, 114–15, 120, 125–7, 130–1, 133, 138, 161
saxatilis L. ssp. laxa (Trautv.) Kalela 77, 97
saxatilis L. ssp. saxatilis 30, 72, 77
scirpoidea Michx. 79, 119
soczavaeana Gorodk. 26, 65, 93
stans Drej. 13, 19, 40, 42, 54, 66, 68, 76–7, 86, 99, 103–5, 108, 112, 115, 118–19, 124–6, 132, 134–5, 150, 168
subspathacea Wormskj. ex Hornem. 13, 88, 132, 141–2
supina Wahlenb. ssp. spaniocarpa (Steud.) Hult. 57, 86

tripartita All. 142
ursina Dew. 13, 118–20, 132
Cassiope D. Don 53–4, 63, 106, 122, 125–6, 129, 135, 138, 141
tetragona (L.) D. Don 3, 21–2, 27, 39, 42, 50, 53–5, 69, 75, 77, 80, 82, 85, 87–8, 93, 96, 105–6, 115, 122, 124–5, 127, 129, 133, 135–6, 138–9
Castilleja Mutis
vorkutensis Rebr. 38
Cerastium L. 16
alpinum L. 80, 97, 118, 127, 130, 139, 154–5, 160–1
arcticum Lge. 97, 124, 154, 156, 167
beeringianum Cham. & Schlecht. ssp. beeringianum Hult. 118, 124
bialynickii Tolm. 111, 165
cerastoides (L.) Britt. 30, 81, 88
regelii Ostenf. 38, 120, 131–2, 135, 140, 148, 155, 159–60, 163–4, 165–6
Chamaenerium Adans.
angustifolium (L.) Scop. 23, 30, 34, 86–7, 131
latifolium (L.) Th. Fries & Lge. 131, 134
Chamaepericlymenum Graebner
suecicum (L.) Graebn. 30, 32
Chosenia Nakai
arbutifolia (Pall.) A. Skv. 14, 24, 59, 62–3, 66
Chrysosplenium L.
alternifolium L. 111–12
rosendahlii Packer 117
tetrandum (Lund) Th. Fries 140
Cirsium Mill.
heterophyllum (L.) Hill 32
Claytonia L.
tuberosa Pall. 54
Claytoniella Jurtz. 16
Cochlearia L.
arctica Schlecht. 38, 118, 120, 140
groenlandica L. 131–2, 140, 157, 160
Colobanthus Bartl.
crassifolius (D'Urv.) Hook. f. 173
quitensis (Kunth) Bartl. 173, 176, 180
Comarum L.
palustre L. 23
Coptis Salisb.
trifolia (L.) Salisb. 86
Crepis L.
nana Richards. 124
Cyperaceae 142
Cystopteris Bernh.
fragilis (L.) Bernh. 130, 139

Dasiphora Rafin.
fruticosa (L.) Rydb. 65
Delphinium L.
elatum L. ssp. cryophilum (Nevski) Jurtz. 35
Dendranthema (DC.) Des Moul.; Tzvelev
mongolicum (Ling) Tzvel. 52
Deschampsia Beauv.
alpina (L.) Roem. & Schult. 72, 81, 97
antarctica Desv. 173, 176–7, 180
borealis (Trautv.) Roshev. 14, 114, 116, 159, 163–5
brevifolia R. Br. s.l. 17, 37, 96–8, 100, 102, 111, 132, 134, 136
caespitosa (L.) Beauv. 30
flexuosa (L.) Trin. (Lerchenfeldia flexuosa (L.) Schur) 23, 30, 32, 34, 86
Dianthus L.
repens Willd. 52
superbus L. 30
Diapensia L.
lapponica L. 30, 76, 79, 80, 85
obovata (F. Schm.) Nakai 47, 65
Draba L. 16, 145–6
allenii Fern. 72, 122
alpina L. 131, 159
bellii Holm 124, 132, 160
cinerea Adams 129, 130, 132, 134
fladnizensis Wulf. 131
glabella Pursh 130
kamtschatica (Ldb.) N. Busch 54
lactea Adams 130–1, 134, 160
macrocarpa Adams 124, 132, 134, 160, 163
micropetala Hook. 131, 157, 163
norvegica Gunn. 72, 154
oblongata R. Br. 91, 134, 155–7, 160, 164, 166–7
pseudopilosa Pohle 110
subcapitata Simm. 53, 91, 128, 132, 139, 156, 160, 167
Dryas L. 53–4, 63, 99, 100, 102, 106, 111, 114, 119, 122, 126, 133, 135, 137
grandis Juz. 54
integrifolia Vahl 3, 21, 69, 77, 79, 80, 93, 113–14, 119, 120, 122–8, 130, 133, 136

Dryas L. – *cont.*
integrifolia Vahl var. canescens Simm. 114
octopetala L. 21–35, 37, 40, 93, 97–101, 103, 114–15, 135, 137, 138
punctata Juz. 12, 21–2, 37, 40, 42, 50–1, 53–5, 87, 93, 101, 104–6, 108, 111, 114–15, 128, 130–1, 145
Dupontia R. Br. 15
Fisheri R. Br. 77, 80, 97, 99, 108, 111, 118–19, 124, 126, 140–1, 160, 167
psilosantha Rupr. 54
Duschekia Opiz
fruticosa (Rupr.) Pouzar 11

Empetraceae 19
Empetrum L.
hermaphroditum (Lange) Hagerup 21, 32, 34, 37, 40, 53, 65, 67–8, 76–7, 79, 82, 85–8, 130
rubrum Willd. 172
Equisetum L.
arvense ssp. boreale (Bong.) Rupr. 131–2, 140
sylvaticum L. 84
variegatum Schleich. 126, 131, 140
Ericaceae 4
Erigeron L.
compositus Pursh 117, 120, 128
eriocephalus J. Vahl 124
muirii A. Gray 70
Eriophorum L. 42
angustifolium Honck. (E. polystachion L.) 13, 40–2, 54, 77, 80, 86, 88, 91, 97, 102–5, 108, 111–12, 115, 118–19, 124–5
brachyantherum Trautv. & Mey. 104, 112
callitrix Cham. 131
medium Anderss. 97, 99, 103, 111
russeolum Fr. 37
scheuchzeri Hoppe 14, 22, 80, 86, 88, 100, 103, 105–6, 112, 118–19, 124, 126, 132, 134, 140, 160, 167
spissum Fern. 66, 69, 75–8, 93
triste (Th. Fries) Löve & Hadač 124, 126, 130, 132, 134–5
vaginatum L. 26, 39, 41–2, 50, 53, 56–7, 61, 63, 65, 66–7, 69, 72, 75, 78, 93, 108, 118
vaginatum L. ssp. spissum (Fern.) Hult. 66, 69, 78
Eritrichium Schrad.
villosum (Ldb.) Bunge 165
Ermania Cham. 16
Erysimum L.
pallasii (Pursh) Fern. 22, 97–8
Eutrema R. Br.
edwardsii R. Br. 22, 80, 131

Festuca L.
altaica Trin. 68
baffinensis Polunin 119, 120
brachyphylla Schult. 77, 86, 105, 111, 118, 123–5, 129, 131, 167
hyperborea Holmen 127, 134
ovina L. 37, 51
Filipendula Mill.
ulmaria (L.) Maxim. 32
Fragaria L.
vesca L. 35

Gastrolychnis Rchb. 16
angustiflora Rupr. 30
apetala (L.) Tolm. & Kozh. 22, 120, 130–1, 134
triflora (R. Br.) Tolm. & Kozh. 132, 134
Geum L.
rivale L. 32
Gnaphalium L.
supinum L. 38, 88
Gorodkovia Botsch. & Karav. 16
jacutica Botsch. & Karav. 54
Gymnocarpium Newman
dryopteris (L.) Newman 84, 86

Harrimanella Cov.
hypnoides (L.) Cov. 37, 79, 86, 88
Helictotrichon Bess.
dahuricum (Kom.) Kitag. 54
krylovii (N. Pavl.) Henrard 12, 63
Hierochloë R. Br.
alpina (Liljebl.) Roem. & Schult. 37, 53, 69, 77, 80, 129, 130–1
pauciflora R. Br. 77, 97, 123, 126
Hippuris L.
vulgaris L. 119, 132
Honkenya Ehrh.
peploides (L.) Ehrh. 118
Hymenophyllaceae 172

Juncus L.
biglumis L. 123, 130, 132, 140, 160, 165, 167–8
triglumis L. 130

Juniperus L.
communis L. 86
sibirica Burgsd. 84

Kobresia Willd. 115
bellardii (All.) Degland 12, 22, 86, 88, 115, 120, 126–7, 130–1, 133, 135
sibirica Turcz. 72
simpliciuscula (Wahlenb.) Mackenzie 22, 130
Koeleria Pers.; Rchb.
pohleana (Domin) Gontsch. 35

Lagotis Gaertn.
minor (Willd.) Standl. 11
Larix Mill.
alaskensis W. F. Wight 70
cajanderi Mayr 24, 47, 52, 56–7, 59, 63, 66
gmelinii (Rupr.) Kuzan. 14, 23–4, 26, 47, 51
laricina (Du Roi) K. Koch 24, 59, 70, 75, 81
sibirica Ldb. 14, 24, 35–6, 38
Ledum L.
decumbens (Ait.) Small 21, 26, 37, 39, 40, 50, 54–6, 62–3, 65–9, 75–7, 79, 85
palustre L. 40
Lesquerella S. Wats.
arctica (Wormskj.) S. Wats. 22, 120, 129, 134
Leymus Hochst.
arenarius (L.) Hochst. 30
mollis (Trin.) Pilger s.l. 118, 134
Liliaceae 19
Linnaea L.
borealis L. ssp. americana (Forb.) Hult. 84, 86
Listera R. Br.
cordata (L.) R. Br. 84
Lloydia Salisb.
serotina (L.) Rchb. 37, 114
Loiseleuria Desv.
procumbens (L.) Desv. 68, 76, 79, 85
Lupinus L.
arcticus S. Wats. 67, 70, 72
Luzula DC.
arcuata (Wahlenb.) Sw. 30, 97
confusa Lindeb. 22, 37, 69, 76–7, 80, 99, 103–5, 107, 111, 116, 120, 123–5, 129–31, 133, 135, 138–9, 160
nivalis Laest. 64, 69, 105, 116, 124–5, 129, 130, 132, 160
Lychnis L.
sibirica L. ssp. villosula (Trautv.) Tolm. 52
Lycopodiaceae 172
Lycopodium L.
complanatum L. (Diphasium complanatum (L.) Rothm.) 51, 84
selago L. ssp. articum (Grossh.) Tolm. 159

Matthioleae O. E. Schulz 15
Mertensia Roth
maritima (L.) S. F. Gray 118
Minuartia L.
macrocarpa (Pursh) Ostenf. 97, 105, 116
rossii (R. Br.) Graebn. 120
rubella (Wahlenb.) Hiern 127–8, 130, 165

Nardosmia Cass.
frigida (L.) Hook. 72, 91, 119
Neuroloma Andrz. ex DC. 15
Novosieversia Bolle 16
glacialis (Adams) Bolle 22, 37, 97–8, 105, 161

Ophioglossaceae 172
Oxycoccus Hill
palustris Pers. ssp. microphyllus (Lge.) Löve & Löve 86
Oxygraphis Bge.
glacialis (Fisch.) Bge. 131
Oxyria Hill.
digyna (L.) Hill 22, 64, 123–5, 130–4, 143, 160–1, 167
Oxytropis DC. 16
arctica R. Br. 119, 120, 124
arctobia Bge. 120
bellii (Britt.) Palib. 78
hudsonica (Greene) Fern. 78
middendorfii Trautv. 114
middendorfii Trautv. ssp. orulganica Jurtz. 54
ochotensis Bge. 64
tschuktschorum Jurtz. 116
wrangelii Jurtz. 117

Papaver L. 16, 145–6
dahlianum Nordh. 10, 139, 160
lapponicum (Tolm.) Nordh. 111
polare (Tolm.) Perf. 10, 111, 148, 155, 162–3, 165
radicatum Rottb. s.l. 10, 80, 111, 119, 120, 124, 127–9, 130, 134, 139, 143, 155–6, 160–7

Parnassia L.
  kotzebuei Cham. & Schlecht. 57
Parrya R. Br. 15
  arctica R. Br. 15, 72, 75
  nudicaulis (L.) Rgl. 72, 114
Pedicularis L.
  capitata Adams 120
  flammea L. 131
  hirsuta L. 126, 129–31, 139
  lanata Cham. & Schlecht. 119
  oederi M. Vahl 11
  sudetica Willd. 77, 123, 126
Phippsia R. Br.
  algida (Soland.) R. Br. 10, 14, 19, 64, 100, 119, 125, 127, 131–2, 134, 142, 145–6, 148, 155, 157, 160, 163–6
  concinna (Th. Fries) Lindeb. 167
Phlojodicarpus Turcz.
  villosus Turcz. 64
Phyllodoce Salisb.
  coerulea (L.) Bab. 4, 32, 76, 79, 85, 88
Picea A. Dietr.
  glauca (Moench) Voss 14, 59, 70, 75, 78, 81
  mariana (Mill.) B.S. & P. 14, 59, 70, 75, 78, 81
  obovata Ldb. 14, 24, 33–5
Pinus L.
  lapponica Mayr 33
  pumila (Pall.) Rgl. 14, 26, 52, 65–6
Pleuropogon R. Br.
  sabinei R. Br. 22, 119, 132, 135
Poa L.
  abbreviata R. Br. 10, 19, 22, 35, 53, 91, 120, 124, 128, 131, 134, 146, 156, 167
  alpigena (Fries) Lindm. 11, 119, 123, 159–60
  alpigena (Fries) Lindm. ssp. colpodea (Th. Fries) Soland. 131–2
  alpina L. 30
  arctica R. Br. 3, 22, 69, 77, 108, 118–19, 123, 129–31, 133
  flabellata (Lam.) Hook. f. 171
  foliosa Hook. f. 171
  glauca M. Vahl 120, 129–33, 135
  hartzii Gandoger 117, 127, 131, 134
  pseudoabbreviata Roshev. 116
  rigens Hartm. 161
  wrangelica Tzvel. 117
Polemonium L.
  boreale Adams 131
Polygonum L.
  bistorta L. 30
  ellipticum Willd. 47
  viviparum L. 77, 85, 88, 105, 120, 123, 125–6, 129–31, 134, 161
Polypodiaceae 19, 172
Populus L.
  balsamifera L. 14, 24, 59, 60, 70, 75
  suaveolens Fisch. 14, 24, 59, 62–3
  tremuloides Michx. 60, 70
Potentilla L. 16
  crantzii (Crantz) Beck 72
  hyparctica Malte 3, 22, 80, 110–11, 120, 130–1, 133, 135, 139, 157, 161, 167
  nivea L. 132
  pulchella R. Br. 131–2
  rubella Th. Sør. 122
  rubricaulis Lehm. 119–20, 124
  vahliana Lehm. 124
  viscosa J. Don 57
  wrangelii Petr. 117
Puccinellia Parl. 16
  andersonii Swallen 120, 134
  angustata (R. Br.) Rand. & Redf. 91, 123–4, 127, 131, 134, 163
  beringensis Tzvel. 64
  bruggemannii Th. Sør. 122, 124, 127
  coarctata Fern. & Weath. 97
  groenlandica Th. Sør. 87, 122
  phryganodes (Trin.) Scribn. & Merr. 13, 88, 107, 118, 120, 132, 141
  poacea Th. Sør. 122
  rosenkrantzii Th. Sør. 87, 122
  svalbardensis Rønning 137
  vahliana (Liebm.) Scribn. & Merr. 132, 134, 160
× Pucciphippsia Tzvel. 15
  vacillans (Th. Fries) Tzvel. 160
Pulsatilla Adans.
  multifida (Pritz.) Juz. 115
  patens (L.) Mill. 12
Pyrola L.
  grandiflora Rad. 42, 79, 85

Ranunculus L. 16
  allenii Robins. 72
  gmelinii DC. 72, 119
  hyperboreus Rottb. 119, 132, 135, 160, 168
  lapponicus L. 77
  nivalis L. 124, 130, 167
  pygmaeus Wahlenb. 16
  sabinei R. Br. 91, 119, 167

spitsbergensis Hadač 140
sulphureus Soland. 37, 110, 124, 130, 134, 140, 142, 160, 164, 166–7
Rhododendron L.
aureum Georgi 65
lapponicum (L.) Wahlenb. 75–7, 79, 85, 130–1
Rosaceae 16, 157
Rubus L.
chamaemorus L. 40, 68, 76–7, 92, 100

Sagina L.
intermedia Fenzl 131–2, 160
Salix L. 77, 108
alaxensis Cov. 45, 47, 67–9, 72, 75, 92
arbusculoides Anderss. 72, 75
arctica Pall. 22, 37, 42, 53, 76–7, 79, 91–3, 98, 104, 119–21, 123–7, 129, 130–3, 135, 144, 152, 158
arctophila Cockerell 77, 86, 91, 125
boganidensis Trautv. 47
fuscescens Anderss. 47, 65, 68, 108
glauca L. 11, 24, 32, 34–5, 37–9, 41–2, 50, 52, 55–7, 68–9, 75, 92, 115
glauca L. ssp. callicarpaea (Trautv.) Böcher 27, 79, 80, 85–8, 93, 124
hastata L. 30, 37
herbacea L. 30, 76, 79, 82, 85–8, 118, 122, 125, 130, 142
krylovii E. Wolf 11, 26, 54
lanata L. 11, 27, 32, 34, 37–8, 40–3, 50, 52, 56–7, 91, 105
lanata L. ssp. richardsonii (Hook.) A. Skv. 27, 67–9, 72, 75, 91, 113, 115, 119, 124–5
lapponum L. 30, 32, 38, 44–5
myrsinites L. 30, 32
myrtilloides L. 65
nummularia Anderss. 37, 53, 92, 103
phlebophylla Anderss. 68–9, 116, 118–9
phylicifolia L. 11, 24, 30, 32, 34, 38, 44, 92
planifolia Pursh 24, 75–7, 92
polaris Wahlenb. 21, 37, 53–4, 92, 97–8, 100, 103–4, 106–8, 110–13, 116, 118, 120–3, 135, 137–9, 140, 142–3, 145–7, 152, 158–9, 160–1, 165, 167
polaris Wahlenb. ssp. pseudopolaris (Flod.) Hult. 72, 93, 96, 105, 112–14, 118
pseudopolaris Flod. 113
pulchra Cham. 27, 36–42, 52–7, 62, 67–9, 72, 91–2, 97, 108, 114, 119
reptans Rupr. 27, 35, 39, 41, 50, 56, 91–2, 98, 105–6, 108, 111, 114
reticulata L. 37, 68, 77, 92, 108, 119, 125–6, 142
rotundifolia Trautv. 114–15
sphenophylla A. Skv. 63, 91, 108
uva-ursi Pursh 72, 79, 80
Saussurea DC.
angustifolia (Willd.) DC. 72
Saxifraga L. 16, 145
aizoides L. 97
caespitosa L. 118, 127–8, 130, 139, 142, 146, 160, 163, 165–6
cernua L. 14, 22, 120, 127, 130–2, 139–40, 142, 157, 160, 162–7
flagellaris Willd. 119, 160, 167
foliolosa R. Br. 111, 131–2, 140, 160, 168
hieracifolia Waldst. & Kit. 30
hirculus L. 120, 142
hyperborea R. Br. 16, 64, 148, 155, 157
nathorstii (Dusén) Hayek 128, 131
nelsoniana D. Don 64
nivalis L. 130, 134, 139, 160, 165, 167
oppositifolia L. 38, 77, 80, 112, 114, 119–20, 124–8, 130, 133–4, 137–9, 142–3, 145, 158, 162–4, 166–7
platysepala (Trautv.) Tolm. 91, 119, 166
rivularis L. 14, 100, 118, 125, 127, 131–2, 140, 142, 160
serpyllifolia Pursh 114, 116, 161
tenuis (Wahlenb.) H. Smith 16, 131, 167
tricuspidata Rottb. 120
Senecio L.
atropurpureus (Ldb.) B. Fedtsch. 97, 98
congestus (R. Br.) DC. 123
frigidus (Richards.) Less. 72
jacuticus Schischk. 64
resedifolius Less. 97
Sibbaldia L.
procumbens L. 86, 88
Silene L.
acaulis (L.) Jacq. 37, 77, 80, 97, 130, 133–4, 139
repens Patr. 12

Sisyrinchium L.
groenlandicum Böcher 87
Solidago L.
lapponica Wither. 32
virgaurea L. 30, 34
Sorbus L.
groenlandica (Schneid.) Löve & Löve 84, 87
Spiraea L.
beauverdiana Schneid. 65, 67–8
Stellaria L.
ciliatosepala Trautv. 124
crassipes Hult. 140
edwardsii R. Br. 130–1, 163–5
humifusa Rottb. 13, 88, 107, 120, 124, 132, 141
longipes Goldie 131
monantha Hult. 123

Taraxacum Wiggers 16
arcticum (Trautv.) Dahlst. 130
arctogenum Dahlst. 122
croceum Dahlst. 86, 88
pumilum Dahlst. 134
Thellungiella O. E. Schulz
salsuginea (Pall.) O. E. Schulz 57
Thymus L.
extremus Klok. 52
Tofieldia Huds.
coccinea Richards. 130
Trichophorum Pers.
caespitosum (L.) Hartm. (Baeothryon caespitosum (L.) A. Dietr.) 76, 80, 86
Trientalis L.
europaea L. 30, 32
Trisetum L.
spicatum (L.) Richt. 3, 130, 133, 135
Trollius L.
apertus Perf. 38
europaeus L. 30

Vacciniaceae 19
Vaccinium L. 14
myrtillus L. 23, 30, 32, 34
uliginosum L 32, 65–6
uliginosum L. ssp. microphyllum Lge. 21, 26–7, 37, 39, 40–1, 50, 53–6, 62, 67–8, 75–7, 79, 80, 82, 85–8, 91, 124–6, 130, 134
vitis-idaea L. 32, 53, 65, 75
vitis-idaea L. ssp. minus (Lodd.) Hult. 34, 39–42, 54–6, 62–3, 67–9, 75–7, 79, 80, 85, 87, 91, 108
Vahlodea Fr.
atropurpurea (Wahlenb.) Fries 30
Valeriana L.
capitata Pall. 91, 119

MOSSES

Andreaea Ehrh. 150, 178
alpestris (Thed.) B.S.G. 138
crassinervis Bruch 125
depressinervis Card. 177
gainii Card. 177
papillosa Lindb. 115, 159
regularis C. Müll. 177–8
rupestris Hedw. 156
Aulacomnium Schwaegr. 88
palustre (Hedw.) Schwaegr. 35, 53, 65, 80, 111, 115
turgidum (Wahlenb.) Schwaegr. 35, 37, 40–2, 50, 54, 56, 65, 85, 101–6, 108, 112, 114–16, 118, 124–5, 129, 133–4, 138, 140, 166

Bartramia Hedw.
ithyphylla Brid. 159
Brachythecium B.S.G.
antarcticum Card. 180
austro-salebrosum (C. Müll.) Par. 176, 180
Bryoerythrophyllum Chen.
recurvirostre (Hedw.) Chen. 183
Bryum Dill. 14, 125, 146, 150–1, 163, 182
algens Card. 180, 182–3, 185–6
antarcticum Hook, f. & Wils. 182–3
argenteum Hedw. 182–3, 185–6
crispulum Hampe 160
filicaule Broth. 189
nitidulum Lindb. 141
obtusifolium Lindb. 134, 140
pallescens Schwaegr. 141
rutilans Brid. 157
ventricosum Hook. & Tayl. 141

Calliergon Kindb. 13, 42, 105, 118
giganteum (Schimp.) Kindb. 126
richardsonii (Mitt.) Kindb. 106
sarmentosum (Wahlenb.) Kindb. 77, 80, 98–9, 111, 115, 132, 140, 142, 160, 180
stramineum (Brid.) Kindb. 88
Campylium (Sull.) Lange & Jensen 14, 150
stellatum (Hedw.) Lange & Jensen 140, 142, 157, 160
zemliae C. Jens. 160

Ceratodon Brid.
grossiretis Card. 180
purpureus (Hedw.) Brid. 182–3, 185
Chorisodontium (Mitt.) Broth. 179
aciphyllum (Hook. f. & Wils.) Broth. 172, 176, 179
Cinclidium Sw.
arcticum (B.S.G.) Schimp. 126

Dicranoweisia Lindb.
crispula (Hedw.) Lindb. 116, 133, 142, 159
grimmiacea (C. Müll.) Broth. 177
Dicranum Hedw. 37, 40, 42, 56
angustum Lindb. 35, 40, 53, 140
bonjeanii De Not. 143
congestum Brid. 40, 56
elongatum Schleich. 35, 65, 99, 103, 104, 111
fuscescens Turn. 85, 129
groenlandicum Brid. 125
scoparium Hedw. 126
Distichium B.S.G.
capillaceum (Hedw.) B.S.G. 151, 160, 163–5
inclinatum (Hedw.) B.S.G. 141
Ditrichum Timm
flexicaule (Schleich.) Hampe 10, 105, 125–6, 139, 146, 148, 151, 154, 156, 163–4, 166
Drepanocladus (C. Müll.) Roth 13, 42, 105, 118, 159
brevifolius (Lindb.) Warnst. 126, 134, 140, 142, 160
exannulatus (B.S.G.) Warnst. 66, 99, 140
fluitans (Hedw.) Warnst. 142
revolvens (Turn.) Warnst. 40, 77, 80, 86, 106, 111, 115, 126, 132, 134, 140
sendtneri (Schimp.) Warnst. 106
uncinatus (Hedw.) Warnst. 68, 85, 103, 110, 137–40, 142–3, 160, 172, 176, 178, 179, 180
vernicosus (Lindb.) Warnst. 111, 142

Encalypta Schreb.
alpina Sm. 156
rhabdocarpa Schwaegr. 151, 156

Grimmia Ehrh.
doniana Sm. 183, 185
plagiopodia Hedw. 183

Hygrohypnum Lindb.
polare (Lindb.) Broth. 156
Hylocomium B.S.G.
alaskanum (Lesq. & James) Kindb. (H. splendens var. alaskanum (Lesq. & James) Limpr.) 41–2, 50, 53–4, 62–3, 96, 98, 100–2, 104–6, 108, 110, 114–15, 125–6, 129, 138–9, 143, 146, 164
splendens (Hedw.) B.S.G. 29, 32, 34, 40, 68
Hypnum Dill.
bambergeri Schimp. 138, 164
revolutum (Mitt.) Lindb. 133

Meesia Hedw.
triquetra (Hook. & Tayl.) Angstr. 77, 126, 134, 142
Myurella B.S.G.
julacea (Schwaegr.) B.S.G. 151

Onchophorus Brid.
wahlenbergii Brid. 104, 116, 138, 150, 166
Orthothecium B.S.G.
chryseum (Schwaegr.) B.S.G. 14, 140, 150, 157, 165–6
killiasii C. Müll. 134

Pleurozium Mitt.
schreberi (Brid.) Mitt. 29, 32, 34, 85
Pogonatum Palis. 118
Pohlia Hedw. 146, 182
cruda (Hedw.) Lindb. 143, 151, 160
cruda (Hedw.) Lindb. var. imbricata (Card.) Bartr. 180
nutans (Hedw.) Lindb. 177, 179–80
obtusifolia (Brid.) L. Koch 157
Polytrichum Dill. 42
alpestre Hoppe 29, 34, 40, 53, 99, 104–5, 111, 176, 178–9
alpinum Hedw. 10, 37, 104–5, 110, 116, 138, 140, 143, 146, 151, 154, 157, 160, 176–80
commune Hedw. 34
hyperboreum R. Br. 38
juniperinum Hedw. 103
piliferum Hedw. 125, 179
Psilopilum Brid.
cavifolium (Wils.) Hag. 151, 157

Rhacomitrium Brid.
canescens (Hedw.) Brid. 139
lanuginosum (Hedw.) Brid. 12, 37, 76–7, 80, 99, 103–4, 115–16,

Rhacomitrium Brid. – *cont.*
125–6, 128–9, 133, 138–9, 143, 150, 156, 166, 172
Rhynchostegium B.S.G.
brachypterygium (Hornsch.) Jaeg. 172
Rhytidium (Sull.) Kindb.
rugosum (Hedw.) Kindb. 50–1, 54, 106

Sacroneurum Bryhn
glaciale (Hook. f. & Wils.) Card. & Bryhn 182–4, 186
Schistidium Brid.
antarctici (Card.) L. Savicz & Z. Smirn. 177, 182–4
apocarpum (Hedw.) B.S.G. 142, 183
gracile (Schleich.) Limpr. 115–16, 128, 142, 159
Sphagnum (Dill.) Ehrh. 40, 62, 150
balticum (Russ.) C. Jens. 65, 103, 140
compactum DC. 40, 56
contortum K. F. Schultz 112, 115
fimbriatum Wils. 91, 99, 103–4, 112, 140
girgensohnii Russ. 34, 86, 112
lenense H. Lindb. 65
nemoreum Scop. 34
obtusum Warnst. 112
squarrosum Pers. 77, 92, 105, 112, 115, 140
subfulvum Sjörs 140
subnitens Russ. & Warnst. 88
subsecundum Nees 105, 115
teres (Schimp.) Ångstr. 80
warnstorfii Russ. 65, 125

Tomenthypnum Loeske
nitens (Hedw.) Loeske 37, 41–2, 50, 53–4, 56, 96, 98, 100–2, 104–6, 112, 114–15, 125, 134, 137–8
Tortula Hedw.
conferta Bartr. 180
robusta Hook. & Grev. 172
ruralis (Hedw.) Crome 126, 133, 143

LIVERWORTS

Anthelia Dum.
juratzkana (Limpr.) Trevis. 38

Barbilophozia Loeske
hatcheri (Evans) Loeske 176, 179, 180

Blepharostoma Dum.
trichophyllum (L.) Dum. 126, 140, 157

Cephaloziella Spruce
arctica Bryhn & Kaal. 10, 146, 148, 155, 157
rubella (Nees) Douin 140
varians (Gottsche) Steph. 176, 179, 180

Gymnomitrium Ångstr.
concinnatum Corda 155
coralloides Nees 38, 125, 151, 156

Lophozia Dum.
alpestris (Scheich.) Evans 157
excisa (Dicks.) Dum. 157
grandiretis (Lindb.) Schiffn. 157

Orthocaulis Buch
quadrilobus (Lindb.) Buch 140

Pleuroclada Spruce
albescens (Hook.) Spruce 38
Ptilidium Nees
ciliare (L.) Hampe 41–2, 53, 56, 68, 85, 101–4, 108, 125–6, 129, 138, 143

Scapania Dum.
calcicola (Arn. & Pears.) Ingham 157
globulifera (G. Jens.) Schljak. 157
Sphenolobus Steph.; Buch
minutus (Crantz) Steph. 104

Tritomaria (Schiffn.) Loeske
quinquedentata (Huds.) Buch 140
scitula (Tayl.) Jørg. 155

LICHENS

Acarospora Mass.
chlorophana (Wahlenb.) Mass. 128
macrocyclos Vain. 178
petalina Golubk. & Savicz 186
Alectoria Ach.
miniuscula (Nyl. ex Arn.) Degel. 182–6
nigricans (Ach.) Nyl. 128, 156, 160
ochroleuca (Hoffm.) Mass. 12, 37, 44, 50, 53, 85, 103–4, 115, 128, 156
pubescens (L.) Howe 115, 139, 156, 159, 177–9

Biatora Th. Fr.
flava (Dodge & Baker) Golubk. & Savicz 185
Biatorella Th. Fr.
antarctica Murr. 178, 184
Blastenia Mass.
sparsa Murr. 184
Buellia de Not. 75, 178
coniops (Wahlenb. ex Ach.) Th. Fr. 178
disciformis (Fries) Mudd. 159
frigida Darb. 182–6
russa (Hue) Darb. 178

Caloplaca Th. Fr. 38, 180
bracteata Jatta 127
cirrochrooides (Vain.) Zahlbr. 178
darbishiri (Dodge & Baker) Cretz. 184
regalis (Vain.) Zahlbr. 178
subolivacea (Th. Fr.) Lynge 157
Candelaria Mass.
concolor (Dicks.) Stein var. antarctica Murr. 184
Catillaria (Ach.) Th. Fr.
corymbosa (Hue) M. Lamb 178
Cetraria Ach. 10, 12, 75–6, 80, 146
crispa (Ach.) Nyl. 53–4, 65, 85, 103, 111, 125, 156, 160, 166
cucullata (L.) Ach. 30, 40, 44, 53–4, 56, 65, 68, 85, 104, 114, 126, 163–4, 166
delisei Th. Fr. 99, 100, 103, 143, 148, 154, 157, 159–60, 163–5, 166
hepatizon (Ach.) Vain. 104, 139, 159
islandica (L.) Ach. 56, 68, 76, 104, 114, 129, 138–9, 143
nivalis (L.) Ach. 32, 53, 66, 76, 88, 99, 114–15, 125–6, 128–9, 134, 139, 156, 160
tilesii Ach. 38
Cladonia (Hill.) Vain. 12, 30, 32, 34, 43–4, 75–6, 80, 151, 180
alpestris (L.) Rabenh. 32, 53
amaurocrea (Flk.) Schaer. 43
bellidiflora (Ach.) Schaer. 160
cornuta (L.) Schaer. 68
elongata (Jack.) Hoffm. 104, 160
furcata (Huds.) Schrad. 176–7, 179
gracilis (L.) Willd. 104, 143
metacorallifera Asah. 176, 179
mitis Sandst. 32, 37, 54, 76, 88, 99, 104, 115, 125, 143
pleurota (Flk.) Schaer. 68
pyxidata (L.) Fr. 139, 143, 151, 157
rangiferina (L.) Web. 32, 40, 53, 65, 68, 76, 99, 104
rangiformis Hoffm. 76
sylvatica (L.) Hoffm. 53, 65, 68
uncialis (L.) Hoffm. 103
vicaria R. Sant. 176–7
Collema (Wigg.) A. Zahlbr. 10, 146, 148, 149
ceranicum Nyl. 155
Collemataceae 166
Cornicularia Ach.
divergens Ach. 12, 37, 44, 50, 53, 66, 104, 115, 125, 156

Dactylina Nyl.
arctica (Hook.) Nyl. 76, 103, 129
Dufourea (Ach.) Nyl.
madreporiformis (Wulf.) Ach. 163

Gasparrinia (Tornab.) Th. Fr.
elegans Stein 182
murorum Tornab. 182

Haematomma Mass.
erythromma (Nyl.) Zahlbr. 178
ventosum (L.) Mass. 139
Himantormia M. Lamb
lugubris (Hue) M. Lamb 177
Hypogymnia Nyl.
apicola (Th. Fr.) Hav. 159

Lecanora Ach. 12, 38, 75, 77, 142, 178
campestris (Schaer.) Hue 156, 157
expectans Darb. 182
polytropa Rabenh. 138, 159
rubina (Vill.) Ach. var. melanophthalma (Ram.) Zahlbr. 184
Lecidea (Ach.) Zahlbr. 12, 38, 75, 77, 142, 178, 180
cancriformis Dodge & Baker 184–5
dicksonii (Gmel.) Ach. 128, 138, 159
physciella Darb. 185
Lepraria Ach.
neglecta (Nyl.) Erichs. 182, 185
Lobaria (Schreb.) Zahlbr.
linita (Ach.) Rabenh. 76, 160

Mastodia Hook. f. & Harv.
tesselata (Hook. f. & Harv.) Hook. f. & Harv. 178, 180
Microglaena Kbr.
antarctica M. Lamb 178

Nephroma Ach.
articum (L.) Torss. 129
expallidum (Nyl.) Nyl. 160

Neuropogon Fw. & Nees
sulphureus (Koenig) Elenk. 151, 156, 182

Ochrolechia Mass. 10, 100, 146, 148–9, 152, 161
frigida (Sw.) Lynge 116, 138, 143, 153, 155–6, 176–9
tartarea (L.) Mass. 153, 155
Omphalodiscus Schol.
antarcticus (Frey & Lamb) Llano 178
decussatus (Vill.) Schol. 182

Pannariaceae 166
Parmelia Ach. 37–8, 77
coreyi Dodge & Baker 185
minuscula Nyl. 159
omphalodes (L.) Ach. 115, 125, 139, 143, 160
saxatilis (L.) Ach. 177
Peltigera Pers.
aphthosa (L.) Willd. 85, 140
canina (L.) Willd. 143
leucophlebia (Nyl.) Gyeln. 76
malacea (Ach.) Funck 85, 140
rufescens (Weis) Humb. 160
spuria (Ach.) DC. 177
Pertusaria DC. 10, 146, 148–9, 152–3, 161
freyi Erichs. 153, 155
glomerata (Ach.) Schaer. 153, 155
octomela (Norm.) Erichs. 153, 155
Physcia (Ach.) Vain. 182
caesioides Golubk. & Savicz 185
muscigena (Ach.) Nyl. 160
Placopsis Müll. Arg.
contortuplicata M. Lamb 178
Protoblastenia (Zahlbr.) Stur
citrina Dodge 182–5
Psoroma Nyl. 180
follmannii Dodge 178, 180
hypnorum (Vahl) S. F. Gray 116, 125, 138
Pyrenodesmia Mass.
mawsonii Dodge 182–3, 185

Ramalina Ach.
terebrata Hook. & Taylor 178
Rhizocarpon (Ram.) Th. Fr. 12, 38, 75, 77, 142
geographicum (L.) DC. 104, 138, 159, 178
Rinodina S. Gray
petermannii (Hue) Darb. 178
turfacea (Wahlenb.) Körb. 138, 182–3

Solorina Ach.
bispora Nyl. 160
crocea (L.) Ach. 38, 160
Sphaerophorus Pers.
fragilis (L.) Pers. 160
globosus (Huds.) Vain. 12, 77, 99, 115, 125, 139, 143, 156, 165, 177–9
Stereocaulon Schreb. 146
alpinum Laur. 37, 77, 85, 111, 114, 116, 126, 129, 138–9, 176–7
fastigiatum Anzi 159
paschale (L.) Hoffm. 32, 143
rivulorum H. Magn. 10, 146, 155–7, 160
vesuvianum Pers. var. depressum (H. Magn.) M. Lamb 156

Thamnolia Ach.
vermicularis (Sw.) Schaer. 3, 105, 114, 116, 125–6, 160, 163, 166
Toninia Th. Fr. 10, 146, 149, 153
coeruleonigricans (Lightf.) Th. Fr. 127
lobulata (Sommerf.) Lynge 153, 161 166

Umbilicaria Hoffm. 12, 37, 75, 142
antarctica Frey & Lamb 177
arctica (Ach.) Nyl. 139, 156
cylindrica (L.) Del. ex Duby 77, 159
decussata (Vill.) Frey 128, 177, 182–6
deusta (L.) Baumg. 160
erosa (Web.) Ach. 139, 159
hyperborea (Ach.) Hoffm. 77, 104, 139, 159
leiocarpa DC. 128
proboscidea (L.) Schrad. 77, 104, 139, 156, 159
propagulifera (Vain.) Llano 177
spongiosa Dodge & Baker 183, 185
Usnea Wigg. 178
acromelana Stirt 182–6
antarctica Du Rietz 176–9, 182–6
fasciata Torr. 177
sulphurea (Koenig) Th. Fr. 151, 182, 184–5

Verrucaria (Wigg.) Th. Fr.
ceuthocarpa Wahlenb. 178
psychrophila M. Lamb 178
tesselatula Nyl. 178

Xanthoria (Fr.) Th. Fr.
candelaria (L.) Th. Fr. f. antarctica (Vain.) Hillm. 178, 182–3
elegans (Link.) Th. Fr. 178, 184
mawsonii Dodge 185
parietina (L.) Beltr. 159

ALGAE

Enteromorpha Link 141

Nostoc Vauch. 141, 149, 181, 183, 186
commune Vauch. 163, 185–6

Prasiola Ag.
crispa (Lightf.) Menegh. 180, 185–6

Fungi (Ascomycetes)

Thyrenectria Seeler
antarctica (Speg.) Seeler var. hyperantarctica D. Hawksw. 180

INDEX

Abbe, E. C. 79
Academy Tundra 113, 116
Acock, A. M. 136
Agapa River 41
Ahti, T. 52
Alaska 26, 28, 57, 64, 66–7, 69–71, 113, 117; subarctic tundra 66–70
Alazeyeva–Kolyma district 56–7, 91, 107
Alekhin, V. V. 51
Aleksandrova, V. D. 1, 8, 9, 14, 18, 24, 42–3, 48, 50–1, 55, 90, 92, 96–100, 109–12, 132, 135, 144–7, 154
Aleutian low pressure 58
Alexandra Land 148, 153–4
Allison, J. S. 175–6, 178–9
America 15, 16, 22, 60, 117
American Arctic 5
Amguema River 61
Amguemsky 64
amphi-atlantic flora 17, 30, 33, 58, 72, 97, 136–7
amphi-pacific flora 33
Amund Ringness Island 166
Anabar basin 4, 6, 129
Anabar–Olenek lowland 52
Anabar River 48, 50
Anabar–Penzhina 26, 65
Anderson River 66, 70, 74
Andreyev, V. N. 6, 9, 11, 24, 33–6, 39, 44, 55–7, 103, 107, 118, 188
Angara 22
Angmagssalik 87
Antarctic 144, 169, 175, 181–2, 187
antarctic convergence 169, 170; frontal system 170; geobotanical regions 174; Peninsula 173–5, 181–2; polar deserts 151, 173–4, 181; zone 174
Antarctica 173
anticyclone, East Siberian 28, 47
Anyusky Range 62–3
apomictic microspecies 16
Arctic 1, 2, 4, 89, 90, 144, 171, 187
arctic belt 7, 89; deserts 144, 162–3, 165; division 1, 2, 5, 7, 72; flora origin 96; islands 71; tundras 4, 7, 20–2, 88, 91–2, 94; vegetation areas 2–8
Arctic Polar Deserts 144, 151–2; Barents province 153; Canadian province 166; Cape Chelyuskin 165; De Long Islands 162; Franz Joseph's Land 154; Nordaustlandet 159; North Novaya Zemlya 158; North Severnaya Zemlya 162; Siberian province 161
arctic–alpine flora 22, 30
Arctic–Subarctic boundary 89, 90
Arctogaea 15
Argentine Islands 179
Arngold' 164
Ary–Mas 14, 27, 47, 51
Asia 15, 22–3, 48, 60, 117
Atlantic Ocean 79
Avramchik, M. N. 39, 44, 48, 50
Axel Heiberg Island 3, 120, 123, 126–7, 166–8

Babb, T. A. 123, 126
Backe Peninsula 120
Baer, K. 98
Baffin Island 71, 79, 80, 82, 120, 124, 126, 129
Baffin Strait 82
Baidaratsky Inlet 39
'baidzharakhs' 44, 48
Banfield, A. W. F. 74
Bangor 'oasis' 181, 186
Banks Island 74, 117, 119, 122–3

Barents province 161
Barents Sea 36, 96, 101, 161; low pressure 36
barren wedge 123, 151
Bathurst Island 123, 166
Bear Island 142
Begichev Island 101
Belsun Bay 139
Bennet Island 162
Bering landbridge 58, 60–1, 116
Bering Sea 58
Beringia 22, 57–8, 70
Beringian element 17; species 60
Beschel, R. E. 3, 123, 126–8, 166–8
Big Lyakovsky Island 97, 106, 109–12
birdcliff vegetation 120, 141
Birulya, A. 165
Bliss, L. C. 67, 123, 126, 168
Boch, M. S. 33, 39, 40, 42, 55, 103
Bochantsev, V. P. 15
Böcher, T. W. 3, 5, 12–13, 24, 81, 83–9, 129
Bogdanovskaya-Gienef 33
Bol'shevik Island 164
Bol'shezemelya Tundra 14
Borden Island 123, 166
boreal belt 7
boreal dwarfshrubs 34
boreal herbs 34
boreal region 31
boreal tundras 31
Borisov, A. A. 47
Borodin 113
botanical–geographical belts 31, 90
boulder fields 115
Brassard, G. R. 126
Braun-Blanquet, J. 2, 5, 15, 137
Breslina, I. P. 31–2
Britton, M. E. 66–7, 70, 118
Brock Island 123
Brooks Range 58, 69, 117
Brown, J. 48, 118
Brown, R. J. E. 48
Bruggemann, P. F. 123
Buks, I. I. 21, 89
Bylot Island 120, 126
Byrranga Foothills 29, 41
Byrranga Mountains 47, 102, 104–5

Camp Tareya 27, 42, 44
Canada 28, 63, 71
Canadian Arctic 5, 26
Canadian Arctic Archipelago 91–2, 113, 117–18, 120, 124–5, 151, 166, 189
Canadian Shield 71–2, 79
Cantlon, J. E. 67
Cape Bathurst 70
Cape Chelyuskin 101, 104, 149, 151–2, 161, 163, 165
Cape Isachsen 167
Cape Ratmanov 101
Cape Svyatoy Nos 107
Cape Ushakov 116
Cape Zhelaniya 158
Carisetum subspathaceae 5
Cassiope tetragona synedrium 138
Cassiopeta 133, 138
'cemetery mounds' 44, 108
characteristics: characterizing 1; diagnostic 1; phytocoenological 2
characterizing traits 15
Chaun 64
Chaun Inlet 62
Chaun Plains 63
Chekanovsky, A. L. 48
Chekanovsky Ridges 48, 52
Cherepanov, S. K. 46, 67, 90
Chernov, Yu. I. 104, 145–6, 149, 152–3, 161, 165–6
Chesterfield Inlet 76
Chevykalov 116
chionophobous species 51
chionophytic associations 108
chironomid larvae 150
Chukoch'ya River 108
Chukotka 26, 28, 57, 61–4, 66, 71, 117
Chukotka–Alaska: geobotanical province 57; subarctic tundras 57, 61
Chukotka Highlands 57–8, 62
Churchill, E. D. 3, 66–7, 69
classification of vegetation 2, 5, 6
Cody, W. J. 74, 76
cold deserts 144
Collembola 150
Collins, N. J. 175–6
coniferous forest floras 60
convergence, antarctic 169, 170
Cordilleras Icesheet 71
Corner, R. W. M. 175–6, 178–9, 180
Corns, I. G. W. 67, 69
Cornwallis Island 124
Crary, A. P. 168
'crooked forests' 26, 35
Crozet Islands 169, 170, 172
cryo-arid epoch 12
cryogenic disruption 34
cryogenic formations 175
cryogenic processes 148
cryogenic relief forms 55

cryogenic soils 171
cryophytes 10, 44
cryophytic associations 88
cryophytic floras 16
cryophytic species 44, 52
cryophytic steppe 52
cryoxeromesophilous meadow-like communities 12
cryoxerophytic vegetation types 28
cushion plants 169, 171, 173–4

DeCandolle, A. 10
Dedov, A. A. 33
'delli' 44, 55
De Long Islands 151, 162
Dennis, J. G. 118
Derviz-Sokolova, T. G. 62–4
Devon Island 120, 123, 126, 168
diagnostic traits 1, 2
Dibner, B. D. 105
disjunct areas: amphi-atlantic 17, 30, 33, 58, 72, 97, 136–7; amphi-pacific 33
Disko Bay 84
Disko Island 84, 86
dispersal 58, 72; amphi-atlantic species 72; Beringian species 72
distribution maps: Betula exilis 46; Betula nana 46; Betula tundrarum 46; Carex bigelowii 95; Carex consimilis 95; Carex ensifolia ssp. arctisibirica 95; Carex lugens 95; Carex miliaris 78; Carex rotundata 78; Eriophorum vaginatum 78; Larix laricina 59; Picea glauca 59; Picea mariana 59; Salix alaxensis 45; Salix lapponum 45; Salix polaris 121
district, definition 17
divergence, subantarctic 169–71
division, botanical-geographical 1
Dobbs, C. G. 136, 140–1
Dostovalov, B. N. 48
Drake Range 173
Drew, J. V. 118
Drury, W. H. 126
dry steppe 127
Dryadion 5, 138
Dryas-heath 3
Dryas-tundras 52, 86
Dudypta River 41, 44, 48
DuRietz, G. E. 5
Dushechkin, V. I. 52–3
dwarfbirch tundra 26, 33–4, 36
dwarfshrubs 21; boreal 34; heath 3; hyparctic 34, 42, 52–3; lichen 32; petrophytic 32; tundra 9

'earth glaciers' 128, 130
East European forest-tundra 9
East European tundras 6
East European–West Siberian tundras 28; East European subprovince 31, 33; Kola subprovince 31; Ural-Pai-Khoy subprovince 31, 35; Yamal-Gydan-West Taimyr subprovince 31, 38
East Greenland Current 82, 87
East Siberian anticyclone 28, 47
East Siberian subarctic tundras 44, 49; Khatanga-Olenek subprovince 48–9; Kharaulakh province 49, 52; map 49; Yana-Indigirka subprovince 49, 54
Eastern Canadian Islands 124
Ebelyakh Inlet 107
ecobiomorphs 9, 21, 169, 173, 187
Edwards, J. A. 175–7
Egedesminde 81
Ekstam, O. 98
element: amphi-atlantic 17; Beringian 17; endemic 17; Siberian 17
elfin-wood 21
Ellef Ringness Island 123, 166–7
Ellesmere Land 120, 125–6, 135, 151, 166–8
Elton, C. S. 135–43, 159
El'veney, Mount 64
Emerald Island 170
Enchytraeidae 150
endemic plant communities 16
endemic taxa 17, 29
endemic tundras 29
endemism in Arctic 15, 16, 122
Enderby Land 182–3
Engelskjön, T. 142
Engler system 16
Eurola, S. 99, 136–41
Europe 28, 30, 33, 47
Euro-Siberian shrub region 7

Faddeyevsky Island 110
Faroe Islands 28, 33
fell-field 3
Festuca-heath 3
field-flur 3
Finegan, E. J. 126

flora 2, 157, 159
floristic zonation 35
florogenesis 12, 17
föhn winds: Antarctic Peninsula 182; Byrranga Mountains 105; Canadian Arctic Archipelago 127; Greenland 12, 81, 85, 88, 120, 128, 133; Peary Land 133; Severnaya Zemlya 164; Spitsbergen 136; Wrangel Island 113, 115–16
forest limit by Larix cajanderi 57
forest-tundra 7
Franz Joseph's Land 16, 144, 148, 151–4, 157, 161, 173, 175, 179
fratria 21, 89, 93
Fredskild, B. 84, 133
freshwater vegetation 3
Fristrup, B. 132

Gelting, P. 84
geobotanical areas 1, 14, 17, 169
geobotanical boundaries 17, 18
geobotanical elements 169
geobotanical fields 7, 188
geobotanical regions 19, 20, 169
geobotanical survey maps 55
Geral'd Island 117, 144
Gerasimenko, T. V. 40
Getty 'oasis' 181
Gimingham, C. H. 175–6
Giterman, R. E. 22, 89
glaciation 30, 58, 71; subsurface 47–8
glycophytes 13, 108
Godthaab 81, 84
Gollerbakh, M. M. 182
Golomyanny Island 163
'gol'tsy' 22, 48, 52–3
Golubkova, N. S. 175
Gorbatsky, G. V. 58
Gorchakovsky, P. L. 36, 38–9
Gorodkov, B. N. 6, 7, 9, 11, 24, 34, 36, 39–41, 62–3, 65, 103–4, 109–16, 144–5, 151
Govorukha, L. C. 149, 154
Govorukhin, V. S. 39, 40
Great Bear Lake 74
Greater Beringia 58, 70
Greene, D. M. 171, 175
Greene, S. W. 170–5, 189
Greenland 3, 5, 12, 23, 26, 28, 33, 70–1, 80–1, 83–4, 86, 91–3, 106–7, 117, 120, 151, 167; föhn winds 120; high pressure levels 120; icefree areas 84; subarctic tundras 83
Gribova, S. A. 24, 33
Grierson 'oasis' 183, 186
Grigor'yev, A. A. 33, 89
ground-ice 48
Gulf Stream 82, 136
Gusinaya Land 98
Gydan Peninsula 39–41, 101

Hadač, E. 5, 33, 82, 136–42
Hallett area 185
halophytic marsh associations 13, 108, 141
Hanson, H. C. 67–8, 118
Hanssen, O. 142, 154
Hare, F. K. 79
Harper, F. 79
Hart, H. C. 126
Hartz, N. 87, 128
Hayes Island 154
Hayes Peninsula 126
Heard Island 170–2
heath 3
Heezen, B. C. 28, 33, 82
hekistotherms 9–12, 14, 172
Henrietta Island 162
herb-field 3
high-antarctic species 173
high-alpine belt 53
high-arctic species 4, 7, 14, 19
high-'gol'tsy' belt 53
Hofmann, W. 5, 136–7, 140
Holarctic Dominion 15, 32
Holdgate, M. W. 175–6, 181
Holmboe, J. 142
Holmen, K. 122, 127, 132–5, 168
Holocene epoch 22, 89
Holtom, A. 175
Holttum, R. E. 83
Hudsonian 72
Hudson's Bay 28, 71–2, 74–5; synclinal 71
Hultén, E. 66, 78, 95, 118
Huntley, B. J. 171–2
Hustich, I. 59, 60, 76, 79
hydrolaccoliths 56
hypantarctic species 173
hyparctic belt 7, 9–11, 19, 21–3, 27, 30–1, 34–5, 41–2, 53, 57, 63–5, 88–9, 92–3
hyperhekistotherms 10, 14, 145

Ice Age 12, 22, 28–9, 33, 47–8, 60, 89, 122
Iceland 13, 28, 33; low pressure 47, 96 levels 47, 96
Ignatenko, I. V. 33, 48, 98, 147

Ignat'yev, G. M. 71, 73
Igoshina, K. N. 36–9
Il'ina, I. S. 21, 39, 40, 89, 93
Indigirka River 55, 95, 106–7, 189
Inglefield 120
Isefjord 136, 142
iso-anomalies 43
isoline 175, 181
isostatic uplift 75
isotaxa 167
isotherm 89, 151, 175, 181

Jeanetta Island 162
Johnson, P. L. 118

Kanin-Pechora subprovince 35
Kara Sea 96, 98, 101, 161
Karamysheva, Z. V. 2, 17, 18
Karavayeva, N. A. 92
Kartushin, V. M. 162
Katenin, A. E. 9, 33–4, 61–3
Kats, N. Ya. 171–2
Keewatin Icesheet 71
Kerguelen Island 170, 172
Kerik, J. 126
Khaarastan 107
Kharaulakh Range 52–4, 57, 64
Kharp, research station 39
Khatanga basin 4, 14, 27, 48, 51–2, 101
Khatanga-Anabar plains 102
Khodachek, E. A. 42
Khrom Inlet 107
Krustal'niy River 113
Kihlman, A. O. 31
King Charles Land 151, 159–60
Kitsing, L. I. 114
Kjellman, F. R. 165
Knapp, R. 4
Knorre, A. V. 48
Kola Peninsula 26, 29, 31–3, 129
Kolguyev Island 33, 35
Kolguyev–Novaya Zemlya Current 96
Kolyma 54, 57, 61–5, 92, 107, 117
Kolyuchinskaya Inlet 62
Komarov, V. L. 11, 55
Komi ASSR 146
Komsomolets Island 151, 162
Korchagin, A. A. 11, 13, 33
Korotkevich, E. S. 132, 144–5, 161–5, 169, 172, 175, 181–6
Kotel'ny Island 109, 112
Kruchinin, Yu. A. 109
krummholz 14, 23–4, 26–7, 47, 51, 90
Kruuse, C. 128
Kuc, M. 136
Kupsch, W. O. 48
Kvitøya 159
Labrador 71, 79, 82
Labrador Icesheet 71
Labradorian 72
Lady Franklin's Fjord 160
Ladyzhenskaya, K. I. 154
Lake Taimyr 44
landbridge: Bering 58, 60–1, 116; Greenland–Europe 33, 82
Larsen, J. A. 74
latitude zonation 18
Laurentian Icesheet 71, 75
Laurentian upland 71, 74
Lavrenko, E. M. 2, 6, 8, 14–15, 32, 82
Lee, H. A. 75
Lena basin 129
Lena River 48, 51, 95
Leskov, A. I. 6, 8, 34, 36, 89, 144
lichen associations 32
lichen heath 3
lichen tundras 37
lichen vegetation 12
Lid, J. 136, 138, 140, 154
Lindsay, D. C. 171, 175–6, 178–9, 181
Lindsey, A. A. 74
Little, E. L. 67
Longton, R. E. 175–6, 179, 182–6
Lougheed Island 166
Löve, Á. 10, 13
Löve, D. 10, 13
low-alpine belt 53
low-arctic region 4
low-'gol'tsy' belt 53
low pressure: Aleutian 58; Barents Sea 36; Iceland 47, 96
Lundager, A. 128
Lynge, B. 153–4

Mackenzie River 58, 66, 71
Mackenzie River Delta 69
Macquarie Island 169–72
Maini, J. S. 74
Malye Karmakuly 97
Malye Ostrova 162
Malyi Yamal 39
Malyshev, L. I. 121
Manakov, K. N. 31, 206
Maria Pronchishcheva Bay 105
marine transgression 28, 75
Marion Island 172
Marr, J. W. 76
marshes: glycophytic 108; halophytic 108, 141
Matochin Shar 97–8

Matveyeva, N. V. 42–3, 104–5, 145–6, 149, 152–3, 161, 165–6
McMurdo area 183–5
Mega-Beringia 47, 72
Meighen Island 123, 151, 166–7
Mel'tser, L. I. 24, 40
Melville Island 123
mesophytes 10
meta-arctic flora 16
Michelmore, A. P. G. 136–7, 140–1
Middle Island 162
Mikhailichenko, V. S. 90, 101–3
Mikhailov, I. S. 109–10, 149, 154
Mikisfjord 87
'miniaturized animals' 150
Minyayev, N. A. 31
Miocene 60
mires: graminoid 124–5; herb-moss 86; hillocky 13, 23, 34–5, 41; homogeneous 13; hyparctic 88; mineral 14, 100, 179; moss 115, 142; oligotrophic 23; palsa 13, 34; peat-mound 66, 88; polygonal 23, 40, 44, 56–7, 62–3, 68, 90, 96, 104, 119; sedge 115; tetragonal 105; tussocky 86
Molenaar, J. C. de 5, 88
Mongolia 12, 63
Morye-Yu 14
Morye-Yu River 34
Mt Berry 116
Mt Gunngjörn 81
Mt Inkala 113
Mt More-Pai 35
Mt Newton 136
Muc, M. 126
Murchison Fjord 159, 161
Murman Current 31
Myggbukta 120

Nakhabtseva, S. F. 24, 55, 66, 107, 188
nano-relief 34, 42
Neilson, A. M. 159
Nematoda 150
Nikolayeva, M. G. 34, 38–9
nival meadow 11, 21
Nordaustlandet 135, 151–4, 159
Nordenskiöld, A. E. 81
Nordvik Bay 106
Norin, B. N. 7, 9, 14, 33, 48, 146
North America 12, 73
North American Quaternary glaciation 73
North Pole 168
North Polar Sea 144, 151
northern limits: Larix gmelinii 51; Picea glauca 59; Picea mariana 59; Populus balsamifera 60; Populus tremuloides 60
Northeastern Greenland 124
Northwestern Greenland 128
Norway 6
Nosova, L. I. 55–6
Nova River basin 48
Novaya Zemlya 17, 90–3, 96–7, 136, 139, 144, 151, 153, 158; amphi-atlantic species 97; 'bora' 96
Novichkova-Ivanova, L. N. 149, 154
Novosiberian Islands 109, 112, 117, 144, 162
nunataks 97

'oases' in Antarctic 181–2
Ob River 39
October Revolution Island 164
Olenek Bay 93
Olenek Inlet 101, 106
Olenek River 44, 48, 51
Omoloy–Indigirka district 55
Oosting, H. J. 128
oro-polar deserts 135, 141–2
Ostenfeld, C. H. 132

Pacific Ocean 58, 71
Pai-Khoy Mountains 29, 35–6, 38
Pakarinen, P. 126–7
paleogeography 17, 47, 58
Paleozoic era 74
Palibin, I. V. 154
Palmer, L. J. 67
Palmer Archipelago 173, 175
palsa mires 13, 34
Parry Archipelago 122
Passarge, S. 144
Peary Land 120, 122, 128, 132, 167–8
Penzhina area 65
Perfil'yev, I. A. 33, 90–1, 98, 145–6
Perfil'yeva, V. I. 44, 55, 57, 92
permafrost 13, 56, 98, 110
Peterson, W. 126
petrophytic associations 32, 99
Petrovsky, V. V. 52, 113–15, 117
Philippi, G. 136–7
phylocoenogenesis 17
physiognomic 5, 6
phytomass 145, 177, 179
'pingos' 56
Pioneer Island 151, 162
'plakor' 8
Pleistocene arctic tundras 89

Pleistocene climate 28
Pleistocene cyclonal activity 29
Pleistocene epoch 12, 63
Pleistocene flora exchange 60
Pleistocene landbridge 33, 58, 60–1, 82, 116
Pleistocene steppe relics 63
Pliocene era 22, 47–8
Pohle, R. 6, 33, 89, 90
Point Barrow 69, 70, 113, 117, 119
Polar Basin 82
polar deserts 4, 10, 20; antarctic 181; arctic 144–5, 151–4, 158–9, 161–2, 165–6; endemism 16; plants 19; polygonal 148, 154, 162, 175
polar forest limit 59, 89
Polar Sea 168
Polar Urals 33, 35–37, 129
Polozova, T. G. 24, 33, 39, 42, 51, 145
Polunin, N. 4, 5, 72, 76–7, 79, 80, 124–6, 135–6, 138
polygonal deserts 148, 154, 162, 175
polygonal mires 27, 44, 62
Popigay River 48, 51
Popov, A. I. 48
Porsild, A. E. 3, 5, 24, 66, 74–8, 83, 95, 119, 120, 123, 126–9, 167
Porsild, M. P. 83
Pospelova, E. B. 41–2
Precambrian 31, 71, 74
Price, L. W. 123
Prince Edward Island 169–70, 172
Prince Patrick Island 123, 166
Prince William Island 123
Pronchishchev Mountain 48, 52
Propashchy Bay 100
Puccinellietalia 5
Pyasina basin 30
Pyasina River 27, 29, 47, 50, 101

Quaternary 55, 70, 73
Quebec Province 79
Queen Elizabeth Islands 168

Rachkovskaya, E. I. 2, 17
radiation balance 89, 181
Rakhmanina, A. T. 33–4, 147
Ramenskaya, M. L. 31
Raup, H. 128
Rebristaya, O. V. 20, 33, 35–6, 151
refugia 122
Regel, K. W. 31, 98
region 8, 15, 20
Reindeer Peninsula 137
relic mounds 48
Reutt, A. T. 62–3, 65
Rhacomitrietum 3, 139, 141
Richardson, D. H. 126
Ritchie, J. C. 74, 76
Rogachev River 90
Roger's Bay 113, 116
Řomanovsky, N. N. 48, 109
Rønning, O. 4, 5, 89, 136–8
Rouse, C. H. 67
Rousseau, J. 79
Rubinshtein, E. S. 43
Rudolph, E. D. 182, 185
Rykova, Yu. V. 92, 108

Safronova, I. N. 107, 164–5
Sagyr River 51
Salazkin, A. S. 31
Sambuk, F. V. 33
Savich-Lyubitskaya, L. I. 182
Savile, D. B. O. 121, 123, 126, 167–8
Scandinavia 28, 33
Scandinavian school 4
Schmithüsen, J. 2
Schofield, W. B. 76
Scholander, P. E. 136, 159–61
Schrenk, A. G. 6, 33, 36
Schwarzenbach, F. H. 88, 128
Schweitzer, H. J. 142
Scotter, G. W. 74
Sdobnikov, V. M. 39, 104
Sedow Archipelago 163
Seidenfaden, G. 128
Sellyak Inlet 107
Semenov, I. V. 164
Serebryakov, I. G. 42
Seslerietalia 5
'settlers from the north' 51
Severnaya Zemlya 144, 151–2, 161–2, 164
Shamurin, V. F. 33, 52–3, 97, 129, 145, 147
Shchelkunova, R. P. 50, 55, 107
Sheludyakova, V. A. 55
Shennikov, A. P. 2
shrub thickets 90
shrub tundra 9, 32
Shukhtina, G. G. 154
Siberia 4, 12, 16, 28, 35, 47, 63–4, 97
Siberian element 17
Signy Island 176–7, 179
Sisko, A. K. 109
Sivaya Maska 146
skeletal soils 37
Skottsberg, C. 171, 175
Skvortsov, A. K. 45, 55, 112, 121

Skvortsov, E. F. 107
Small Islands 162
Smirnova, Z. N. 33, 182
Smith, R. I. L. 171–2, 175–6, 178–80
Snaddvika 161
Sob' River 36
Sochava, V. B. 2, 4, 6, 7, 9, 15, 18, 21, 24, 28, 36–7, 48, 50, 65, 82, 89, 90, 93, 106, 129, 145
sodium chloride 13
soil formation 149
Sokolovsky, V. 98
solifluction lobes 130
Solonevich, N. G. 33
Soper, J. D. 79, 80
Sørenson, Th. 128–30
South America 173
South Georgia 170–2
South Orkney Islands 173, 175–6
South Polar Sea 169
South Sandwich Islands 173, 175
South Shetland Islands 173, 175, 178–9
Southampton Island 77
Soviet Arctic 7, 25; latitudinal zonation 25
Soviet Far North 7
Soviet Union 3–8, 26, 188
Soviniy River 113
Soykudakh Mountains 53
Spitsbergen 5, 95, 99, 135–6, 168; amphi-atlantic species 136–7
spotty tundras 4, 41, 44, 55, 90, 98, 101, 105–6, 108, 111, 146
Stannovoye Foothills 121
Steffen, H. 98
steppe 12, 23, 44, 63–4
Stolbovy Island 109
Størmer, P. 154
Storøya 159
subalpine belt 52
subantarctic divergence 169–71
subarctic tundras 7, 21–3, 44; boundary 7; Canadian province 70–81; Chukotka–Alaska province 57–70; East European-West Siberian province 11, 29, 30; East Siberian province 44–57; Greenland province 26, 81–88; middle belt 26, 27, 62; northern belt 27; southern belt 24–26
subendemics 16, 122
sub-'gol'tsy' belt 52, 65
Sukachev, V. N. 39
Sumina, O. I. 105, 109
Summerhayes, V. S. 135–43, 159, 160
Sunding, P. 141–2
Svatkov, N. M. 113, 116
Sverdrup syncline 166
Svoboda, J. 126–7
synclinal, Hudson's Bay 71
synusia 8, 9, 10
Syroyechkovsky, Ye. Ye. 182

taiga trees 60
Taimyr 14, 27, 30, 38, 41, 90, 101–2
Takhtadzhyan, A. L. 1
Tanfil'yev, G. I. 33
Tareya, research station 27, 42, 44
Taylor, B. W. 171
Taylor 'oasis' 181
Tazov Peninsula 39
Tedrow, J. C. F. 121, 123, 126, 149
Telfer, E. S. 74
temperature inversion 127
Terent'yev, P. V. 1
Tertiary Arctic 15, 16
Thannheiser, D. 5, 74, 123–4
Tharp, M. 28, 33, 82
thermokarst 48, 108–9
Thomas, J. H. 66, 118–9
Tikhaya Bay 154
Tikhomirov, B. A. 52–3, 90, 104, 165
Tiksy Bay 52–3, 97
Timan 6
Timan Ridge 29
Timofeyev, D. A. 7, 89
Tokarevskikh, S. A. 14, 33–4
Tolchel'nikov, Yu. S. 40
Tolmachev, A. I. 9, 14, 22, 33–4, 42, 44, 66, 97–8, 154, 158, 162, 188
Trapnell, G. O. 83
Trautvetter, E. R. 6, 33, 89
Triloff, E. G. 136
Troll, C. 171
Trotsenko, G. V. 38–9
Tsareva, 55
Tsinzerling, Yu. D. 31
Tuktoyaktuk Peninsula 69
tundra 6, 10, 20, 64, 83; alpine 48, 53–4, 63–4; arctic 6, 7, 21, 88–9, 152; associations 8, 9, 19; belt 37; boundaries 6; Cassiope 53, 122, 125, 129, 135, 138, 141; Dryas 53, 68, 86, 102, 111, 114, 119, 122, 126, 133, 137; dwarfshrub 4, 8, 32, 34, 36–8, 53, 56, 64, 68, 76; glycophytic 13; graminoid 64, 111, 115; grass-moss 111; halophytic 13; herb 110, 142; hillocky 13; hummocky 4, 41, 56, 103, 119, 124;

hyparctic 63, 68, 70, 91, 108, 152; lichen 8, 37, 53–4, 64; moss 8, 115; nival 37; oro-arctic 52; polygonal 90, 111, 124; sedge 64; spotty 4, 41, 44, 55, 90, 98, 101, 105–6, 108, 111, 146; steppe 12, 57, 63–4, 86; subarctic 6, 7, 21, 23–4, 76; swampy 57; tussocky 26, 38–9, 41–2, 50, 56–7, 61–3, 65, 67, 69, 70, 86; underground structure 110; willow-moss 53, 111
Tyrrell Sea 28, 75
Tyrtikov, A. P. 55
Tyulina, L. N. 14, 48, 50–1

Umiat region 69
underground organs 110, 148
Ungava Peninsula 79
Upernivik 120
Urals 6, 29, 35–6, 38
Ural-Hercynian complex 96
Usnea-Andreaea association 177
USSR 3–8, 26, 188
USSR tundra zone 6, 7

Vaigach 17, 33, 35, 96–9
Vanger 'oasis' 183
vascular plants: antarctic islands 172; arctic islands 97, 109, 112, 117–19, 157, 159, 162
Vasil'yev, V. N. 65
Vasilyevich, V. I. 1, 42
vegetation 2; petrophytic 99
vegetation areas 5
vegetation classification 2, 6
vegetation complexes 5
vegetation maps 6, 33
vegetation types 5, 6
Verkhoyansk Mountains 52, 54, 95
Vestfold 'oasis' 181
Victoria Island 74, 117, 119–20, 122–3
Victoria Land 182–6
Victoria 'oasis' 181
Viereck, L. A. 67
Vikhireva-Vasil'kova, V. V. 65, 147
Vil'kitsky Island 162
Vinogradova, A. N. 39, 41–2
Vitt, D. H. 126–7
Vollosovich expedition 107
Vul'f, E. V. 171

Walton, D. H. W. 171–2
Walton, G. F. 123
Walton, J. 136, 141
Warming, E. 3, 83, 87, 128, 145
West Greenland Current 82, 84
West Greenland tundras 83–4
West Novaya Zemlya Current 96
West Siberia 28, 30, 38
West Spitsbergen 17, 136–7
West Taimyr 41
White Sea 33
Wiggins, I. L. 66, 118–19
willow thickets 27, 32
Wisconsin Icesheet 71, 73
Wrangel Island 12, 17, 93, 109, 112–13, 116–17
Wright 'oasis' 181

xerophytic 22, 86

Yakutia 54
Yamal 4, 38–9, 40, 90, 101–2
Yana River 189
Yegorova, T. V. 95
Yenisey River 30, 38
Yesipov, V. K. 98
Young, S. B. 20, 89, 90, 93, 151, 159
Yugorsky Shar 35
Yurtsev, B. A. 2, 7, 8, 10–12, 15–16, 20–2, 27–8, 31–2, 36, 38, 44, 47–8, 51–4, 58, 60, 62–5, 67, 70, 72, 82, 89, 90, 114, 116, 121, 151

Zhadrinskaya, N. G 109
Zharkova, Yu. G. 41–2
Zhitkov, B. M. 6, 39
Zhokhov Island 162
Zhukova, A. L. 149, 154
zonal belts 17
zonation 18, 35
Zubakov, V. A. 48
Zubkov, A. I. 98–9, 158